The Use and Management of Invasive Plants

The Use and Management of Invasive Plants

Collection Editors

Danijela Poljuha
Barbara Sladonja

MDPI

Basel • Beijing • Wuhan • Barcelona • Belgrade • Novi Sad • Cluj • Manchester

Collection Editors
Danijela Poljuha
Institute of Agriculture and Tourism Poreč
Poreč
Croatia

Barbara Sladonja
Institute of Agriculture and Tourism Poreč
Poreč
Croatia

Editorial Office
MDPI AG
Grosspeteranlage 5
4052 Basel, Switzerland

This is a reprint of the Topical Collection, published open access by the journal *Plants* (ISSN 2223-7747), freely accessible at: https://www.mdpi.com/journal/plants/topical_collections/Ecosystem_Phytochemicals.

For citation purposes, cite each article independently as indicated on the article page online and as indicated below:

Lastname, A.A.; Lastname, B.B. Article Title. *Journal Name* **Year**, *Volume Number*, Page Range.

ISBN 978-3-7258-5517-9 (Hbk)
ISBN 978-3-7258-5518-6 (PDF)
https://doi.org/10.3390/books978-3-7258-5518-6

Cover image courtesy of Danijela Poljuha

Contents

About the Editors

Danijela Poljuha

Danijela Poljuha graduated with a degree in molecular biology in the Faculty of Science, University of Zagreb, Croatia, where she also holds a PhD degree in natural sciences in the field of biology. She has worked as a researcher in the Faculty of Science, University of Zagreb, The Institute of Agriculture and Tourism Poreč, and The Materials Research Centre METRIS Pula. She has participated in numerous national and international projects and has published over 50 scientific papers. She is the founder of two laboratories and a Biotechnical Department. Her research interests are focused on invasive plants' phytochemistry, plant genetics, and molecular markers' application in the conservation of plant genetic resources. She is also involved with the popularization of science.

Barbara Sladonja

Barbara Sladonja graduated with a degree in biology-ecology from the Faculty of Science, University of Zagreb, Croatia. She earned her PhD in Natural Sciences through a joint program at the University of Udine, Pisa and the University of Florence in Italy. She is employed in the Institute of Agriculture and Tourism, Poreč, Croatia. She organized major international conferences in Poreč, has an extensive publication record in peer-reviewed journals and substantial experience in environmental education and outreach. Her work exemplifies the integration of rigorous scientific research, environmental stewardship, and community engagement, significantly contributing to biodiversity conservation and sustainable agricultural practices.

Preface

The Reprint of this topical collection, entitled "The Use and Management of Invasive Plants", emerges from a growing recognition that invasive plant species, often regarded solely as ecological threats, may also represent untapped resources of scientific, environmental, and economic value. Traditionally, invasive plants have been approached primarily from the perspective of control and eradication. However, contemporary research increasingly supports a more integrated view—one that seeks not only to mitigate their negative impacts but also to investigate their potential applications in fields such as phytopharmacy, bioenergy, and sustainable agriculture.

Danijela Poljuha and Barbara Sladonja
Collection Editors

plants

MDPI

Editorial

The Use and Management of Invasive Plants

Danijela Poljuha *,† and Barbara Sladonja †

Department of Agriculture and Nutrition, Institute of Agriculture and Tourism, Karla Huguesa 8, 52440 Poreč, Croatia; barbara@iptpo.hr
* Correspondence: danijela@iptpo.hr
† These authors contributed equally to this work.

Invasive plant species pose significant challenges to biodiversity, ecosystem stability, and economic sustainability, such as disrupting native flora, altering soil chemistry, and creating ecological imbalances. However, recent research highlights the potential of these species as resources. Their secondary metabolites, allelopathic properties, and adaptability offer avenues for novel applications, particularly in phytopharmacy, energy production, and agriculture. This topical collection presents key advancements in both the control and utilization of invasive plant species, exploring their chemical properties, ecological impact, potential applications, and alternative management strategies.

Several studies in this collection investigate the bioactive compounds of invasive plants and their potential medicinal applications. One study focused on *Solidago canadensis* L., an invasive species in Europe, to investigate its total phenolics, flavonoids, non-flavonoids, and species-specific phenolic compounds. The findings indicated that phenolic acids and flavonoid-rich extracts from the invasive *S. canadensis* exhibit strong antioxidant properties, suggesting their potential in addressing oxidative-stress-related diseases [1]. In addition to the phytochemical properties of invasive alien plant species, their potential as antimicrobial agents is also being explored. Species such as *Ailanthus altissima*, *Ambrosia artemisiifolia*, *Conyza canadensis*, *Dittrichia viscosa*, *Erigeron annuus*, and *Xanthium strumarium* have shown significant antimicrobial activity. In one study, among all tested plant extracts, some effectively inhibited bacterial and fungal growth while exhibiting lower cytotoxicity compared to a positive control, with the exception of *X. strumarium* [2]. These findings suggest that invasive plants may be valuable sources of natural antimicrobials and pharmaceuticals. Further analysis of their phytochemicals also revealed potential cytotoxic effects against human cancer cells. *Acacia melanoxylon*, another invasive species, was studied to determine its chemical properties and cytotoxicity. The research identified five major compounds in *A. melanoxylon* extracts, including lupeol derivatives, which showed potential cytotoxic activity against colorectal cancer cells [3]. Another study showed that extracts of the highly invasive *Reynoutria japonica* rhizome exhibit promising activity against multidrug-resistant bacteria, such as *Mycobacterium smegmatis*. This plant is also an important source of stilbenes and anthranoids. The investigation concentrated on extraction methods, revealing that simple, food-grade solvent extractions can efficiently concentrate bioactive compounds, making them more viable for pharmacological use. The study highlighted *Reynoutria japonica*'s potential as an antimicrobial agent, suggesting alternative management strategies that incorporate its bioactive properties rather than focusing solely on eradication [4].

Understanding the biochemical strategies that enhance the invasiveness of plant species is crucial for effective management. Research in this area investigates how flavonoids and other secondary metabolites contribute to the competitive advantages of invasive plants. These compounds interact with soil nutrients and microbial communities, influencing root development and plant–microbe interactions. Some invasive species, such as *S. canadensis*, form nitrogen-containing complexes that regulate plant growth, even when released at very low concentrations, potentially altering local ecosystems and promoting further invasion [5]. Managing invasive species often involves costly eradication efforts with limited success. This collection presents alternative management strategies, including biomass utilization and microbial-based control. The highly toxic species *Jacobaea vulgaris* has been studied to determine the energetic potential of its biomass through biomethanization, demonstrating that ensiling can significantly reduce its toxic alkaloid content, making this invasive species safer for potential use in bioenergy production. Biomethanization experiments revealed that the biomasses of some invasive plants, though limited in terms of energy potential, could still provide sustainable alternatives for waste management and renewable energy [6].

The review articles in this collection provide a broader context for managing invasive plants, integrating biological control, chemical utilization, and novel eradication techniques. *Senecio madagascariensis* is an exotic invasive species that poses a major threat to natural and modified ecosystems around the world. The review presented in [7] synthesizes the currently available information on the biology, distribution, and management options of *S. madagascariensis*'s, as well as on its ecological impact. It also discusses integrated management strategies, including biological control and competitive planting, as part of a systematic approach. Another review on *Fallopia japonica* provides an overview of its high therapeutic potential, highlighting its antioxidant, antimicrobial, anti-inflammatory, and anticancer effects. Among the key biological compounds found in this species are resveratrol, emodin, and polydatin. The review also explores its potential for honey production, offering an innovative perspective on its economic value [8]. Other studies explore microbial-based control methods, emphasizing the role of plant–microbe interactions in suppressing invasive growth while supporting native plant species. One review highlights the increased research into microbial-mediated phytochemical production in invasive plants, microbial consortia that reduce the competitiveness of invasive plants, microbial-mediated increases in the herbicidal tolerance of native plants, and the increased susceptibility of invasive plants to plant pathogens [9]. Despite numerous novel control and eradication methods, classic mechanical methods can still be valuable, especially if key factors for their successful implementation are identified. Managing invasive exotic plant species is a complex challenge, particularly for Asian knotweeds (*Reynoutria* spp.). Tarping is a common but poorly documented method of managing this species, and the review presented in [10] emphasizes the importance of site-specific conditions and long-term monitoring for its successful implementation. Moreover, based on the review's bibliography and survey work, practical recommendations are proposed.

This topical collection demonstrates the complex nature of invasive plant research, emphasizing the importance of viewing these species not only as ecological threats but also as potential resources. The work reported in this collection of papers will contribute to a growing body of evidence that invasive plants can be managed in ways that promote sustainability, whether through pharmaceutical applications, bioenergy production, or alternative control strategies. Future research should continue to explore these dual perspectives, ensuring that ecological balance is maintained while maximizing the potential benefits of invasive species.

Author Contributions: All authors contributed to the design, writing, and editing of this Editorial article. All authors have read and agreed to the published version of the manuscript.

Funding: This work was supported by the Croatian Science Foundation under project number HRZZ-IP-2020-02-6899.

Conflicts of Interest: The authors declare no conflicts of interest.

References

1. Uzelac Božac, M.; Poljuha, D.; Dudaš, S.; Bilić, J.; Šola, I.; Mikulič-Petkovšek, M.; Sladonja, B. Phenolic Profile and Antioxidant Capacity of Invasive *Solidago canadensis* L.: Potential Applications in Phytopharmacy. *Plants* **2025**, *14*, 44. [CrossRef]
2. Poljuha, D.; Sladonja, B.; Šola, I.; Šenica, M.; Uzelac, M.; Veberič, R.; Hudina, M.; Famuyide, I.M.; Eloff, J.N.; Mikulic-Petkovsek, M. LC–DAD–MS Phenolic Characterisation of Six Invasive Plant Species in Croatia and Determination of Their Antimicrobial and Cytotoxic Activity. *Plants* **2022**, *11*, 596. [CrossRef] [PubMed]
3. Alves, D.; Duarte, S.; Arsénio, P.; Gonçalves, J.; Rodrigues, C.M.P.; Lourenço, A.; Máximo, P. Exploring the Phytochemicals of *Acacia melanoxylon* R. Br. *Plants* **2021**, *10*, 2698. [CrossRef] [PubMed]
4. Alperth, F.; Melinz, L.; Fladerer, J.-P.; Bucar, F. UHPLC Analysis of *Reynoutria japonica* Houtt. Rhizome Preparations Regarding Stilbene and Anthranoid Composition and Their Antimycobacterial Activity Evaluation. *Plants* **2021**, *10*, 1809. [CrossRef] [PubMed]
5. Likhanov, A.; Oliinyk, M.; Pashkevych, N.; Churilov, A.; Kozyr, M. The Role of Flavonoids in Invasion Strategy of *Solidago canadensis* L. *Plants* **2021**, *10*, 1748. [CrossRef] [PubMed]
6. Müller, J.; Wiedow, D.; Chmit, M.S.; Beuerle, T. Utilization of Biomasses from Landscape Conservation Growths Dominated by Common Ragwort (*Jacobaea vulgaris* Gaertn.) for Biomethanization. *Plants* **2022**, *11*, 813. [CrossRef] [PubMed]
7. Wijayabandara, K.; Campbell, S.; Vitelli, J.; Shabbir, A.; Adkins, S. Review of the Biology, Distribution, and Management of the Invasive Fireweed (*Senecio madagascariensis* Poir). *Plants* **2022**, *11*, 107. [CrossRef]
8. Cucu, A.-A.; Baci, G.-M.; Dezsi, Ş.; Nap, M.-E.; Beteg, F.I.; Bonta, V.; Bobiş, O.; Caprio, E.; Dezmirean, D.S. New Approaches on Japanese Knotweed (*Fallopia japonica*) Bioactive Compounds and Their Potential of Pharmacological and Beekeeping Activities: Challenges and Future Directions. *Plants* **2021**, *10*, 2621. [CrossRef] [PubMed]
9. Shahrtash, M.; Brown, S.P. A Path Forward: Promoting Microbial-Based Methods in the Control of Invasive Plant Species. *Plants* **2021**, *10*, 943. [CrossRef] [PubMed]
10. Dusz, M.-A.; Martin, F.-M.; Dommanget, F.; Petit, A.; Dechaume-Moncharmont, C.; Evette, A. Review of Existing Knowledge and Practices of Tarping for the Control of Invasive Knotweeds. *Plants* **2021**, *10*, 2152. [CrossRef] [PubMed]

plants

MDPI

Article

Phenolic Profile and Antioxidant Capacity of Invasive *Solidago canadensis* L.: Potential Applications in Phytopharmacy

Mirela Uzelac Božac [1,†], Danijela Poljuha [1,*,†], Slavica Dudaš [2], Josipa Bilić [3], Ivana Šola [4], Maja Mikulič-Petkovšek [5] and Barbara Sladonja [1]

1 Department of Agriculture and Nutrition, Institute of Agriculture and Tourism, Karla Huguesa 8, 52440 Poreč, Croatia; mirela@iptpo.hr (M.U.B.); barbara@iptpo.hr (B.S.)
2 Agricultural Department, Polytechnic of Rijeka, Karla Huguesa 6, 52440 Poreč, Croatia; sdudas@veleri.hr
3 METRIS Research Centre, Istrian University of Applied Sciences, Zagrebačka 30, 52100 Pula, Croatia; jbilic@iv.hr
4 Department of Biology, Faculty of Science, University of Zagreb, 10000 Zagreb, Croatia; ivana.sola@biol.pmf.unizg.hr
5 Department of Agronomy, Biotechnical Faculty, University of Ljubljana, Jamnikarjeva 101, 1000 Ljubljana, Slovenia; maja.mikulic-petkovsek@bf.uni-lj.si
* Correspondence: danijela@iptpo.hr; Tel.: +385-52-408-336
† These authors contributed equally to this work.

Abstract: Canadian goldenrod (*Solidago canadensis* L.), an invasive plant in Europe, is known for its allelopathic activity and is rich in bioactive compounds like flavonoids and phenolic acids, with significant pharmacological potential. This study presents the LC-MS phenolic profiles of leaf and flower extracts from *S. canadensis*, an invasive alien plant in the Istria region (Croatia). Total phenolics (TP) (45.78–110.68 mg GAE/g DW) and non-flavonoids (TNF) (28.38–72.20 mg GAE/g DW) were found to be more abundant in ethanolic than in methanolic extracts. The antioxidant capacity (AC), as measured by ABTS, DPPH, and FRAP assays, was higher in flower extracts compared to leaf extracts. A non-targeted metabolomics approach was used, and 41 phenolic compounds in leaves and 36 in flowers were identified, with hydroxycinnamic acids and flavonols being the most abundant. 5-caffeoylquinic acid was quantitatively predominant in the leaf extracts, while quercetin-3-rutinoside dominated the flower extracts. Five leaf-specific compounds were identified (dicaffeoylquinic acid 2, 4-*p*-coumaroylquinic acid 1, *p*-hydroxybenzoic acid, quercetin-3-rhamnoside, and quercetin acetylhexoside 1), suggesting targeted extraction for different pharmacological applications. This study highlights the therapeutic potential of *S. canadensis* and underscores the need for further research on the bioavailability, efficacy, and safety of its compounds, potentially transforming this ecological threat into a valuable resource for drug development.

Keywords: Canadian goldenrod; flower; invasive plants; leaf; plant extracts

1. Introduction

Plants have a long history of use in folk herbal medicine as food sources and remedies for treating various diseases. Their bioactive compounds, such as alkaloids, terpenoids, coumarins, nitrogen-containing and organosulfur compounds, and phenolics, are currently being researched to discover and develop new pharmaceuticals [1]. Phenolics, in particular, are significant due to their role as effective antioxidants and antibacterials [2]. They are synthesized through the shikimate and phenylpropanoid pathways, which convert primary metabolites like phenylalanine into biologically active phenolics such as quercetin, rutin, and chlorogenic acid. The shikimate pathway transforms simple carbohydrates into aromatic amino acids like phenylalanine and produces precursors like chorismate, which leads to phenolic acids such as caffeic acid, chlorogenic acid, and ferulic acid. The phenylpropanoid pathway converts phenylalanine into *p*-coumaric acid, a key intermediate for synthesizing flavonoids and polyphenols like quercetin, rutin, and epigallocatechin [3]. These compounds are known for their strong antioxidant, anti-inflammatory and antimicrobial activities [2]. Additionally, plant phenolics exhibit significant antitumor activity, inhibiting cancer initiation, progression, and metastasis in both in vitro and in vivo studies. Phenolics exert their effects by modulating cellular pathways such as growth factor-receptor interactions and signaling cascades (e.g., kinases and transcription factors), leading to cell cycle arrest, apoptosis, and reduced cell survival [4].

Phenolics in invasive alien plants are especially intriguing because some studies suggest they produce higher quantities of these compounds than native species [5] or the same species in their native ranges [6]. One theory is that the novel biochemistry of invasive plants in new habitats represents their response to new environments and climate conditions [7,8].

Canadian goldenrod, *Solidago canadensis* L., a perennial herb native to North America, was introduced in Europe in the early 17th century as an ornamental plant [9]. The species is a member of the *Solidago* genus (*Compositae, Asterales*), commonly referred to as "goldenrods" [10]. It quickly naturalized, spreading along abandoned fields and riverbanks. Nowadays, *S. canadensis* is widespread across Europe and is listed on the EPPO list of invasive alien plants [11]. Its invasive nature is enhanced by the fast vegetative propagation through long rhizomes, high growth rate, and allelopathy, which leads to the transformation of soil properties and plant communities [7,12,13]. *S. canadensis* exhibits strong allelopathic effects by releasing allelochemicals that inhibit the growth, germination, and development of neighboring plants, thereby reducing biodiversity and outcompeting native species. These allelochemicals are released through root exudates, plant residue decomposition, and leachates, which alter soil microbial communities and can degrade arbuscular mycorrhizal fungi, vital for nutrient and water uptake in native plants [6]. The phytotoxicity of *S. canadensis* is linked to oxidative stress and damage to cell membranes, leading to electrolyte leakage, reduced chlorophyll content, and impaired photosynthesis. Some of the key phenolic allelochemicals in this species that contribute to its allelopathic impact are chlorogenic acid, rutin (quercetin-3-O-rutinoside), kaempferol-3-O-D-glucoside, and quercitrine [6]. Phenolic allelochemicals interfere with several enzymes and the major physiological processes, such as phytohormone activity, mineral uptake, plant water balance and stomatal function, photosynthesis, respiration, organic synthesis of certain compounds, and flow of carbon, contributing to the phytotoxicity [14,15]. These phenolic compounds enhance species invasiveness but may offer potential as natural herbicides. The beneficial aspects of *S. canadensis* have been recognized for centuries. The herbal material of this invasive species has been used in European phytotherapy for a long time to treat urinary and genital diseases [16]. This species is well-known in traditional medicine due to its complex composition of specialized metabolites, such as polyphenolics, which contribute to its antioxidant, antimicrobial, and anti-inflammatory activities [16–20].

In our recent comprehensive review of *S. canadensis* phytochemicals [21], we highlighted the existing knowledge gaps, and the present study builds on this by providing the site-specific comparison of phenolic profiles between leaf and flower extracts of this species, further advancing our understanding of its potential for phytopharmaceutical applications. In this paper, we selected *S. canadensis*, commonly found in anthropogenically disturbed areas and along abandoned riverbanks in Istria, Croatia, for the first site-specific phytochemical screening to explore its potential for phytopharmaceutical applications.

This choice was based on the understanding that phytochemical profiles are influenced by ecological and climatic conditions [22]. This study aimed to identify the main phenolic compounds and the antioxidant capacity of leaf and flower extracts of *S. canadensis* from Istria. For the first time, we provide a novel comparison of the phytochemical profile and antioxidant activity of individual parts of the plant—leaf and flower—offering a detailed insight into the significant differences between them. This analysis not only establishes a foundation for understanding the local biological activity of *S. canadensis* but also sets the stage for proposing a site-specific model of its exploitation as a provider of new ecosystem services. Alongside future ecological and biological analyses, we expect such a model to trigger similar approaches in broader geographical contexts.

For this purpose, we (1) spectrophotometrically measured the content of different groups of bioactive compounds (total phenolics, flavonoids, and non-flavonoids) and their antioxidant capacity using three standard assays (DPPH, ABTS, and FRAP); (2) identified and quantified the main phenolic compounds using the LC-DAD-MS method; and (3) statistically determined the influence of different factors, such as plant part (leaf and flower) and solvent (70% ethanol and 80% methanol), on measured variables using two-way ANOVA and Tukey's test ($p \leq 0.05$ and 0.01).

2. Results and Discussion

2.1. Phenolic Content and Antioxidant Capacity

The type of solvent had a significant effect on total phenolic (TP) and total non-flavonoids (TNF) content in tested extracts, with values being higher for ethanol (EtOH) than for methanol (MeOH) (Figure 1, Table S1). The plant part significantly affected TNF content, which was higher in leaf than in flower extracts, but only when the EtOH was used as a solvent. Neither the plant part nor the solvent significantly influenced the total phenolic (TP), total flavonoid (TF), or total non-flavonoid (TNF) content. The highest TP values were found in the ethanolic extracts of leaves and flowers, measuring 110.68 and 110.77 mg of gallic acid equivalent (GAE)/g of dry weight (DW), respectively. The TNF values were higher in the ethanolic extract for both leaves and flowers, with 72.20 mg of GAE/g DW (in leaves) and 64.70 mg of GAE/g DW (in flowers), compared to the methanolic extracts, which contained 29.68 and 28.38 mg of GAE/g DW in leaves and flowers extracts, respectively. This could be attributed to the solvent's efficiency in solubilizing specific non-flavonoid compounds. Previous studies reported that the choice of solvent depends on the extraction method and the type of phenolic compounds desired, with methanol being optimal for maceration and microwave extraction of hydrolyzable tannins and polyphenols, while ethanol is best for infusion and the extraction of condensed tannins across most methods [23].

The TF values were similar across solvents and plant parts, ranging from 29.27 to 34.14 mg of catechin equivalent (CE)/g DW.

The TP and TF contents of *S. canadensis* leaf extract obtained in our study were higher than in the findings of Du et al. [24], who analyzed samples collected in China, where this species is also invasive. Furthermore, flower extracts exhibited significantly higher TP and TF content compared to the values reported by Shelepova et al. [25]. Their study determined TP and TF contents of 105.36 ± 1.45 mg GAE/100 g and 58.23 ± 0.17 mg quercetin equivalent (QE)/100 g, respectively, in 96% EtOH and 80% MeOH extracts. The higher levels of detected flavonoids can be attributed to variations in climatic conditions and vegetative phase [26,27]. In clay soils, plant fixation of NH_4^+ increases with increasing pH [28]. In soils with a high pH, the roots of *S. canadensis* secrete more compounds from the flavonoid group compared to other phenolic groups. This aligns with the fact that our samples were collected along riverbanks in Istria with alkaline swamp-clay soil on a sunny day in August.

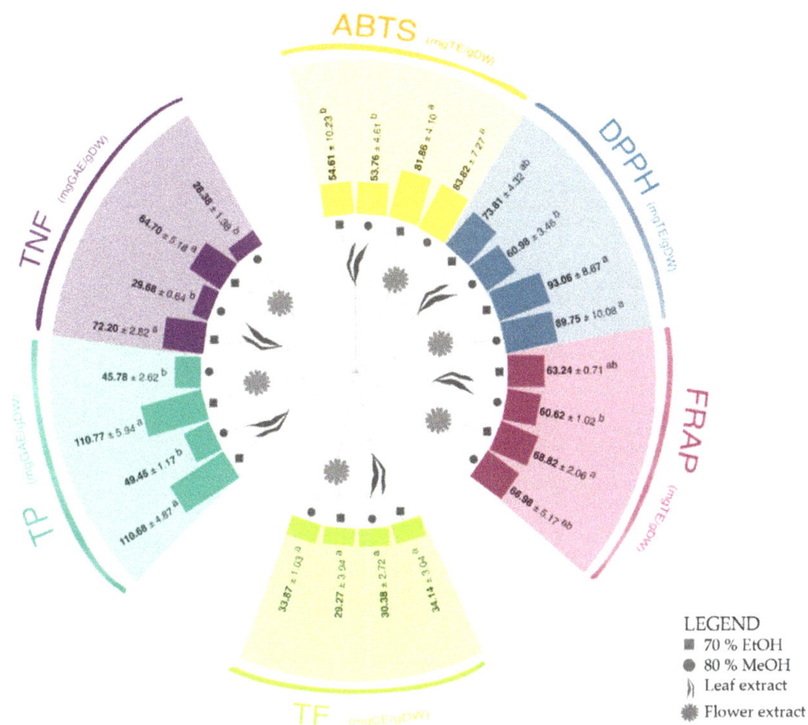

Figure 1. The content of total phenolics (TP), total non-flavonoids (TNF), and total flavonoids (TF), and antioxidant capacity measured by ABTS, DPPH, and FRAP assays in *S. canadensis* leaf and flower extracts in 70% ethanol and 80% methanol. Different letters (a–b) in the same section indicate significant differences between the measured values (two-way ANOVA, Tukey's test, $p \leq 0.01$). All values are also shown in Table S1.

The antioxidant capacity (AC) of *S. canadensis* flower extracts, measured across all three assays, was higher than that of the leaf extracts, indicating that the plant part had a more significant effect on the AC than the solvent used. AC of *S. canadensis* was also tested by other authors, who proved the antioxidant capacity of aerial parts [17,19,20,29–31] and root [20,32] extracts of this species. However, comparing the obtained values is challenging due to the differences in the extraction methods, solvents used, and the tests themselves. Deng et al. [19], for example, showed that the AC and contents of TP, TF, and total tannins depended on the ripeness stage, tissue type, and extraction method. Furthermore, a comprehensive evaluation of antioxidant potential requires multiple in vitro assays due to the diverse mechanisms by which antioxidants operate. Direct comparisons between methods are thus challenging [33].

In all the extracts, highly significant ($p \leq 0.01$) total ($0.9 < r < 1.0$) and very strong positive ($0.7 < r < 0.9$) correlations, according to the Roemer-Orphal scale, were observed between AC values determined by ABTS, DPPH, and FRAP assays, as well as between TP and TNF values (Figure 2). There were no other significant correlations (Table S2).

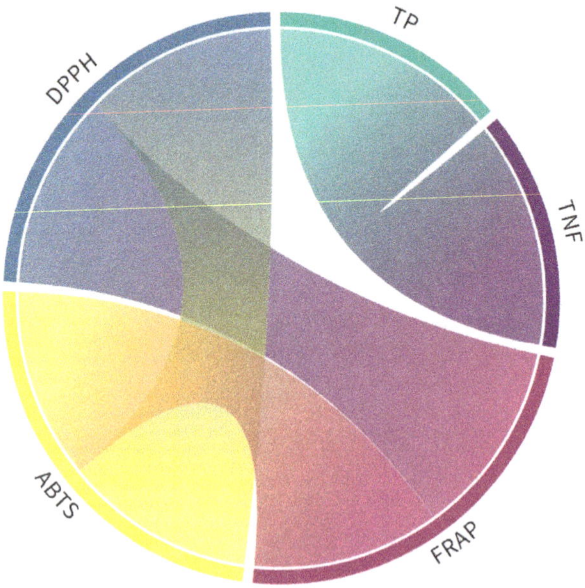

Figure 2. Pearson's correlation coefficients of total phenolics (TP) and non-flavonoids (TNF) contents and antioxidant capacities in extracts in both solvents, determined by ABTS, DPPH, and FRAP assays. Only significant correlations ($p \leq 0.01$; Table S2) are shown. The outer ring represents the variables (TP, TNF, DPPH, ABTS, and FRAP), while the connecting bands show the strength of correlations between the variables: thicker bands represent stronger correlations.

2.2. Identification and Quantification of Phenolic Compounds

We identified 41 phenolic compounds in leaf and 36 in flower extracts, belonging to the groups of phenolic acids (hydroxycinnamic and hydroxybenzoic acid derivatives) and flavonoids (flavonols) (Table 1, Figure 3). In leaf extracts, flavonols dominated, accounting for 55% of the total detected phenolics in ethanolic and 60% in methanolic extracts (Figure 3A). Similarly, the flower extracts had the highest share of flavonols (76% in EtOH and 67% in MeOH) (Figure 3B).

Table 1. Phenolic compounds (mg/g of dry weight (DW)) of *S. canadensis* leaf and flower extracts in 70% ethanol and 80% methanol extracts identified by LC-MS. Values represent the mean ± SD of three replicates. The total phenolic contents of the main phenolic groups are given in bold. Different letters (a–d) in the same row indicate significant differences (ANOVA, Tukey's honest significant difference, p-value ≤ 0.01; * p-value ≤ 0.01); n.d.—not detected.

Phenolic Compounds	*Solidago canadensis* Leaf		*Solidago canadensis* Flower	
	70% EtOH	80% MeOH	70% EtOH	80% MeOH
3-caffeoylquinic acid 1	1.393 ± 0.092 a	1.270 ± 0.124 a	0.433 ± 0.161 b	0.617 ± 0.071 b
4-caffeoylquinic acid 1	0.542 ± 0.098 a	0.653 ± 0.123 a	0.383 ± 0.096 a	0.477 ± 0.121 a
5-caffeoylquinic acid 1	17.748 ±1.558 a	20.423 ± 0.737 a	9.837 ± 0.685 c	13.600 ± 0.667 b
5-caffeoylquinic acid 2	0.920 ± 0.053 a	0.797 ± 0.078 a	0.732 ±0.200 ab	0.413 ± 0.088 b
Caffeic acid	0.008 ± 0.000 c	0.007 ± 0.000 c	0.172 ± 0.047 a	0.097 ± 0.021 b
Caffeic acid hexoside 1	0.652 ± 0.013 a	0.293 ± 0.017 b	0.766 ± 0.128 a	0.532 ± 0.088 ab
Caffeic acid hexoside 2	0.006 ± 0.001 a	0.006 ± 0.001 a	0.008 ± 0.001 a	0.009 ± 0.002 a
Dicaffeoylquinic acid 1	0.349 ± 0.023 c	0.225 ± 0.014 c	0.930 ± 0.337 ab	1.538 ± 0.244 a
Dicaffeoylquinic acid 2	2.412 ± 0.058 b	2.868 ± 0.018 a	n.d.	n.d.
Dicaffeoylquinic acid 3	0.149 ± 0.014 a	0.130 ± 0.013 a	0.237 ± 0.117 a	0.275 ± 0.015 a

Table 1. *Cont.*

Phenolic Compounds	*Solidago canadensis* Leaf		*Solidago canadensis* Flower	
	70% EtOH	80% MeOH	70% EtOH	80% MeOH
Dicaffeoylquinic acid 4	0.344 ± 0.042 [c]	0.463 ± 0.024 [c]	0.965 ± 0.124 [b]	1.247 ± 0.088 [a]
p-coumaric acid	0.486 ± 0.007 [a]	0.169 ± 0.006 [b]	0.124 ± 0.040 [bc]	0.069 ± 0.011 [c]
p-coumaric acid hexoside 1	0.261± 0.026 [a]	0.114 ± 0.013 [b]	0.069 ± 0.011 [c]	0.048 ± 0.008 [c]
p-coumaric acid hexoside 2	0.173 ± 0.010 [a]	0.188 ± 0.012 [a]	0.001 ± 0.000 [b]	0.001 ± 0.000 [b]
3-*p*-coumaroylquinic acid	1.694 ± 0.554 [a]	0.468 ± 0.080 [b]	0.752 ± 0.071 [b]	0.119 ± 0.024 [b]
4-*p*-coumaroylquinic acid 1	0.293 ± 0.018 [a]	0.088 ± 0014 [b]	n.d.	n.d.
5-*p*-coumaroylquinic acid 1	0.556 ± 0.036 [a]	0.572 ± 0.046 [a]	0.264 ± 0.062 [b]	0.200 ± 0.051 [b]
5-*p*-coumaroylquinic acid 2	0.290 ± 0.033 [a]	0.381 ± 0.012 [a]	0.273 ±0.138 [a]	0.234 ± 0.039 [a]
3-feruloylquinic acid	0.081 ± 0.014 [ab]	0.092 ± 0.019 [a]	0.034 ± 0.009 [c]	0.043 ± 0.011 [bc]
5-feruloylquinic acid 1	1.013 ± 0.095 [ab]	1.225 ± 0.060 [a]	0.805 ± 0.056 [bc]	0.750 ± 0.130 [c]
Ferulic acid	0.002 ± 0.000 [b]	0.002 ± 0.001 [b]	0.191 ± 0.045 [a]	0.145 ± 0.037 [a]
Hydroxycinnamic acid derivatives	**27.749 ± 1.081 [a]**	**28.594 ± 3.694 [a]**	**17.272 ± 1.480 [b]**	**20.498 ± 0.944 [b]**
p-hydroxybenzoic acid	0.591 ± 0.023 [a]	0.213 ± 0.027 [b]	n.d.	n.d.
Syringic acid	0.713 ± 0.021 [b]	0.417 ± 0.009 [c]	0.901 ± 0.002 [a]	0.224 ± 0.033 [d]
Protocatechuic acid	0.023 ± 0.002 [b]	0.021 ± 0.002 [b]	0.526 ± 0.024 [a]	0.535 ± 0.008 [a]
Hydroxybenzoic acid derivatives	**1.328 ± 0.020 [a]**	**0.645 ± 0.020 [c]**	**1.427 ± 0.023 [b]**	**0.759 ± 0.023 [d]**
Quercetin pentoside 1	0.045 ± 0.005 [a]	0.036 ± 0.003 [a]	0.041 ± 0.014 [a]	0.037 ± 0.001 [a]
Quercetin pentoside 2	0.324 ± 0.021 [a]	0.210 ± 0.013 [b]	0.056 ± 0.020 [c]	0.092 ± 0.015 [c]
Quercetin-3-rutinoside	14.465 ± 0.265 [bc]	11.317 ± 0.559 [c]	30.702 ± 4.158 [a]	19.653 ± 2.434 [b]
Quercetin-3-galactoside	0.816 ± 0.106 [a]	0.454 ± 0.072 [b]	0.734 ± 0.160 [ab]	0.699 ± 0.056 [ab]
Quercetin-3-glucoside	0.710 ± 0.018 [c]	0.684 ± 0.058 [c]	4.881 ± 0.442 [a]	4.020 ± 0.223 [b]
Quercetin-3-rhamnoside	0.127 ± 0.005 b	0.157 ± 0.016 a	n.d.	n.d.
Quercetin acetylhexoside 1	3.210 ± 0.258 [b]	3.957 ± 0.017 [a]	n.d.	n.d.
Quercetin pentosylhexoside	0.578 ± 0.090 [b]	0.252 ± 0.069 [b]	1.064 ± 0.171 [a]	1.242 ± 0.162 [a]
Isorhamnetin hexoside	0.827 ± 0.000 [a]	1.226 ± 0.677 [a]	0.382 ± 0.144 [b]	0.159 ± 0.018 [b]
Isorhamnetin pentosylhexoside	0.395 ± 0.044 [b]	0.438 ± 0.046 [b]	0.537 ± 0.137 [b]	1.210 ± 0.062 [a]
Isorhamnetin-3-rutinoside	3.765 ± 0.327 [b]	3.781 ± 0.502 [b]	6.034 ± 0.452 [a]	6.548 ± 0.393 [a]
Isorhamnetin acetylhexoside	1.489 ± 0.151 [b]	1.590 ± 0.128 [b]	2.436 ± 0.332 [a]	2.123 ± 0256 [a]
Kaempferol rhamnosylhexoside 1	2.275 ± 0.058 [a]	2.100 ± 0.058 [a]	1.514 ± 0.137 [b]	1.248 ± 0.069 [c]
Kaempferol-3-galactoside	0.749 ± 0.089 [a]	0.601 ± 0.051 [ab]	0.450 ± 0.157 [b]	0.408 ± 0.016 [b]
Kaempferol-3-rutinoside	7.921 ± 0.340 [a]	6.792 ± 0.465 [a]	5.411 ± 0.555 [b]	4.189 ± 0.254 [c]
Kaempferol-3-glucoside	2.120 ± 0.203 [a]	1.430 ± 0.207 [b]	0.937 ± 0.354 [bc]	0.391 ± 0.044 [c]
Kaempferol acetylhexoside 1	4.595 ± 0.306 [a]	4.277 ± 0.450 [a]	1.851 ± 0.102 [b]	1.982 ± 0.154 [b]
Flavonols	**43.580 ± 1.392 [b]**	**38.686 ± 1.699 [b]**	**57.028 ± 4.649 [a]**	**44.000 ± 3.774 [b]**
TOTAL	**73.063 ± 1.553 [ab*]**	**67.924 ± 3.701 [ab*]**	**75.723 ± 5.836 [a*]**	**65.257 ± 4.702 [b*]**

Leaf extracts exhibited the highest compound diversity within the hydroxycinnamic acids (HCA) group (21 compounds), followed by flavonols (17 compounds) (Table 1, Figure 3A). The flower extracts showed a similar distribution of phenolic groups; HCA was represented by 19 compounds, followed by flavonols, represented by 15 compounds (Table 1, Figure 3B).

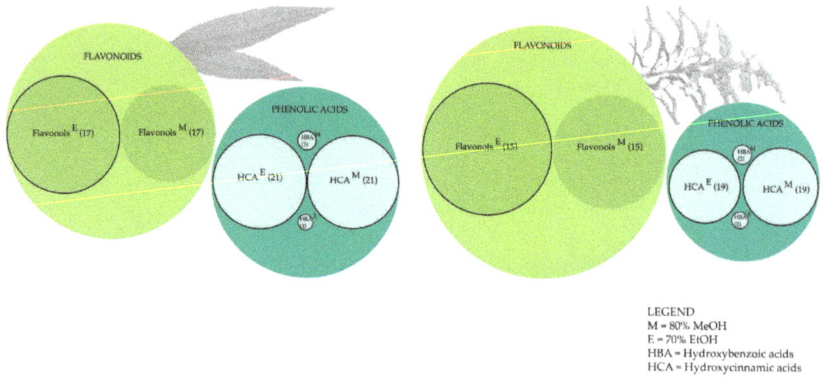

(**A**) (**B**)

Figure 3. The main phenolic groups obtained by the LC-DAD-MS method in the leaf (**A**) and flower (**B**) extracts. The ratio of the circles' sizes corresponds to the total concentration of individual phenolic groups (mg/g of dry weight (DW)), and the numbers in parentheses indicate the numbers of identified individual compounds within each phenolic group. The circle border indicates statistically significant differences between concentrations of individual phenolic groups in the plant part (two-way ANOVA, Tukey's test, $p \leq 0.01$).

These findings are consistent with the results reported in our recent study on *S. canadensis* phytochemicals, which provides a more detailed comparison of the phytochemical composition of this species and highlights hydroxycinnamic and hydroxybenzoic acids, as well as flavonoids, as the primary polyphenolic constituents [21]. As noted in several studies, hydroxycinnamic acids, particularly chlorogenic acid, are the most prevalent phenolic acids, with flavonols, including quercetin and kaempferol, being the dominant flavonoids found in various plant parts [6,31]. These compounds were also noted in the flowers and leaves of *S. canadensis*, with significant differences in the glycoside composition—rutin being the most abundant in flowers and hyperoside in leaves [34]. Our study similarly identifies a rich diversity of hydroxycinnamic acids in both leaves and flowers, supporting the conclusions of Woźniak et al. [31], who reported a high concentration of chlorogenic acid derivatives in aerial parts. Notably, we also observe that *S. canadensis* contains a broad spectrum of flavonoids, with flavonols being the predominant group, consistent with previous reports [6,17]. Additionally, our data on the comparison of different plant parts complements the findings by Woźniak et al. [31], where the underground parts were enriched with hydroxycinnamic acid conjugates, highlighting distinct biochemical profiles between the aerial and underground components of *S. canadensis*. These observations further underline the potential for *S. canadensis* to be explored for its diverse bioactive compounds in both ecological and pharmacological contexts. Five compounds were leaf-specific (dicaffeoylquinic acid 2, 4-*p*-coumaroylquinic acid 1, *p*-hydroxybenzoic acid, quercetin-3-rhamnoside, and quercetin acetylhexoside 1), while we found no flower-specific compounds. Our results differ from the results of Zekič et al. [35], who detected dicaffeoylquinic acid in the *S. canadensis* inflorescence from Slovenia.

Various factors significantly affected the content of different phenolic groups ($p \leq 0.01$), whereas for total phenolics, the significance was observed at the $p \leq 0.05$ level. The plant part had a significant impact on the HCA content, while both the solvent and plant part affected the total hydroxybenzoic acids (HBA) content. HCAs might be more concentrated in certain plant parts due to the specific biosynthetic pathways or storage mechanisms in those parts. The higher concentration of HCAs in leaves compared to flowers can be explained by several factors; leaves are often more exposed to herbivores, pathogens, and environmental stress (like UV radiation) than flowers. HCAs, which have antioxidant, antimicrobial, and UV-absorbing properties, play a crucial role in protecting leaves from these threats. Plants may, therefore, synthesize and accumulate more HCAs in leaves as a protective measure [36]. Leaves are the primary site of photosynthesis, a process that generates reactive oxygen species (ROS) as byproducts. With their antioxidant properties, HCAs help neutralize ROS, protecting the leaf tissues from oxidative damage. The high metabolic activity in leaves may necessitate higher levels of HCAs [37]. The biosynthesis of HCAs occurs through the phenylpropanoid pathway, which is active in many plant parts but is particularly prominent in leaves due to their role in photosynthesis and defense. This pathway is closely linked to the production of lignin, flavonoids, and other phenolics, which are often more concentrated in leaves [38,39]. Both the plant part and the solvent significantly influenced the flavonol content (two-way ANOVA and Tukey's test at p-value ≤ 0.01). One of the reasons is assumed to be their structure. Namely, flavonols (a type of flavonoid) are more structurally complex than HCAs and HBAs. They contain multiple hydroxyl groups attached to a polycyclic ring system, making them more reactive to changes in extraction conditions, such as the solvent used and the plant part. They often bind to sugars, proteins, or other cell wall components, forming complex molecules such as glycosides. The ability of a solvent to break these bonds can vary significantly, affecting how much flavonol is extracted from different plant parts [36,40]. In our case, the influence of the solvent on the flavonoid content was significant only in the flower extracts, with deviations shown for individual compounds. HCAs and HBAs tend to be less bound to complex matrices and are more likely to exist in free forms or as esters that are easier to extract. As a result, solvent choice may not have as significant an impact on their extraction [41]. Flavonols are often unevenly distributed among different plant parts. For instance, leaves tend to have higher concentrations due to their role in protecting the plant from UV radiation and oxidative stress. This variation in distribution among plant parts means the part chosen for extraction can significantly impact the flavonol content [42,43]. HCAs and HBAs are more evenly distributed across various plant tissues, being involved in general plant metabolism and cell wall structure. Therefore, the specific plant part used for extraction might not result in such pronounced differences in HCA or HBA content compared to flavonols [39]. Flavonols are part of the flavonoid biosynthesis pathway, which can be more sensitive to environmental factors such as light exposure, stress, and plant part-specific metabolic activity. This means that the location within the plant and the extraction conditions can have a more significant impact on flavonol levels [44,45]. HCAs and HBAs, on the other hand, are produced through the shikimate pathway, which tends to be more stable across different plant parts and less influenced by environmental conditions, leading to relatively consistent levels across plant parts and solvents [39,46].

In the leaf extracts, 5-caffeoylquinic acid 1 was the major compound in both ethanolic and methanolic extracts (17.75 mg/g DW and 20.42 mg/g DW, respectively) (Table 1). The 5-*O*-caffeoylquinic acid (5-CQA) (Figure 4), also known as chlorogenic acid, is one of the major chlorogenic acids present in many fruits, vegetables, and herbs [47]. This compound exhibits antioxidant activity against oxidative stress-mediated liver injury, as well as anti-inflammatory and antimicrobial capacity [47,48]. These properties make 5-CQA potentially useful as a preservative in the food, pharmaceutical, and cosmetic industries. Marksa et al. [18] discovered that chlorogenic acid was the critical component responsible for the antioxidant properties in the leaves and inflorescences of the sibling species *S. gigantea*. In contrast, in *S. canadensis*, the primary antioxidant component was 3,5-dicaffeoylquinic acid. Their study revealed that di-caffeoylquinic acids have a stronger radical scavenging effect than mono-caffeoylquinic acids. Similar major compounds, including chlorogenic acid, quercitrin, and rutin, were identified in 70% ethanolic extracts of the leaves and inflorescences of *S. canadensis* [49]. Chlorogenic acid, quercetin, and kaempferol rutinosides were identified as the main compounds in both aerial and underground parts in 70% methanol *S. canadensis* and *S. gigantea* extracts [31]. Furthermore, Shelepova et al. [25] identified phenolic acids such as chlorogenic, caffeic, and ferulic acids as the main components in the *S. canadensis* flower extracts.

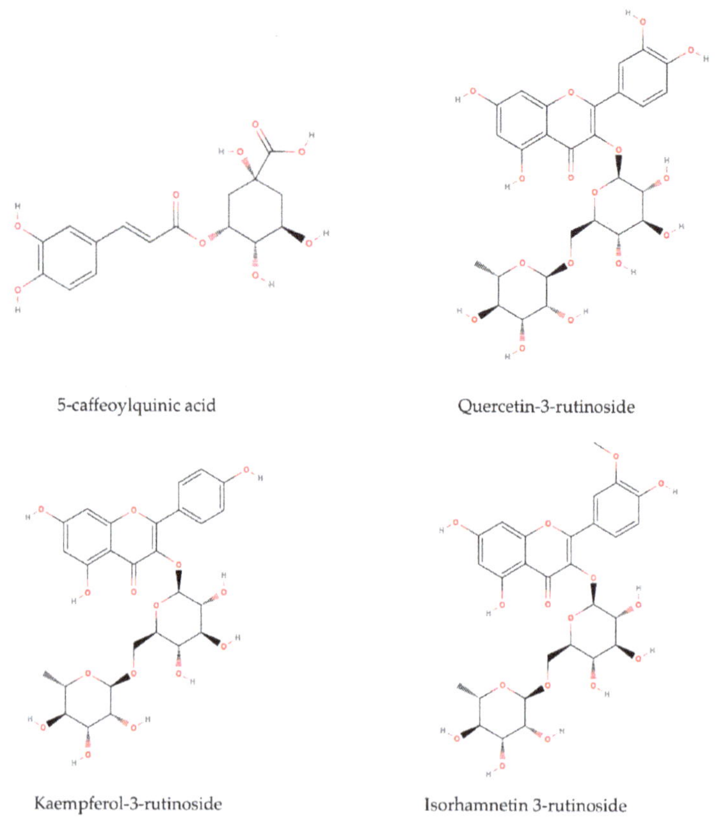

Figure 4. The structural molecules of four main compounds found in leaf and flower extracts in *S. canadensis* (MolView. Available online: https://molview.org/, accessed on 11 August 2024).

Quercetin-3-rutinoside, also known as rutin (Figure 4), was the primary flavonoid identified in flower *S. canadensis* extracts, with concentrations of 30.70 mg/g DW in 70% ethanol and 19.65 mg/g DW in 80% methanol (Table 1). These findings align with those of Zekič et al. [35], who reported a rutin content of 27.62 ± 0.45 mg/g DW in 70% methanol extracts of inflorescences. Furthermore, Shelepova et al. [17] identified rutin as a major compound in aerial parts of *S. canadensis* extracted with methanol, ethanol, and acetone (200.45–211.20 mg/g). The solubility of flavonoids and organic acids in alcoholic extracts is primarily influenced by solvent polarity. Apati et al. [50] demonstrated that 70% ethanol yielded the highest concentration of this compound at 572.5 mg/L. Quercetin-3-rutinoside showed significant antibacterial activity and can be used as a natural antibiotic to treat different infectious diseases [51]. Quercetin glycosides, such as rutin, can interact with ammonium while entering the rhizosphere through plant exudates. As a result, new compounds, such as the phenol-ammonia complex, are formed. At low concentrations (20 µg/mL), these substances promote the development of lateral and adventitious roots in plants, and in concentrations higher than 100 μg/mL, inhibit them [52]. The goldenrods are also known for the rapid uptake of NO_3–N and phosphorus, which can halt the development of co-occurring plants [53].

Furthermore, the flower extract was rich in flavonoid isorhamnetin-3-rutinoside, known as narcissin (Figure 4). It is naturally synthesized in plants via the phenylpropanoid pathway, and its production is triggered by environmental stressors such as UV radiation [54]. Studies have shown that isorhamnetin exhibits a wide range of pharmacological effects on cardiovascular diseases [55] and various types of tumors [56]. Additionally, it holds the potential to prevent neurodegenerative diseases such as Alzheimer's disease [57]. This compound has garnered significant attention recently due to its widespread availability, affordability, high effectiveness, low toxicity, and minimal side effects [58].

Kaempferol-3-rutinoside, also known as nicotiflorin (Figure 4), was identified in both leaf and flower extracts. The measured concentrations were 7.92 mg/g DW in ethanolic and 6.79 mg/g DW in methanolic leaf extracts, while in flower extracts, the values were 5.41 mg/g DW for ethanolic and 4.19 mg/g DW for methanolic extracts. Kaempferol is a low molecular weight flavonoid that plants use to stimulate and regulate their growth as well as for defense purposes [59]. The bioassay results demonstrated that flavonoid nicotiflorin has a neuroprotective effect [60], protective effects on reducing memory dysfunction [61], and antioxidant activities [62]. It is often referred to as a compound with multifaceted therapeutic potential, known for its anti-inflammatory properties, hepatoprotective effects, anti-cancer activity, wound healing capabilities, and cardioprotective benefits [63].

Furthermore, leaf (0.816 ± 0.106 and 0.454 ± 0.072 mg/g DW) and flower (0.734 ± 0.160 and 0.699 ± 0.056 mg/g DW) samples in both solvents contained quercetin-3-galactoside, known as hyperoside. This compound is recognized by the European Pharmacopoeia as a standard used for calculating flavonoid content in *Solidaginis* herba. Similarly, in the research of Avertseva et al. [34], this compound was predominant in leaves in 50% ethanol (8.39 ± 0.60 mg/g). The difference in content can be explained by the difficulty of extracting this compound due to its chemical complexity, interactions with other plant matrix components, solubility challenges, and sensitivity to environmental conditions [64].

During the preparation of leaf and flower samples, foam formation was observed, suggesting the potential presence of saponin compounds. The extraction of saponins can be challenging due to their high polarity and molecular weight [65]. Given the established abundance of saponins in *S. canadensis* extracts [31,66] and their recognized role in plant defense against herbivory and pathogens [67], further investigation into the isolation and characterization of these compounds from this species is needed.

Given the pharmaceutical industry's ongoing search for novel phytochemicals, *S. canadensis* emerges as a promising candidate. The plant's substantial antioxidant capacity provides a strong foundation for developing high-value phytopharmaceutical products. Moreover, its widespread distribution across Europe presents a considerable resource for exploitation [68]. Phenolic compounds and essential oils are recognized for their antioxidant, antimicrobial, and antifungal properties, making them key constituents in developing phytotherapeutic drugs for chronic disease management [69,70]. The leaf extracts of *S. canadensis* were declared to have promising potential in the green synthesis of gold nanoparticles used in medicine [71]. Mariychuk et al. [71] showed that the extract of *S. canadensis* with its specialized metabolites can act both as a reducing agent and as a stabilizing agent for noble metal nanoparticles.

Most studies on *S. canadensis* focus on the phytochemical content or biological activity of its underground and aerial parts. Our study provides a novel comparison of leaves and flowers (separated from the inflorescence) in terms of phenolic content and antioxidant activity, representing the first targeted analysis of this kind. Our study reveals significant differences in phytochemical profiles between flowers and leaves, providing new insights into the species' biological activity.

We present the site-specific phenolic profile of this invasive species in the context of the particular potential use of this species. By assessing the antioxidant capacity of extracts from different plant parts and evaluating various extraction solvents, our findings establish a basis for future research into its potential applications locally and beyond.

3. Materials and Methods

3.1. Plant Material

The leaves and inflorescences of *S. canadensis* were collected during the vegetation year 2021, from June to September, in the Istria region (Croatia). The 15 samples were gathered from three distinct locations (5 per each location), spanning latitudes from N 45.4073056 to N 44.8461944. The plant material underwent air-drying in the dark at room temperature after harvest. Before the grinding and extraction process, we separated the individual flowers from the larger clusters, known as inflorescences. To ensure clarity, we used the term "flower" extracts throughout this paper.

3.2. Extraction Procedure

The dry plant material of each sample (250 g) was pooled to create a representative sample for the study area. The material was minced using the Grindomix GM 200 knife mill, programmed at 10,000× /30 s (Retsch, Haan, Germany).

To spectrophotometrically determine phenolic content, extracts were prepared in quadruplicate in three repetitions following a standardized protocol of Bilić et al. [72]. Within the tubes, 0.06 g of plant material was dissolved in 2 mL of solvent (70% EtOH and 80% MeOH). These prepared solutions were sonicated for 30 min in an ultrasonic bath (40 kHz, 300 W ultrasound power, 400 W heater power, Holon, Israel), followed by centrifugation at 12,000 rpm/10 min (Jouan MR23i, Jouan S.A., Saint-Herblain, France) and filtration through 0.20 µm polytetrafluoroethylene filters (Macherey-Nagel, Düren, Germany) before being stored at +4 °C until analysis.

3.3. HPLC-DAD-MS Analysis of Phenolic Compounds in Leaf and Flower Extracts

The extraction of phenolic compounds for identification via HPLC-DAD-MS was performed following the protocol of Mikulič-Petkovšek et al. [73]. The 0.2 g of dried plant tissue was extracted with 6 mL of 70% EtOH and 80% MeOH containing 3% (*v*/*v*) formic acid in a cooled ultrasonic bath for 60 min. Extracts were centrifuged for 10 min at 10,000× *g* and filtered through 20 µm polytetrafluoroethylene (PTFE) filters (Macherey-Nagel, Düren, Germany). All extracts were subjected to LC-DAD-MS analysis to identify and quantify specific phenolic compounds. Analyte separation was performed using HPLC (Dionex UltiMate 3000, Thermo Fisher Scientific, San Jose, CA, USA) with a DAD detector, maintaining a column temperature (Gemini C18, Phenomenex, Torrance, CA, USA) at 25 °C. Compounds were detected at wavelengths of 280 and 350 nm. Two mobile phases were used for the separation of phenolic compounds: mobile phase A consisted of double-distilled water/acetonitrile/formic acid (96.9/3/0.1, *v*/*v*/*v*), and mobile phase B was double-distilled water/acetonitrile/formic acid (3/96.9/0.1, *v*/*v*/*v*). The elution followed a linear gradient: from 5% to 20% B over the first 15 min, from 20% to 30% B in the next 5 min, then an isocratic mixture for 5 min, followed by a gradient from 30% to 90% B in 5 min, and an isocratic mixture for 15 min before reverting to initial conditions, as per the method described by Mikulic-Petkovsek et al. [74].

The sample injection volume was 20 µL, with a mobile phase flow rate of 0.6 mL/min. Individual metabolites were identified using mass spectrometry (LTQ XL Linear Ion Trap

Mass Spectrometer, Thermo Fisher Scientific, San Jose, CA, USA) with electrospray ionization (ESI) in a negative scanning mode, following modified parameters from the study of Mikulic-Petkovsek et al. [73]. The scanning range was from m/z 110 to 1700—a data-dependent full scan. Phenolic compounds were confirmed based on fragmentation products, retention times of corresponding standards, and the spectral comparison of individual peaks with those of the standards. The share of each phenolic compound was determined by analyzing the peak areas of the samples against the corresponding standard curves and was expressed in mg/g DW. The following external standards were used: caffeic acid, apigenin-7-glucoside, ferulic acid, quercetin-3-O-rhamnoside, neochlorogenic (3-caffeoylquinic) acid, naringenin, ellagic acid, gallic acid, chlorogenic acid, and rutin (quercetin-3-O-rutinoside); (-)epicatechin, quercetin-3-O-galactoside, quercetin-3-O-glucoside, *p*-coumaric acid, procyanidin B1, and kaempferol-O-glucoside; quercetin-3-O-xyloside and quercetin-3-O-arabinopyranoside; and isorhamnetin-3-O-glucoside.

3.4. Total Phenolic, Flavonoid, and Non-Flavonoid Content and Antioxidant Capacity

Total phenolics (TP) were determined using the method of Singleton and Rossi [75], while total non-flavonoids (TNF) were measured using the method described by Ough and Amerine [76]. Both methods rely on the reduction in the Folin–Ciocalteu (FC) reagent in the presence of phenolics, resulting in the formation of molybdenum-tungsten blue, which is then measured spectrophotometrically at 765 nm. In the TP assay, 0.1 mL of extract was mixed with 1.5 mL of distilled water and 0.1 mL of the Folin–Ciocalteu (FC) reagent. The reaction was allowed to proceed for 5 min at room temperature. Following this, 1.5 mL of 20% sodium carbonate solution was added to the mixture, and the reaction was allowed to develop for 30 min at room temperature. The formation of the molybdenum-tungsten blue complex was measured spectrophotometrically at 765 nm. In the TNF assay, a 0.1 mL sample of the extract was mixed with 1.5 mL of distilled water and 0.1 mL of the FC reagent. The reaction was allowed to develop for 5 min at room temperature. Following this, 1.5 mL of 20% sodium carbonate solution was added and incubated for 30 min at room temperature. The resulting blue complex was quantified spectrophotometrically at 765 nm. The results were calculated according to the calibration curve for gallic acid ($y = 0.004x$, $R^2 = 0.991$ for 70% ethanol extracts and $y = 0.009x$, $R^2 = 0.996$ for 80% methanol extracts, where y is the absorbance at 765 nm and x is the concentration of gallic acid mg/L) and expressed as mg of gallic acid equivalents (GAE) per g of dry weight (DW).

Total flavonoid (TF) content was measured following the protocol described by Martins et al. [77]. In the TF assay, to a 1 mL sample of the extract, 1 mL of 2% aluminum chloride ($AlCl_3$) in ethanol was added. After 15 min of incubation at room temperature, the absorbance of the reaction mixture was measured spectrophotometrically at 430 nm. The results were calculated according to the calibration curve for catechin ($y = 0.0024x$, $R^2 = 0.993$ for 70% ethanol extracts and $y = 0.0022x$, $R^2 = 0.994$ for 80% methanol extracts, where y is the absorbance at 765 nm and x is the concentration of gallic acid mg/L) and expressed as mg of (+)-catechin equivalents (CE) per g of dry weight (DW).

The antioxidant capacity (AC) of the extracts was determined spectrophotometrically using standard DPPH, ABTS, and FRAP assays following the methods described in the study of Poljuha et al. [78]. The DPPH assay was performed by mixing 0.1 mL of the extract with 3.9 mL of DPPH radical solution (0.1 mM in methanol). After 30 min of incubation in the dark at room temperature, the absorbance of the mixture was measured at 517 nm. The results of DPPH analysis were calculated against a Trolox calibration curve ($y = 44.991x$, $R^2 = 0.972$ for 70% ethanol extracts; $y = 47.786x$, $R^2 = 0.992$ for 80% methanol extracts) and expressed as mg of Trolox equivalents (TE) per g of dry weight (mg TE/g DW). The FRAP assay was carried out by mixing 0.1 mL of extract with 3 mL of FRAP reagent (containing

10 mM 2,4,6-tripyridyl-s-triazine in 40 mM HCl, 20 mM $FeCl_3$, and 300 mM acetate buffer, pH 3.6). The mixture was incubated for 30 min at room temperature, and the absorbance was measured at 593 nm. FRAP values were calculated using a Trolox calibration curve (y = 1.638x, R^2 = 0.993 for 70% ethanol extracts; y = 1.586, R^2 = 0.991 for 80% methanol extracts) and as mg of Trolox equivalents (TE) per g of dry weight (mg TE/g DW). For the ABTS assay, the ABTS radical cation was generated by reacting ABTS solution (7 mM) with potassium persulfate (2.45 mM) and incubating the mixture in the dark for 12 h at room temperature. The radical cation was then diluted with ethanol to obtain an absorbance of 0.70 ± 0.02 at 734 nm. A 100 μL aliquot of the extract was mixed with 3.9 mL of the ABTS solution, and after 6 min, the absorbance was measured at 734 nm. ABTS values were calculated using a Trolox calibration curve (y = 46.137x, R^2 = 0.981 for 70% ethanol extracts; y = 41.432, R^2 = 0.968 for 80% methanol extracts) and as mg of Trolox equivalents (TE) per g of dry weight (mg TE/g DW).

All measurements were performed in triplicate using a NanoPhotometer P300 spectrophotometer (Implen GmbH, München, Germany) adjusted to a 2 mL cuvette volume.

3.5. *Statistical Analysis*

Two-way analysis of variance (ANOVA) with post hoc Tukey's test was conducted to determine the significance of the factors and to assess the significance of differences between the extracts ($p \leq 0.05$ and 0.01). Pearson's correlation coefficients were calculated to assess the interaction between bio-compounds and antioxidant capacity. Data were statistically analyzed using the software Statgraphics Plus 4.0 (Manugistics, Inc., Rockville, MD, USA) and IBM SPSS 23 (Chicago, IL, USA). Visualization (Figures 1–3) was generated using Flourish Studio 1.0.0.3 (Canva, Sidney, Australia) and Sketch version 101 (Eindhoven, The Netherlands).

4. Conclusions

This study presents a comprehensive site-specific analysis of the phenolic profile and antioxidant capacity of *Solidago canadensis* leaf and flower extracts. The results reveal that leaf extracts exhibit higher diversity in hydroxycinnamic acids (21 compounds) and flavonols (17 compounds), while flower extracts demonstrate a similar distribution, with 19 hydroxycinnamic acids and 15 flavonols identified. Key compounds include 5-caffeoylquinic acid, quercetin-3-rutinoside, kaempferol-3-rutinoside, and isorhamnetin-3-rutinoside, with notable organ-specific variations in concentration. Ethanolic extracts were more efficient than methanolic extracts in extracting total phenolics (TP) and non-flavonoids (TNF), demonstrating higher yields across both plant parts. Antioxidant capacity, assessed through ABTS, DPPH, and FRAP assays, was higher in flowers than leaves, suggesting significant plant part-specific antioxidant potential. Strong positive correlations were observed between TP, TNF, and antioxidant capacities, further validating the contribution of these compounds to the biological activity of *S. canadensis*.

The findings underscore the potential of S. *canadensis* as a valuable source of bioactive phenolics, particularly for phytopharmaceutical applications. Moreover, the differentiation in phenolic composition between plant parts offers insights into optimizing extraction for specific bioactive compounds. These results open opportunities for further research into using invasive species as a source of antioxidants and therapeutic agents, offering a balance between ecological management and potential pharmaceutical applications.

Supplementary Materials: The following supporting information can be downloaded at https://www.mdpi.com/article/10.3390/plants14010044/s1, Table S1: The concentrations of total phenolics (TP), total non-flavonoids (TNF), total flavonoids (TF), and antioxidant capacity (obtained by DPPH, ABTS, and FRAP assays) in *Solidago canadensis* L. leaf and flower extracts in two solvents; Table S2:

Pearson's correlation coefficients (two-tailed) between total phenolic (TP), total non-flavonoids (TNF), and total flavonoids (TF) contents and antioxidant capacity (obtained by DPPH, ABTS, and FRAP assays).

Author Contributions: Conceptualization, D.P. and B.S.; methodology, D.P., M.M.-P., I.Š., J.B. and S.D.; validation, S.D. and M.M.-P.; formal analysis, M.U.B., I.Š., J.B., S.D. and M.M.-P.; investigation, M.U.B., B.S. and D.P.; resources, D.P., M.M.-P. and J.B.; data curation, M.U.B., S.D., D.P., J.B. and B.S.; writing—original draft preparation, M.U.B., D.P. and I.Š.; writing—review and editing, M.U.B., D.P., I.Š., B.S., J.B., M.M.-P. and S.D.; visualization, D.P. and M.U.B.; project administration, D.P.; funding acquisition, D.P. All authors have read and agreed to the published version of the manuscript.

Funding: This research was funded by the Croatian Science Foundation under the project "NATURe as an ALLY: Alien Invasive Plants as Phytopharmaceuticals—NATURALLY" (IP-2020-02-6899) and by the Slovenian Research and Innovation Agency (ARIS) under the Horticulture program No. P4-0013-0481. Mirela Uzelac Božac's work was supported by the Croatian Science Foundation "Young Researchers' Career Development Project–Training New Doctoral Students" (DOK-2021-02-3094).

Data Availability Statement: Data are presented in the manuscript.

Acknowledgments: We especially thank Slavko Brana and the Istrian Botanical Society for the help in the fieldwork and availability.

Conflicts of Interest: The authors declare no conflicts of interest. The funders had no role in the design of the study; in the collection, analyses, or interpretation of data; in the writing of the manuscript; or in the decision to publish the results.

References

1. Fotsing, Y.S.F.; Kezetas, J.J.B.; El-Saber, B.G.; Ali, I.; Ndjakou, B.L. Extraction of Bioactive Compounds from Medicinal Plants and Herbs. In *Natural Medicinal Plants*; El-Shemy, H., Ed.; IntechOpen: London, UK, 2022. [CrossRef]
2. Li, A.N.; Li, S.; Zhang, Y.J.; Xu, X.R.; Chen, Y.M.; Li, H.B. Resources and Biological Activities of Natural Polyphenols. *Nutrients* **2014**, *6*, 6020–6047. [CrossRef] [PubMed]
3. Corso, M.; Perreau, F.; Mouille, G.; Lepiniec, L. Specialized Phenolic Compounds in Seeds: Structures, Functions, and Regulations. *Plant Sci.* **2020**, *296*, 110471. [CrossRef] [PubMed]
4. Wahle, K.W.; Brown, I.; Rotondo, D.; Heys, S.D. Plant Phenolics in The Prevention and Treatment of Cancer. *Adv. Exp. Med. Biol.* **2010**, *698*, 36–51. [CrossRef] [PubMed]
5. Kim, Y.O.; Lee, E.J. Comparison Of Phenolic Compounds And The Effects of Invasive and Native Species in East Asia: Support for The Novel Weapons Hypothesis. *Ecol. Res.* **2011**, *26*, 87–94. [CrossRef]
6. Kato-Noguchi, H.; Kato, M. Allelopathy and Allelochemicals of *Solidago canadensis* L. and *S. altissima* L. for Their Naturalization. *Plants* **2022**, *11*, 3235. [CrossRef]
7. Abhilasha, D.; Quintana, N.; Vivanco, J.; Joshi, J. Do Allelopathic Compounds in Invasive *Solidago canadensis* s.l. Restrain the Native European Flora? *J. Ecol.* **2008**, *96*, 993–1001. [CrossRef]
8. Batish, D.R.; Singh, H.P.; Priyanka, S.K.; Kohli, R.K. Novel Weapon Hypothesis for the Successful Establishment of Invasive Plants in Alien Environments. In *Invasive Plant Ecology*; Jose, S., Singh, H.P., Batish, D.R., Kohli, R.K., Eds.; Taylor and Francis Group: Abingdon, UK, 2013; pp. 19–28.
9. Weber, E. The Dynamics of Plant Invasions: A Case Study of Three Exotic Goldenrod Species (*Solidago* L.) in Europe. *J. Biogeogr.* **1998**, *25*, 147–154. [CrossRef]
10. Semple, J.C.; Beck, J.B. Revised Infrageneric Classification of *Solidago* (Asteraceae ASTEREAE). *Phytoneuron* **2021**, *10*, 1–60.
11. EPPO Global Database. *Solidago canadensis*. Available online: https://gd.eppo.int/taxon/SOOCA/distribution (accessed on 7 October 2024).
12. Oitsius, L.V.; Volovyk, H.P.; Doletskyi;, S.P.; Lysytsya, A.V. Distribution of Adventive Species *Solidago canadensis*, *Phalacroloma annuum*, *Ambrosia artemisiifolia*, *Heracleum sosnowskyi* in Phytocenoses of Volyn' Polissya (Ukraine). *Biosyst. Divers.* **2020**, *28*, 343–349. [CrossRef]
13. Bielecka, A.; Borkowska, L.; Królak, E. Environmental Changes Caused by the Clonal Invasive Plant *Solidago canadensis*. *Ann. Bot. Fenn.* **2020**, *57*, 33. [CrossRef]
14. Li, Z.H.; Wang, Q.; Ruan, X.; Pan, C.D.; Jiang, D.A. Phenolics and Plant Allelopathy. *Molecules* **2010**, *15*, 8933–8952. [CrossRef] [PubMed]

15. Rosell, R.; Dayan, F.; Nanayakkara, N.; Romagni, J. *Ecophysiology and Potential Modes of Action for Selected Lichen Secondary Metabolites, In Allelopathy—Chemistry and Mode of Action of Allelochemicals*; Macias, F.A., Galindo, J.C.G., Molinillo, J.M.G., Eds.; CRC Press: Boca Raton, FL, USA, 2003. [CrossRef]
16. Apáti, P.; Szentmihályi, K.; Kristó, S.T.; Papp, I.; Vinkler, P.; Szoke, É.; Kéry, Á. Herbal Remedies of Solidago—Correlation of Phytochemical Characteristics and Antioxidative Properties. *J. Pharm. Biomed. Anal.* **2003**, *32*, 1045–1053. [CrossRef] [PubMed]
17. Shelepova, O.; Vinogradova, Y.; Vergun, O.; Grygorieva, O.; Brindza, J. Assessment of Flavonoids and Phenolic Compound Accumulation in Invasive *Solidago canadensis* L. in Slovakia. *Potravin. Slovak J. Food Sci.* **2020**, *14*, 587–594. [CrossRef] [PubMed]
18. Marksa, M.; Zymone, K.; Ivanauskas, L.; Radušienė, J.; Pukalskas, A.; Raudone, L. Antioxidant Profiles of Leaves and Inflorescences of Native, Invasive and Hybrid *Solidago* Species. *Ind. Crops Prod.* **2020**, *145*, 112123. [CrossRef]
19. Deng, Y.; Zhao, Y.; Padilla-Zakour, O.; Yang, G. Polyphenols, Antioxidant and Antimicrobial Activities of Leaf and Bark Extracts of *Solidago canadensis* L. *Ind. Crop Prod.* **2015**, *74*, 803–809. [CrossRef]
20. Quinty, V.; Nasreddine, R.; Colas, C.; Launay, A.; Nehmé, R.; El-Khiraoui, A.; Piot, C.; Draye, M.; Destandau, E.; Da Silva, D.; et al. Antioxidant and Anti-Lipase Capacities from the Extracts Obtained from Two Invasive Plants: *Ambrosia artemisiifolia* and *Solidago canadensis*. *Food Biosci.* **2023**, *55*, 103069. [CrossRef]
21. Poljuha, D.; Sladonja, B.; Uzelac Božac, M.; Šola, I.; Damijanić, D.; Weber, T. The Invasive Alien Plant *Solidago canadensis*: Phytochemical Composition, Ecosystem Service Potential, and Application in Bioeconomy. *Plants* **2024**, *13*, 1745. [CrossRef]
22. Pietta, P.; Minoggio, M.; Bramati, L. Plant Polyphenols: Structure, Occurrence and Bioactivity. Stud. *Nat. Prod. Chem.* **2003**, *28*, 257–312. [CrossRef]
23. Rhazi, N.; Oumam, M.; Hannache, H.; Sesbou, A.; Charrier, B.; Pizzi, A.; Charrier—El Bouhtoury, F. Comparison of the impact of different extraction methods on polyphenols yields and tannins extracted from Moroccan *Acacia mollissima* barks. *Ind. Crops Prod.* **2015**, *70*, 245–252. [CrossRef]
24. Du, L.; Liu, H.; Yan, M.; Li, J.; Li, J. Individual Plasticity of the Shade Response of the Invasive *Solidago canadensis* in China. *PLoS ONE* **2017**, *12*, e0170049. [CrossRef]
25. Shelepova, O.; Vinogradova, Y. Phytochemistry and Inflorescences Morphometry of Invasive *Solidago* L. (Goldenrods) Species—Valuable Late Autumn Mellifers. *Agrobiodiver. Improv. Nutr. Health Life Qual.* **2021**, *5*, 209–214. [CrossRef]
26. Rosłon, W.; Osińska, E.; Mazur, K.; Geszprych, A. Chemical Characteristics of European Goldenrod (*Solidago virgaurea* L. subsp. *virgaurea*) From Natural Sites In Central And Eastern Poland. *Acta Sci. Pol. Hortorum Cultus* **2014**, *13*, 55–65.
27. Karlová, K. Accumulation of Flavonoid Compounds in Flowering Shoots of *Achillea collina* Becker ex. Rchb. Alba During Flower Development. *Hortic. Sci.* **2006**, *33*, 158–162. [CrossRef]
28. Stevenson, F.J.; Nommik, H.; Vahtras, K. Retention and Fixation of Ammonium and Ammonia in Soils. *Agron. Monogr.* **1982**, *22*, 123–171. [CrossRef]
29. Suleymanova, F.; Nesterova, O.; Matyushin, A. HPLC Quantification of Hydroxycinnamic and Organic Acids of Canadian Goldenrod (*Solidago canadensis* L.). *Pharmacogn. J.* **2019**, *11*, 400–404. [CrossRef]
30. Vinogradova, Y.; Vergun, O.; Grygorieva, O.; Brindza, J. Antioxidant Activity of *Solidago* L. Complex. *Acta Hortic.* **2021**, *1324*, 373–380. [CrossRef]
31. Woźniak, D.; Ślusarczyk, S.; Domaradzki, K.; Dryś, A.; Matkowski, A. Comparison of Polyphenol Profile and Antimutagenic and Antioxidant Activities in Two Species Used as Source of *Solidaginis herba*—Goldenrod. *Chem. Biodivers.* **2018**, *15*, e1800023. [CrossRef]
32. Radušienė, J.; Karpavičienė, B.; Vilkickytė, G.; Marksa, M.; Raudonė, L. Comparative Analysis of Root Phenolic Profiles and Antioxidant Activity of Five Native and Invasive *Solidago* L. Species. *Plants* **2024**, *13*, 132. [CrossRef]
33. Munteanu, I.G.; Apetrei, C. Analytical Methods Used in Determining Antioxidant Activity: A Review. *Int. J. Mol. Sci.* **2021**, *22*, 3380. [CrossRef]
34. Avertseva, I.N.; Suleymanova, F.S.; Nesterova, O.V.; Reshetnyak, V.Y.; Matveenko, V.N.; Zhukov, P.A. Study of Polyphenolic Compounds in Extracts from Flowers and Leaves of Canadian Goldenrod and Dwarf Goldenrod (*Solidago canadensis* L. and *Solidago nana* Nitt.). *Mosc. Univ. Chem. Bull.* **2020**, *75*, 47–51. [CrossRef]
35. Zekič, J.; Vovk, I.; Glavnik, V. Extraction and Analysis of Flavonoids and Phenolic Acids from Canadian Goldenrod and Giant Goldenrod. *Forests* **2021**, *12*, 40. [CrossRef]
36. Liu, S.; Jiang, J.; Ma, Z.; Xiao, M.; Yang, L.; Tian, B.; Yu, Y.; Bi, C.; Fang, A.; Yang, Y. The Role of Hydroxycinnamic Acid Amide Pathway in Plant Immunity. *Front. Plant Sci.* **2022**, *13*, 922119. [CrossRef] [PubMed]
37. Cheynier, V.; Comte, G.; Davies, K.M.; Lattanzio, V.; Martens, S. Plant Phenolics: Recent Advances on Their Biosynthesis, Genetics, and Ecophysiology. *Plant Physiol. Biochem.* **2013**, *72*, 1–20. [CrossRef] [PubMed]
38. Grace, S.C.; Logan, B.A. Energy Dissipation And Radical Scavenging By The Plant Phenylpropanoid Pathway. *Philos. Trans. R. Soc. Lond. B Biol. Sci.* **2000**, *355*, 1499–1510. [CrossRef]
39. Vogt, T. Phenylpropanoid Biosynthesis. *Mol. Plant* **2010**, *3*, 2–20. [CrossRef]

40. Crozier, A.; Jaganath, I.B.; Clifford, M.N. Dietary Phenolics: Chemistry, Bioavailability And Effects on Health. *Nat. Prod. Rep.* **2009**, *26*, 1001–1043. [CrossRef]
41. Manach, C.; Scalbert, A.; Morand, C.; Rémésy, C.; Jiménez, L. Polyphenols: Food Sources And Bioavailability. *Am. J. Clin. Nutr.* **2004**, *79*, 727–747. [CrossRef]
42. Agati, G.; Azzarello, E.; Pollastri, S.; Tattini, M. Flavonoids as Antioxidants in Plants: Location and Functional Significance. *Plant Sci.* **2012**, *196*, 67–76. [CrossRef]
43. Treutter, D. Significance of Flavonoids in Plant Resistance and Enhancement of Their Biosynthesis. *Plant Biol.* **2005**, *7*, 581–591. [CrossRef]
44. Winkel-Shirley, B. Biosynthesis of Flavonoids and Effects of Stress. *Curr. Opin. Plant Biol.* **2002**, *5*, 218–223. [CrossRef]
45. Agati, G.; Tattini, M. Multiple Functional Roles of Flavonoids in Photoprotection. *New Phytol.* **2010**, *186*, 786–793. [CrossRef] [PubMed]
46. Herrmann, M.K.; Weaver, M.L. The Shikimate Pathway. *Annu. Rev. Physiol. Plant Mol. Biol.* **1999**, *50*, 473–503. [CrossRef] [PubMed]
47. Bajko, E.; Kalinowska, M.; Borowski, P.; Siergiejczyk, L.; Lewandowski, W. 5-O-Caffeoylquinic Acid: A Spectroscopic Study And Biological Screening For Antimicrobial Activity. *LWT—Food Sci. Technol.* **2015**, *65*, 471–479. [CrossRef]
48. Ruano González, A.; Pinto, A.; Chinchilla, N.; Palma, M.; Barbero, G.; Carrera, C.; Vázquez Espinosa, M. Determination of Caffeoylquinic Acids Content by UHPLC in *Scolymus hispanicus* Extracts Obtained through Ultrasound-Assisted Extraction. *Plants* **2023**, *12*, 2340. [CrossRef] [PubMed]
49. Li, J.; Ye, Y.; Huang, H.; Dong, L. Kaempferol-3-*O*-β-D-glucoside, a Potential Allelochemical Isolated From *Solidago canadensis*. *Allelopath. J.* **2011**, *28*, 259–266.
50. Apati, P.; Szentmihályi, K.; Balázs, A.; Baumann, D.; Hamburger, M.; Kristó, T.S.; Szőke, É.; Kéry, Á. HPLC Analysis of the Flavonoids in Pharmaceutical Preparations from Canadian Goldenrod (*Solidago canadensis*). *Chromatographia* **2002**, *56*, S65–S68. [CrossRef]
51. Al-Shabibi, M.H.S.; Al-Touby, S.S.J.; Hossain, M.A. Isolation, Characterization and Prediction of Biologically Active Glycoside Compounds Quercetin-3-rutinoside From the Fruits of *Ficus sycomorus*. *Carbohydr. Res.* **2022**, *511*, 108483. [CrossRef]
52. Likhanov, A.; Oliinyk, M.; Pashkevych, N.; Churilov, A.; Kozyr, M. The Role of Flavonoids in Invasion Strategy of *Solidago canadensis* L. *Plants* **2021**, *10*, 1748. [CrossRef]
53. Zhang, C.B.; Wang, J.; Qian, B.Y.; Li, W.H. Effects of the Invader *Solidago canadensis* on Soil Properties. *Appl. Soil Ecol.* **2009**, *43*, 163–169. [CrossRef]
54. Gómez-Zorita, S.; Trepiana, J.; Laskibar, I.; Macarulla, M.; Eseberri, I.; Laura, A.G.I.; Merino-Valdeolmillos, R.; McGeoch, I.; Fernández-Quintela, A.; Portillo, M. Isorhamnetin: Current Knowledge And Potential Benefits For Disease Management. In *Handbook of Dietary Flavonoids*; Xiao, J., Ed.; Springer: Cham, Switzerland, 2023; pp. 1–61.
55. Zhao, Z.; Liu, Y. Cardiovascular Protective Effect of Isorhamnetin. *Med. Recapitul.* **2008**, *14*, 2321–2323.
56. Li, J.; Wang, G.; Du, S. Research progress on antitumor effect and mechanism of isorhamnetin. *Shanxi Med. J.* **2011**, *12*, 1215–1217.
57. Gong, G.; Guan, Y.Y.; Zhang, Z.L.; Rahman, K.; Wang, S.J.; Zhou, S.; Luan, X.; Zhang, H. Isorhamnetin: A Review of Pharmacological Effects. *Biomed. Pharmacother.* **2020**, *128*, 110301. [CrossRef] [PubMed]
58. Lei, J.; Yang, J.; Bao, C.; Lu, F.; Wu, Q.; Wu, Z.; Lv, H.; Zhou, Y.; Liu, Y.; Zhu, N.; et al. Isorhamnetin: What is the in Vitro Evidence for its Antitumor Potential and Beyond? *Front. Pharmacol.* **2024**, *15*, 1309178. [CrossRef] [PubMed]
59. Havsteen, B.H. The Biochemistry and Medical Significance of the Flavonoids. *Pharmacol. Ther.* **2002**, *96*, 67–202. [CrossRef] [PubMed]
60. Li, R.P.; Guo, M.L.; Zhang, G.; Xu, X.F. Neuroprotection of Nicotiflorin in Permanent Focal Cerebral Ischemia and in Neuronal Cultures. *Biol. Pharm. Bull.* **2006**, *29*, 1868–1872. [CrossRef]
61. Huang, J.L.; Fu, S.T.; Jiang, Y.Y.; Cao, Y.B. Protective Effects of Nicotiflorin on Reducing Memory Dysfunction, Energy Metabolism Failure and Oxidative Stress in Multi-Infarct Dementia Model Rats. *Pharmacol. Biochem. Behav.* **2007**, *86*, 741–748. [CrossRef]
62. Abreu, P.M.; Braham, H.; Ben Jannet, H.; Mighri, Z.; Matthew, S. Antioxidant Compounds from *Ebenus pinnata*. *Fitoterapia* **2007**, *78*, 32–34. [CrossRef]
63. Patel, D.K. Medicinal Importance, Pharmacological Activities and Analytical Aspects of a Flavonoid Glycoside 'Nicotiflorin' in the Medicine. *Drug Metab. Bioanal. Lett.* **2022**, *15*, 2–11. [CrossRef]
64. Xu, S.; Chen, S.; Xia, W.; Sui, H.; Fu, X. Hyperoside: A Review of Its Structure, Synthesis, Pharmacology, Pharmacokinetics and Toxicity. *Molecules* **2022**, *27*, 3009. [CrossRef]
65. Dobjanschi, L.; Zdrinca, M.; Mureşan, M.; Vicas, S.; Antonescu, A. The Thin Layer Chromatography Analysis of Saponins Belonging to *Solidago* Species. *Ann. Univ. Oradea* **2014**, *11*, 56–60.
66. Takahashi, Y.; Shiojiri, K.; Yamawo, A. Aboveground Plant-to-Plant Communication Reduces Root Nodule Symbiosis and Soil Nutrient Concentrations. *Sci. Rep.* **2021**, *11*, 12675. [CrossRef] [PubMed]

67. Szakiel, A.; Pączkowski, C.; Henry, M. Influence of Environmental Biotic Factors on the Content of Saponins in Plants. *Phytochem. Rev.* **2011**, *10*, 493–502. [CrossRef]
68. Global Biodiversity Information Facility. *Solidago canadensis*. Available online: https://www.gbif.org/species/5389029 (accessed on 7 October 2024).
69. Huang, Y.; Bai, Y.; Wang, Y.; Kong, H. Allelopathic Effects of the Extracts From an Invasive Species *Solidago canadensis* L. on *Microcystis aeruginosa*. *Lett. Appl. Microbiol.* **2013**, *57*, 451–458. [CrossRef] [PubMed]
70. Elshafie, H.S.; Gruľová, D.; Baranová, B.; Caputo, L.; De Martino, L.; Sedlák, V.; Camele, I.; De Feo, V. Antimicrobial Activity and Chemical Composition of Essential Oil Extracted from *Solidago canadensis* L. Growing Wild in Slovakia. *Molecules* **2018**, *24*, 1206. [CrossRef]
71. Mariychuk, R.; Gruľová, D.; Grishchenko, L.M.; Linni, R.P.; Lisnyak, V.V. Green Synthesis of Non-spherical Gold Nanoparticles Using *Solidago canadensis* L. Extract. *Appl. Nanosci.* **2020**, *10*, 4817–4826. [CrossRef]
72. Bilić, J.; Svorcina, M.; Poljuha, D. Antioxidant Capacity of Fruit Species Characteristic for Gardens in Istria (Croatia). In Proceedings of the 9th International Congress of Food Technologists, Biotechnologists and Nutritionists, Zagreb, Croatia, 3–5 October 2018.
73. Mikulic-Petkovsek, M.; Stampar, F.; Veberic, R.; Sircelj, H. Wild *Prunus* Fruit Species as a Rich Source of Bioactive Compounds. *J. Food Sci.* **2016**, *81*, C1928–C1937. [CrossRef]
74. Mikulic-Petkovsek, M.; Koron, D.; Rusjan, D. The Impact of Food Processing on the Phenolic Content in Products Made from Juneberry (*Amelanchier lamarckii*) Fruits. *J. Food Sci.* **2020**, *85*, 386–393. [CrossRef]
75. Singleton, V.L.; Rossi, J.A. Colorimetry of Total Phenolics with Phosphomolybdic-Phosphotungstic Acid Reagents. *Am. J. Enol. Vitic.* **1965**, *16*, 144–158. [CrossRef]
76. Amerine, A.; Ough, C.S. Methods for Analysis of Musts and Wines. *J. Inst. Brew.* **1981**, *87*, 223–224. [CrossRef]
77. Martins, D.; Barros, L.; Carvalho, A.M.; Ferreira, I.C.F.R. Nutritional and in Vitro Antioxidant Properties of Edible Wild Greens in Iberian Peninsula Traditional Diet. *Food Chem.* **2011**, *125*, 488–494. [CrossRef]
78. Poljuha, D.; Šola, I.; Bilić, J.; Dudaš, S.; Bilušić, T.; Markić, J.; Rusak, G. Phenolic Composition, Antioxidant Capacity, Energy Content and Gastrointestinal Stability of Croatian Wild Edible Plants. *Eur. Food Res. Technol.* **2015**, *241*, 573–585. [CrossRef]

plants

MDPI

Article

Utilization of Biomasses from Landscape Conservation Growths Dominated by Common Ragwort (*Jacobaea vulgaris* Gaertn.) for Biomethanization

Jürgen Müller [1,*], Denny Wiedow [1], Mohammad Said Chmit [2] and Till Beuerle [2]

[1] Faculty of Agricultural and Environmental Sciences, University of Rostock, Justus-von-Liebig-Weg 6, 18059 Rostock, Germany; denny.wiedow@uni-rostock.de

[2] Institute of Pharmaceutical Biology, Technical University of Braunschweig, Mendelssohnstr. 1, 38106 Braunschweig, Germany; s.chmit@tu-bs.de (M.S.C.); t.beuerle@tu-bs.de (T.B.)

* Correspondence: juergen.mueller3@uni-rostock.de; Tel.: +49-381-498-3143

Abstract: The highly toxic species common ragwort (*Jacobaea vulgaris* Gaertn.) prefers to migrate into protected dry grassland biotopes and limits the use of the resulting biomass as animal feed. There is an urgent need for a safe alternative use of the contaminated biomass apart from landfill disposal. We investigated the optional utilization of biomethanization of fresh and ensiled common ragwort biomasses and evaluated their energetic potentials by estimation models based on biochemical characteristics and by standardized batch experiments. The fresh and ensiled substrates yielded 174 $L_N \cdot kg^{-1}$ oDM methane and 185 $L_N \cdot kg^{-1}$ oDM, respectively. Ensiling reduced the toxic pyrrolizidine alkaloid content by 76.6%; a subsequent wet fermentation for an additional reduction is recommended. In comparison with other biomasses from landscape cultivation, ragwort biomass can be ensiled readily but has a limited energy potential if harvested at its peak flowering stage. Considering these properties and limitations, the energetic utilization is a promising option for a sustainable handling of *Senecio*-contaminated biomasses in landscape conservation practice and represents a safe alternative for reducing pyrrolizidine alkaloid entry into the agri-food sector.

Keywords: pyrrolizidine alkaloids; *Jacobaea vulgaris*; *Senecio jacobaea*; fermentation characteristics; PA degradation; energetic conversion; wet co-digestion; biogas yield

1. Introduction

Although an autochthonous species in Europe [1], the increased spread of common ragwort (*Jacobaea vulgaris* Gaertn., syn. *Senecio jacobaea* L., Asteraceae) into the landscapes of northern Germany since the 1990s has invasive features [2]. The probable causes of this phenomena, namely climate change [3]; the increased spread of diasporas due to human activities [4]; and the establishment of diaspora banks [5], especially on fallow land [6], cannot be eliminated in the short term. Common ragwort (syn. tansy ragwort) is ubiquitous in ruderal habitats and on road verges [7], where it acts as a pollen supplier [8], provides host for many invertebrates [9], and fulfills landscape aesthetic functions via its flowering aspect. Despite these ecosystem functions, the species is also well known for its toxic 1,2-Dehydro-pyrrolizidine alkaloids (PAs), including their corresponding *N*-oxides (PANOs) [10–12]. PAs are widely known and well-studied liver toxins [13–15] that include a proven genotoxic potential [16] and are carcinogenic to mammals including humans [17,18]. The intrinsic deterrence effect of these secondary plant constituents against herbivores is problematic when the common ragwort migrates into farmland areas intended for livestock feeding [19,20]. Furthermore, beyond the apparent toxicologically relevant feed issue, high population densities of ragwort are today regarded as critical, since this represents an entry path for PAs into the food chain via co-harvesting of PA weeds together with food, medicinal herbs, or via nectar and pollen into bee-related foodstuffs [21–24].

Common ragwort prefers to migrate into extensively used grasslands with low nitrogen fertilization [25]. This constellation applies to the majority of nature conservation grasslands; therefore, these preservation areas suffer in particular from common ragwort mass infestations [26]. To prevent this noxious weed from outcompeting forage species and rare plants in such habitats, regulation measures are necessary [20] or even obligate [27]. Although effective [28,29], the use of herbicides is in general not an option for preservation areas; instead, twofold mowing is the recommended standard alternative measure [30]. According to Siegrist-Maag et al. [31] mowing should take place when half of the plants start anthesis. During this phenological stage, *J. vulgaris* has already accumulated a considerable aboveground biomass, including its toxic PAs/PANOs. In ragwort-dominant stands, up to 100 dt ha^{-1} may occur. Hence, removing the ragwort-infested biomass is necessary to avoid sward degradation and to enhance the biodiversity of the infested grasslands. However, the traditional use of conservation growths, such as forage for livestock, is often not possible due to its PA contamination [32–34]. Therefore, there is an urgent need for a sustainable and safe use of the contaminated biomass apart from disposing of it as landfill. In addition, as was recently demonstrated, careless landfill disposal may cause subsequent environmental contamination and distribution via leaching into surface waters [35].

As was recently shown, composting represents a disposal alternative, bearing the potential to safely degrade the PAs/PANOs [36,37]. However, this is a rather expensive alternative for larger quantities of biomass. Professional composting plants are rare in rural areas, and high standards for the quality and continuity of the feedstock are required [38]. These needs are rarely met by the naturally discontinuously accumulating amounts of conservation growths. In contrast to composting plants, biogas plants exist in Europe in high spatial density [39], and the produced "green" energy has a secure market acceptance. Hence, the utilization of ragwort biomass for biogas production seems to be a possible alternative. Concerning energy recovery, according to van Meerbeek et al. [40], biomethanization is the preferred option for summer growth of herbaceous plants. However, the exploitation of this bioenergy potential cannot always be realized right away using the fresh biomass. In order to process large quantities of short-term-available biomass, these landscape conservation growths are frequently ensiled before use [41]. Although there have been some ensiling experiments involving ragwort infested grassland biomass, those studies were mainly aimed at determining the rate of alkaloid degradation [32,42,43]. The results of these degradation studies did not show a consistent pattern or generally valid endpoints, and a large number of possible factors are discussed [32,33]. Thus far, the question of whether a low-loss conservation of common ragwort as a biogas substrate is possible has not yet been addressed. In addition, there are no studies on the economic efficiency of the methane yields or on the question of whether the fresh-cut substrate or the ensiled substrate is more favorable for biomethanization. Furthermore, it is very likely that the ontogenetic development stage of ragwort may influence the biogas yield, as is known from other herbaceous plants [44]. Hence, relevant for a practical implementation, the question of the most favorable cutting time is of great importance, not only for regulating ragwort, but also for biomethanization.

Therefore, the purpose of this study is to examine the potential use/safe disposal of common ragwort-dominated biomass for methane production in different states/conditions.

In particular, we focus on the following key points:

(I) What are the substrate characteristics of a biomass from common ragwort at the recommended stage of cutting?
(II) Can substrates dominated by common ragwort readily be ensiled without the addition of a carbohydrate-rich biomass?
(III) To what extent does ensiling contribute to the reduction of pyrrolizidine alkaloids?
(IV) Which specific methane yield can be attained by the common wet fermentation technique?
(V) Which stage of ragwort development is best for controlling spreading of the plant, yet maintains the best option for efficient biomethanization?

These questions were addressed on the basis of two separate plant material collections. The first ragwort biomass material was used to investigate the fermentation process of lab-scale ensiling and biomethanization. The second collection was a time-series, conducted to study the influence of the phenological developmental stage of ragwort on its biochemical composition, which was necessary to estimate the stage-dependent bioenergy potential.

2. Results

2.1. Characterization of the PA Pattern

The pattern and relative abundance of individual PAs of the *J. vulgaris* plant material were analyzed by GC-MS. Not surprisingly, all relevant PAs were 1,2-Dehydro-PAs of the senecionine-type (1,2-unsaturated, retroncine-, cyclic diester-, 12-membered ring-type) [11]. Figure 1 shows that the local chemotype of *J. vulgaris* was strongly dominated by the occurrence of erucifoline (79.5%), while the remaining 20.5% was distributed over five minor PAs ranging from 2.4 to 8.3%.

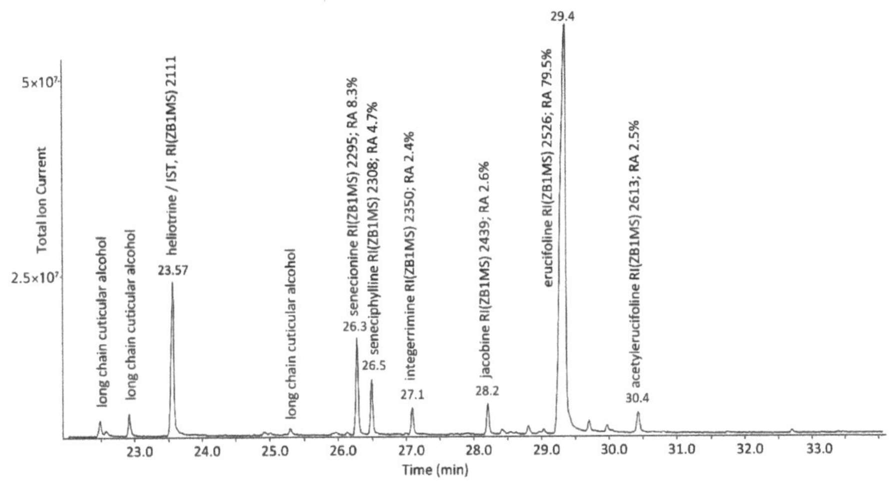

Figure 1. Total ion chromatogram (TIC) of a GC-MS analysis for identification and determination of the relative abundance (RA) of the PA-bouquet of *J. vulgaris* plant material utilized in this study.

2.2. Further Biochemical Properties of the Ragwort's Biomass

2.2.1. Collection I—Fermentation

The mean crude ash (CA) content of the feedstock collected for fermentation purposes was relatively low (51.09 g kg^{-1} DM, sd = 0.52). The average development stage of early flowering was reflected by a high crude fiber (CF) content of 284.03 (sd = 10.77) g kg^{-1} of dry matter (DM). The sum of the cell wall components (neutral detergent fiber, NDF) was 457.93 (sd = 18.83) g kg^{-1} DM, of which the acid detergent fiber (ADF) accounted for 366.98 (sd = 14.10) g kg^{-1} DM. The average amount of 404.75 (sd = 19.78) g kg^{-1} DM of the organic matter was enzyme-insoluble (EISOM). Further biochemical properties relevant to ensiling success are listed in Table 1.

The crude protein (CP) fraction is responsible for the bulk of the buffering substances that counteract the decrease in the pH value during ensiling. Therefore, the moderate CP content of 10.87% DM did not indicate a large substrate-inherent resistance to lactic acid-initiated acidification. At the beginning of flowering, the substrate showed sufficient water-soluble carbohydrates (WSCH, 10.2% DM), which are the food sources of the lactobacilli. The WSCH:CP-ratio combines both characteristics into a single parameter that indicates moderate ensilability; i.e., sufficient sugar supply for the lactobacilli is countered by a notable buffering capacity in the substrate. Nevertheless, the pH value of 4.5 showed that under the laboratory conditions and at a high substrate moisture of 77.8% FM, the desired vigorous microbial conversion was readily achieved. The relatively high proportion of acetic acid, as well as the occurrence of ethanol and butyric acid, indicate a rather heterogeneous and non-homofermentative lactic acid fermentation. The ammonia content of 0.34 g kg^{-1} FM corresponds to a mean NH_3-N content of 11.29 (sd = 7.62)%, both parameters indicates significant protein decomposition during ensiling.

Table 1. Ensiling characteristics of biomass from landscape conservation growths dominated by common ragwort (*J. vulgaris*).

Status of Ragwort Biomass	Feedstock Characteristics	Means (sd)
Chopped starting material before ensiling	Parameters of Ensilability	$n = 3$
	Dry matter content at start (DMC, g kg^{-1})	224.3 (3.79)
	Crude protein (CP, g kg^{-1} DM)	108.7 (0.41)
	Water-soluble carbohydrates (WSCH, g kg^{-1} DM)	102.3 (11.72)
	WSCH:CP-ratio	0.94 (0.12)
Substrate after 92 days of lab-scale ensiling	Fermentation patterns	
	Dry matter content after ensiling (DMC, g kg^{-1})	222.5 (0.12)
	pH-value	4.50 (0.00)
	Lactic acid (g kg^{-1} DM)	34.40 (4.48)
	Acetic acid (g kg^{-1} DM)	8.65 (0.83)
	Ethanol (g kg^{-1} DM)	1.60 (0.01)
	Butyric acid (g kg^{-1} DM)	<0.50
	Ammonia (g kg^{-1} FM)	0.34 (0.01)

2.2.2. Collection II—Growth Stages

Collection II includes biomass stocks from common ragwort at the typical growth stages that match the cutting time schemes of *J. vulgaris*-infested grasslands. The ontogenetic development stages, sampling dates, matching grassland use options, and two metric traits characterizing the average plant development at the sampling dates are given in Table 2. The growth stage significantly influenced the average plant height and the dry matter content. The biochemical characteristics of these four stages are presented in Table 3. The ontogenetic stage affected all biochemical characteristics significantly. The direction of characteristic increase or decrease aligned with increasing plant development and was always consistent during the sampling period from BBCH 44 to BBCH 66.

Table 2. Ontogenetic characterization of the developmental stages of *J. vulgaris* at different cuttings during the season.

Stage of Development (BBCH) [1]	Vegetative Parts Reached ca. 40% of Final Size (44)	First Flowers Visible, Largely Still Closed (55)	Inflorescence Fully Emerged (59)	Flowering Nearly Finished (66)	Effect of Stage [3]
Sampling Date	**1 June 2021**	**8 June 2021**	**16 June 2021**	**5 July 2021**	
Corresponding grassland use	cut for silage	cut for hay	late cut for hay	conservation cut	
Averaged plant height (cm) [2]	33.70 (3.91)	56.80 (0.43)	84.33 (4.41)	99.33 (3.26)	$p < 0.001$ ***
Dry matter content (g kg^{-1}) [2]	116.79 (3.70)	156.98 (0.88)	192.57 (7.34)	267.26 (6.88)	$p < 0.001$ ***

[1] General BBCH-scale according to Meier (2001); [2] sample mean with standard deviation in brackets; [3] results of the one-way-ANOVA; *** significant at p-values less than 0.001.

While the fiber-related characteristics (CF, NDF, ADF, HC) increased with plant growth, a decrease in the energy-related CP, CL, and WSCH parameters was observed. In accordance, the crude ash decreased as a result of C-dilution via the increase in the plant biomass. It is noteworthy that within the flowering period, there was a notable increase in the proportion of cellolytically insoluble substance (EISOM).

Table 3. Constituent characteristics of common ragwort (*J. vulgaris*) biomass depending on its averaged phenological stage. The means are presented with their standard deviations in brackets.

Stage of Development (BBCH) [1]	BBCH 44	BBCH 55	BBCH 59	BBCH 66	Effect of Stage [2]
Crude ash (CA, g kg^{-1} DM)	100.73 (2.37)	76.13 (0.90)	55.60 (3.22)	45.30 (1.34)	$p < 0.001$ ***
Crude protein (CP, g kg^{-1} DM)	155.92 (4.32)	118.92 (6.20)	93.04 (3.17)	79.95 (4.48)	$p < 0.001$ ***
Crude fiber (CF, g kg^{-1} DM)	161.28 (7.02)	220.28 (10.11)	293.31 (14.93)	340.30 (7.78)	$p < 0.001$ ***
Crude lipid (CL, g kg^{-1} DM)	23.93 (1.35)	22.58 (2.15)	18.35 (1.02)	18.02 (0.53)	$p = 0.002$ **
Neutral detergent fiber (aNDF$_{OM}$, g kg^{-1} DM)	284.39 (8.79)	388.12 (16.04)	479.97 (22.89)	537.45 (12.12)	$p < 0.001$ ***
Acid detergent fiber (ADF$_{OM}$, g kg^{-1} DM)	224.26 (6.11)	299.65 (12.96)	378.60 (14.20)	428.86 (7.08)	$p < 0.001$ ***
Hemicellulose (HC, g kg^{-1} DM)	60.13 (6.26)	88.46 (4.15)	101.37 (8.69)	108.59 (5.06)	$p < 0.001$ ***
Water-soluble carbohydrates (WSCH, g kg^{-1} DM)	106.15 (3.34)	93.00 (8.15)	77.58 (1.71)	42.15 (4.71)	$p < 0.001$ ***
Enzyme-insoluble organic matter (EISOM, g kg^{-1} DM)	131.94 (8.63)	262.70 (14.22)	386.38 (32.71)	534.36 (14.54)	$p < 0.001$ ***

[1] General BBCH-scale according to Meier (2001); [2] results of the F-test (one-way-ANOVA) or Kruskal–Wallis test—depending on data distribution; ** significant at p-values less than 0.01; *** significant at p-values less than 0.001.

2.3. PA Degradation during Ensiling Process

2.3.1. Results of the Ensiling Experiment

A significant degradation of the sum of the total PA content (sum of PA/PANOs) after 92 days of lab-scale ensiling was detected (Figure 2). The extent of PA degradation was a remarkable 76.6% on average for the three laboratory silages. Beginning with a mean of 7962 mg in the starting material, an average of 1860 mg was still found per kg of dry matter in the silage after fermentation.

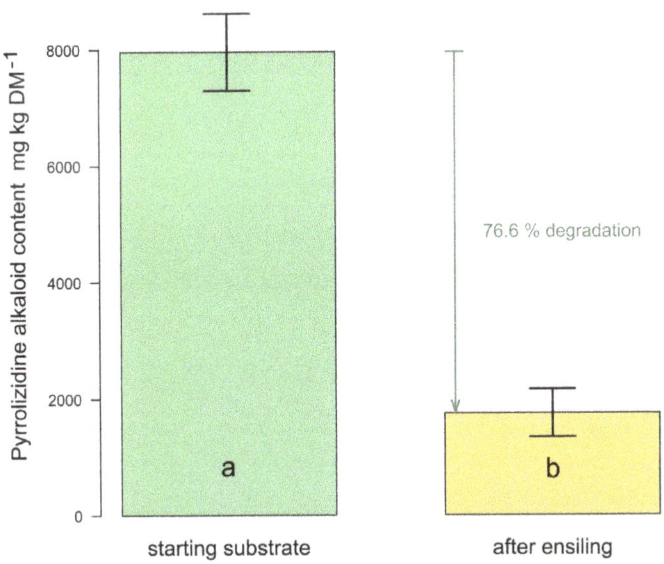

Figure 2. Extent of degradation of the total amount of PAs/PANOs after 92 days of ensiling. Different letters indicate significant differences in PA/PANO content (Dunn-Test, $p < 0.001$). Error bars represent the standard deviations of the means.

2.3.2. Extent of PA-Degradation in the Context of Similar Studies

To evaluate this result, we compared our data to similar experiments where the PA degradation of a biomass containing species of the genus Senecio during ensiling were studied. The studies of Berendonk and Hünting [45], Candrian et al. [46], Gottschalk et al. [32], and Klevenhusen et al. [47] fulfilled the similarity requirements (see chapter 4.6 for details) and were used for the layout in Figure 3. The data used for this ordination are additionally listed in Appendix A Table A1.

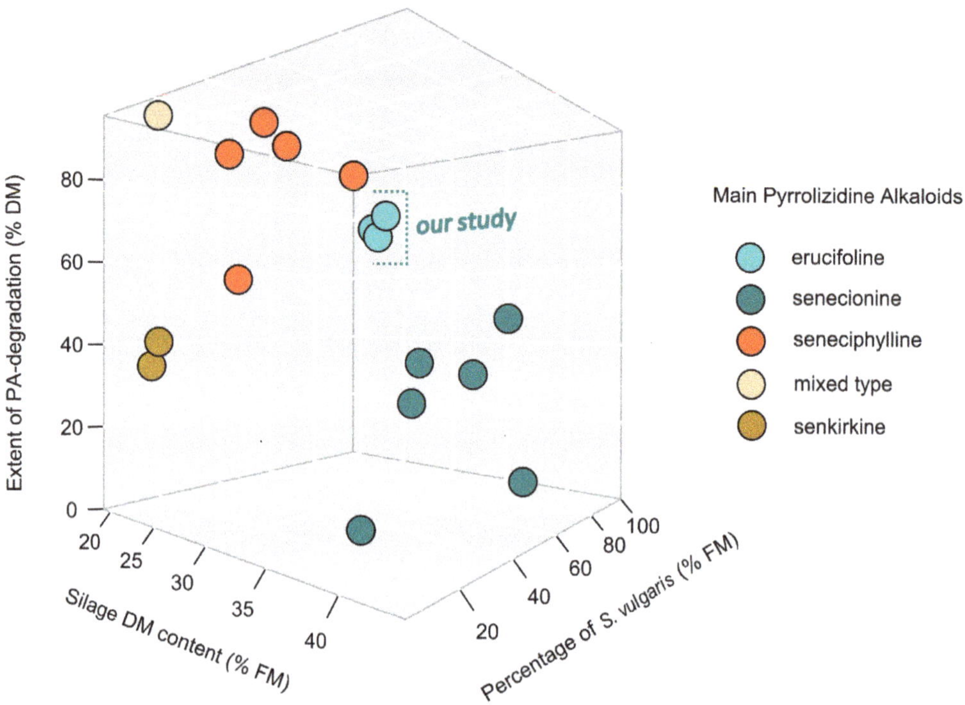

Figure 3. Extent of PA/PANO-degradation during ensiling of our study and comparable studies with PA-containing *Senecio* species. Results are ordinated according to the dry matter content of their silage substrates and the percentage of the *S. vulgaris* resp. *Senecio* biomass in the ensiled feedstock. The main types of the pyrrolizidine alkaloids are color coded (see legend for details).

With the exception of the PA-type senkirkine-dominated substrates, a gradient of decreasing degradation rates with decreasing moisture contents is apparent (Figure 3). Furthermore, there seems to be a trend towards an increase in the degradation rate with an increasing proportion of Senecio in the feedstock.

2.4. Methane Yields

2.4.1. Results of the Batch Wet Co-Fermentation Test

The specific methane yields (SMY) of the fresh and ensiled biomasses from *J. vulgaris* were 173.6 $L_N \cdot kg^{-1}$ oDM and 185.1 $L_N \cdot kg^{-1}$ oDM, respectively. They were not significantly different ($p > 0.05$), despite a tendency of slightly higher SMY-values combined with lower SMY-variations of the ensiled feedstock (Figure 4). The distance between the mean values (top of the bars) and estimates of the methane potential based on the biochemical composition of the substrates (dots) is considerable for both ragwort substrates. In order to show a comparison to the most common gramineous-accompanying vegetation associated with J. vulgaris, the results of Meserszmit [48] have been integrated in Figure 4. These are *Arrhenatherum elatius* and *Festuca rubra*, the dominant grasses of the *Arrhenatheretum elatioris* grassland association, which is especially prone to common ragwort invasion. The feedstock of the companion grasses match that of the Senecio substrates, not only in terms of the plant sociological context, but also in terms of the time of harvest (July). In addition, the authors followed the same guidelines for the experimental determination of specific methane yields (VDI 4630).

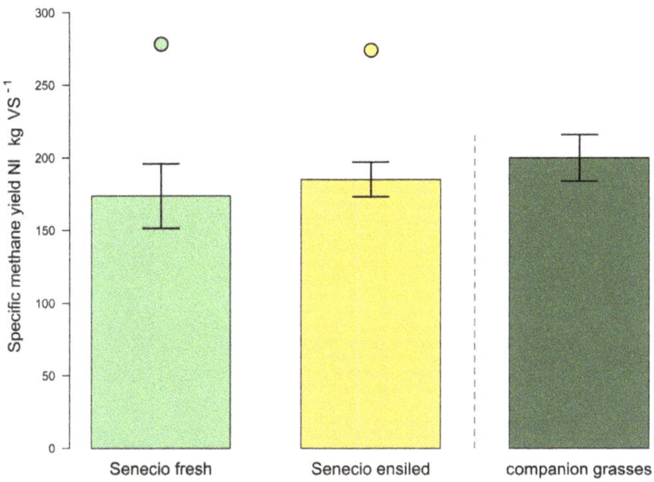

Figure 4. Specific methane yields of fresh and ensiled biomasses from *J. vulgaris*. Presented are the means and the corresponding standard deviations as error bars. Dots mark the amount of specific methane yield potential according to Weißbach (2008). For comparison, the yield of the prevalent companion grasses according to Meserszmit et al. (2019) is additionally shown.

The analysis of daily methane production rates provides insight into how methane yields are generated. Figure 5 compares the daily methane formation rates of the two different ragwort feedstock conditions, fresh (*J. vulgaris* fresh) and ensiled (*J. vulgaris* ensiled), throughout the whole experimental period of 30 days. Both variants reached the peak of daily methane formation, approx. 5 $L_N \cdot kg^{-1}$ oDM, at the same time, that is after 11 days. The dynamics of CH_4-formation up to this point, however, are quite different: the fresh material started with a progressive formation rate, while the silage required a significantly extended start-up time. Conversely, methane production did not collapse quite as strong in the silage feedstock with an increasing lack of fermentable substance in the downstream phase.

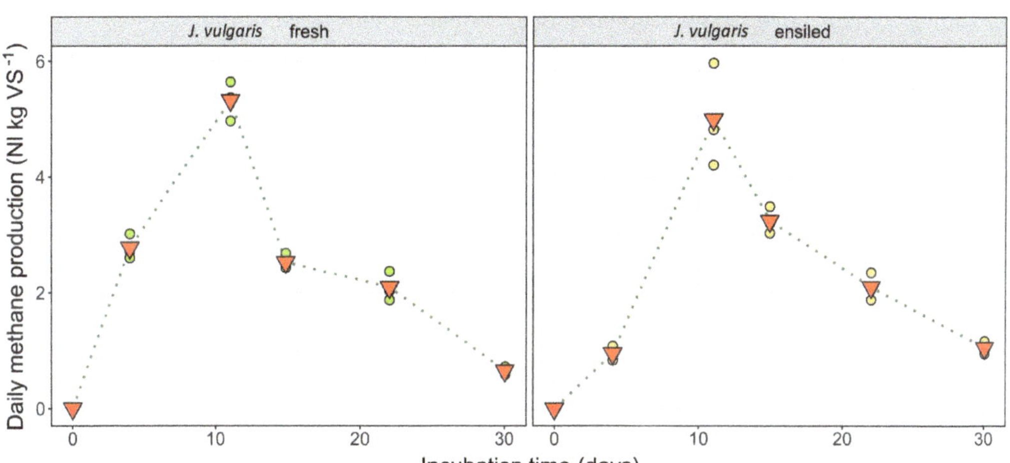

Figure 5. Daily methane production (expressed as norm liters per kg volatile matter) throughout the batch trial period of 30 days. Red triangles present the mean methane yield of the respective measurement event. Dots indicate the single results of the three batch vessels that served as replicates.

When looking at Figure 6, it becomes obvious that the methane yields of the initial phase were mainly due to high gas quantities at still low methane concentrations. With overall comparable carbon dioxide concentrations between the two feedstock conditions, the methane concentrations of the fresh substrate seem to be superior to those of the silage. From the fifth to the tenth day after starting the batch test, a stagnation of the two measured gas concentrations in the silage treatment was observed, which suggests a stronger presence of other non-measured gases (such as O_2, NH_3, NO_x, H_2S, i.a.).

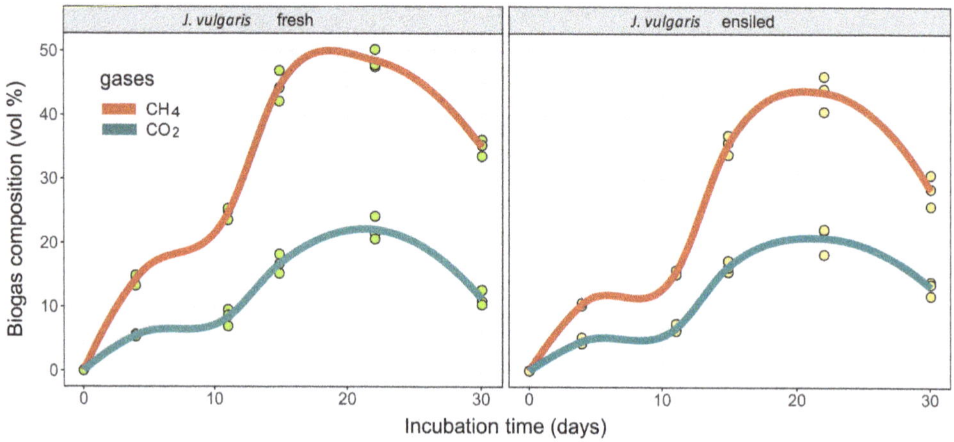

Figure 6. Volume percentages of the two main biogas components methane (CH_4) and carbon dioxide (CO_2) during the course of the batch wet fermentation period. Dots indicate the measurement results of the individual batch vessels in triplicate. Trendlines were constructed using local regression approaches (loess).

2.4.2. Potential Methane Yields at Different Growth Stages

The potential methane yields (PMY) of Collection II (see Table 2) are shown in Figure 7. There was a significant effect of the developmental stage on PMY (p = 0.015 *). It is obvious that the potential of the substrate for methane production decreases with increasing phenological development. The difference between the cut for silage harvest and the late landscape conservation cut was considerable (106 $L_N \cdot kg^{-1}$ oDM, corresponding to 31%). Even the lowest PMY at the late developmental stage of BBCH 66 was higher than the measured specific methane yield in the batch test.

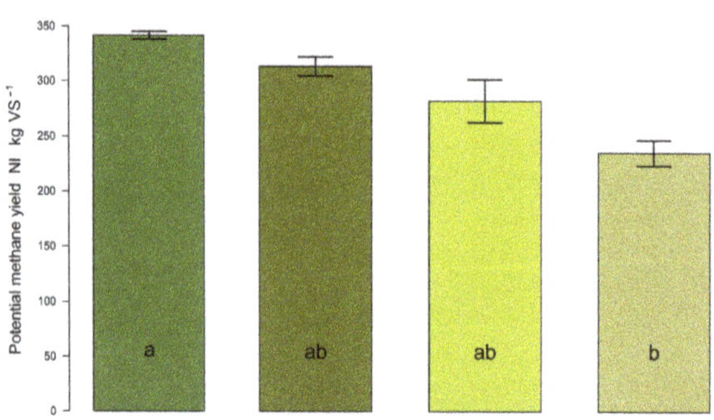

Figure 7. Potential specific methane yields (expressed as norm liters per kg volatile substrate) of common ragwort feedstock harvested at different developmental stages. Different letters inside the bars indicate statistically significant differences between the four stages ranging from BBCH 44 to BBCH 66. Error bars represent standard deviations of the mean.

3. Discussion

3.1. Characteristics of Common Ragwort Biomass as Feedstock for Ensiling and Biomethanization

Apart from the discussion about the inhibiting effect of plants containing toxins in the biomethanization process [49], the ingredients of the substrate determine its utilization efficiency [50]. The knowledge of the composition of the lignocellulosic feedstock can help in choosing appropriate technical measures to achieve better hydrolysis which, in turn, would result in higher biogas yield and conversion efficiency [51]. Lignification of the substrate plant inhibits hydrolytic fermentation by hampering the microbes from attacking the cell wall carbohydrates. Therefore, many authors base their assessment of the biogas potential of unknown plant substrates on the lignin content as a key characteristic [52,53]. Nevertheless, choosing lignin content as the determining characteristic for biogas yield prediction is error-prone, as not only the amount of lignin, but also the incrustation of the polymeric lignin (crystallization), is relevant for microbial degradability [52,54]. In addition, the incorporation of the lignin into the plant tissue matrix and the binding via phenolic acids are of relevance [55]. Therefore, we decided to use enzyme insolubility (EISOM) as a key feature for estimating biogas potential, instead of lignin, as suggested by Weißbach [56].

The determined EISOM contents of *J. vulgaris* varied more (ranging from 132 to 534 g kg^{-1} DM) than those observed for the grass-dominated grassland biomass under comparable environmental conditions (range: 180–488 g kg^{-1} DM [57]). The enzyme-insoluble organic matter of the ragwort feedstock correlated with the phenological stage of development. Hence, the usefulness of *J. vulgaris* as a feedstock for bioenergy via the methanization pathway is strongly dependent on the average development stage of ragwort at the time of harvest. To our knowledge, there is no information on the enzyme insolubility of the organic matter (EISOM) of the species *J. vulgaris*. Only Frost et al. [58] and Purcell et al. [59] have provided information on the cell wall components of *J. vulgaris* depending on the developmental stage. According to Frost et al. [58], NDF-content increased from 37.6% DM at the rosette stage to 49.2% DM at flowering. Purcell et al. [59] did not specify developmental stages, but approximations can be drawn from the collection times given. Accordingly, plant material collected in early spring (in Ireland) had a cell wall content (NDF) of only 19.4% DM, whereas, at the end of August, the NDF content increased to 44.4% DM. The corresponding ADF contents in this study were 15.5% DM for the early collection and 34.8% DM for the late summer collection. ADF includes the fractions of lignin and cellulose and thus represents the recalcitrant part of a biomass for the microbiome involved in methanogenesis. The ADF contents in our study varied between 22.4 and 44.8% DM, representing a higher level. Those deviations could be explained by the fact that the spring sampling of the Irish survey took place very early in the season. Alternatively, the late summer sampling was probably the regrowth of an already pre-harvested area. Both conditions would lead to a biomass that is dominated by leaves and depleted in stems, which in turn would lead to lower levels of structural substances, such as NDF and ADF. In this light, these data are quite consistent with our results, confirming the overriding influence of the ontogenetic developmental stage on ragwort biomass composition, even across different environments.

It is known from other herbaceous plants that their CP content responds more strongly to the N supply of the growing site [60]. Nevertheless, in the case of *J. vulgaris*, similar crude protein contents have been found across North America and Europe at comparable developmental stages. Debessai [61] reported a CP level of 13.1% DM in a not ontogenetically classified *J. vulgaris* collection, which was apparently collected at early flowering stages in Oregon (U.S.). Frost et al. [58] recorded higher CP contents (15.5% DM) in the rosette stage at sites in Idaho (U.S.), which corresponds well with the 15.4% CP that Purcell et al. [59] found in Ireland, also at an early developmental stage. The CP contents in our study fit very well with those of Frost et al. [58] and indicate a decrease in potentially buffering substances with the increasing maturity of the plants.

Surprisingly, we observed the highest contents of water-soluble carbohydrates in the stage of vegetative development in Collection II. This is in contrast to Purcell et al. [59], who reported quite low WSCH contents in the spring-harvested ragwort (52 g per kg DM) and a considerably high level of 144 g per kg DM in the late summer collection. Both factors, different morphology (leaf/stem ratio and growth duration), and weather influences (solar radiation before collection) are possible explanations for these deviations. In any case, the WSCH content proves to be the most uncertain parameter for estimating the ensilability of ragwort biomasses. DM contents below 25–30% lead to the formation and release of silage effluent (McDonald et al. [62]). The DM content of the freshly harvested ragwort reached this threshold only at the end of flowering. Therefore, all earlier harvests need to be wilted before ensiling to avoid environmental threats, like the PA-leach-out risk, as well as additional energy losses associated with silage effluent run offs. In our study, the jars used for fermentation did not allow such effluent losses. Hence, under these ideal lab conditions, only gaseous ensiling losses of 0.23% DM were observed. Thus, in our experiments, the substrate characteristics of the fresh substrate differed only slightly from those of the silage, leading to comparable potential methane yields (shown as dots in Figure 4). Under practical conditions, however, this might be quite different.

3.2. *Degradation of Pyrrolizidine Alkaloids and Their N-Oxides*

The idea of using the process of ensiling as a means of detoxification of biomasses containing PAs was first conceived by forage scientists. Debessai [61] reported a reduction in PA content from an original 2454 to 584 μg kg^{-1} (−76%) after a 30-day experimental storage of ragwort-contaminated feedstock with the addition of urea preparations. In this study, grinding of the plant material was beneficial in reducing final PA contents as well. However, the alkaline environment achieved by urea ammonization, does not correspond to a lactic acid fermentation ensilaging process, which leads to an acidic environment. Therefore, we only compared our results with studies that also investigated lactic acid fermentation-based ensiling with regard to its effects on PA degradation. In their studies involving *Senecio alpinus*, Candrian et al. [46] found a decline in PA degradation from 92 to 54% with a decrease in the proportion of *Senecio* biomass in the silage substrate ranging from 100 to 3.5%, respectively. In the discussion of this finding, it was disregarded that the replacement of *Senecio* by a drier grass component was accompanied by an increase in the dry matter content while factor grading took place. Since the dry matter content significantly influences the overall microbial activity in silage fermentation processes [63], it may be that the observed decrease in PA degradation was also an effect of the overall reduction in microbial conversion activity too. Our results of the percentage PA degradation fit well with those of other studies if they are placed in an orientation framework (Figure 3) that takes both aspects, namely the ragwort percentage and the substrate moisture, into account. A number of authors who have experimentally studied the fermentative detoxification of PA-containing substrates, emphasize the importance of the specific PA type in the discussion of their results [32,43,47]. With the exception of the apparently recalcitrant otonecine-type PA senkirkine (a major PA in *Senecio vernalis* [47]), no striking differences in the degradation behavior of the general dominating senecionine-type PAs, such as erucifoline, senecionine, or seneciphylline, can be identified under comparable experimental conditions. This is not surprising since all these PAs share a similar basic chemical senecionine-type structure (1,2-unsaturated, retroncine-, cyclic diester-, 12-membered ring-type). In contrast to most other studies, our detection method was based on the detection of the core base structure of PAs and comprised all 1,2-unsaturated retronecine and heliotridine-type ester PAs [64]. This means that it would also cover for unknown toxic PA-metabolites that could be formed during the degradation process but still bear the general toxic principle of 1,2-unsaturated ester PAs.

Klevenhusen et al. [47] observed that the addition of homofermentative lactic acid bacteria did not increase the PA degradation rate but that the addition of molasses did. The addition of homofermentative lactic acid-producing inoculants results in a faster pH-drop, which strongly limits further microbial conversions, thereby shortening the fermentation-intensive phase of ensiling. Presumably, ensiling conditions that promote heterofermentative processes at the beginning, and allow lactic acid formation to stabilize the ensiled material by lowering the pH value a little later, are more conducive to PA degradation. Further ensiling experiments would also be desirable in this area, e.g., with staggered buffering of otherwise identical PA-containing and PA-type classified substrates.

3.3. *Specific Methane Yields of Ragwort Biomasses*

In a previous study, we were able to show that the combined process of ensiling/biomethanization leads to a significant and reliable reduction of PAs/PANOs by 91–99% of the starting material [37]. These findings are only useful for practical implementation if the *S. vulgaris* biomass can be converted into green bioenergy with economic success. This, in turn, requires solid knowledge of the substrate-specific methane formation potential, as well as of the behavior in the biogas process [52]. We found specific methane yields of only 173.6 $L_N \cdot kg^{-1}$ oDM (fresh) and 185.1 $L_N \cdot kg^{-1}$ oDM (silage) in the standard batch digestion test. These were significantly lower yields than can be achieved by herbaceous cultivated crops, such as catch crops (250–350 $L_N \cdot kg^{-1}$ oDM) [50]. Even grass-dominated and extensively used grassland growths have higher specific methane yields of 197–221 $L_N \cdot kg^{-1}$ oDM [48]. Consequently, the usability of ragwort-dominated growths harvested in the flowering stage has to be considered critical, even if the prime costs of the biomass production were not credited to the biogas generation. There was a greater discrepancy between the potential energy yield and the yield measured in the batch tests (Figure 4) than the usually reckoned 15%. This is an indication that the microbiome in the test did not work as expected and thus seems to confirm that PAs could disrupt the complex and sensitive microbiological conversion processes of the methanogenesis [49]. However, Chmit et al. [37] could identify neither a pronounced lag phase nor a correlation between PA levels and methane formation. Therefore, other ingredients, such as inhibitors or even an insufficiently adapted microbiome of the inoculum, must be responsible for the reduced methanogenesis. The inoculum we used originated from a biogas plant based on renewable raw materials. In order to prove or exclude the aspect of inhibition, the adaptability of the microorganisms would have to be examined in continuous tests over a longer period of time. In this way, the microorganisms could adapt to the substrate, and the accumulation of inhibitory substances could also be better detected.

The development of the gas formation in the course of the batch test can provide information about the process dynamics. As shown in Figure 5, the methane formation kinetics of fresh material and ensiled material differed somewhat, despite a comparable basic pattern. The higher acidity of the silages, in combination with a lack of buffering substances, was probably responsible for the delayed increase in the gas formation rate in the first week [65]. In contrast, the degression in the gas formation after the consumption of the microbially more-easily-accessible carbohydrate sources was not as strong in the silage variants after almost two weeks, most likely due to a certain pre-digestion of the cell wall components by the enzymatic activity of the silage microflora [66]. This is also indicated through the somewhat lower EISOM values of the silages compared to the fresh substrates. The non-ensiled biomasses tended to have higher methane contents in the biogas, although, according to Herrmann et al. [67], the ethanol and butyric acid formed during ensiling should lead to higher methane contents in the ensiled substrates. However, both fermentation products could only be detected in limited quantities.

The strong dependence of the potential methane yield on the developmental stage of the plants from Collection II is not surprising and is also reported for a number of other plant species [44,68,69]. While the energy yield per ha in the case of cultivated bioenergy crops plays a decisive role in the determination of the time of harvest [69], the compatibility of landscape conservation objectives with biomass usability has priority in the case of conservation growths. When ragwort-dominated areas are cut at the time of full flowering in order to suppress the noxious weed, the resulting biomass can hardly be used for methane production in an economically viable way. The challenge for the bioenergetic use of *J. vulgaris* is to schedule the time of harvest in such a way that the regulating effect for the containment of this invasive species is compatible with the requirements on substrate quality for efficient methane production. If this succeeds, an invasive burden can be turned into a bioenergetic opportunity. Already, at the recommended cutting time of early flowering [31] and with a growth height of about 57 cm, only marginal methane yields are achieved. Therefore, we recommend harvesting such growths somewhat earlier to ensure sustainable energy production with simultaneous detoxification at a development stage not earlier than BBCH 48, but not later than BBCH 56. Cuts taken too early do not provide enough biomass to justify the harvesting costs, are more difficult to ensile, and, according to Berendonk et al. [42], also possess lower PA degradation rates during ensiling.

3.4. Synthesis

We have investigated the usefulness and behavior of ragwort-dominated biomasses for the fermentation processes of ensiling and biomethanization based on a collection of a larger amount of uniform plant material (referred to as Collection I). This corresponded in its phenological development to the sensitive stage of full flowering, recommended by weed researchers, when cutting most severely affects the vitality of the invasive species. In addition, we have investigated the expected behavior of the ragwort feedstock at different development stages on the basis of a second time/development-based collection of plant material (referred to as Collection II). Evaluation of ensilability was done by means of biochemical characteristics and, in the case of biomethanization, with the help of a validated model approach based on enzymatic degradability. In the preceding sections of this discussion, the results of both collections and aspects were discussed in order to comply with the complex character of the matter. The final conclusions for the practical significance in landscape management and biomass utilization will be summarized in Section 5.

Further research on the topic would be desirable. In particular, the measurement of methane yields in a continuous-flow-digester, instead of the batch tests, would generate additional useful information.

4. Materials and Methods

4.1. Chemicals and Reagents

All chemicals used were purchased from Roth (Karlsruhe, Germany) and Sigma-Aldrich (Seelze, Germany) and were of HPLC-grade purity or redistilled before use. Lithium aluminum hydride solution (1 M) in THF and pyridine, both in AcroSeal quality, were acquired from ACROS Organics (Geel, Belgium). Strong cation-exchange solid-phase extraction cartridges (SCX-SPE) were obtained from Phenomenex (Aschaffenburg, Germany). Isotopically labeled internal standard (IST) 7-*O*-9-*O*-dibutyroyl-[9,9-$^2H_2$]-retronecine was synthesized in our laboratory [64].

4.2. Plant Material

The plant material was harvested in two collections of already investigated J. vulgaris populations near Rostock (northeastern Germany). Collection I (fermentation collection) was collected on 5 July 2019 at the main developmental stage of BBCH 57 (beginning of the flowering period), according to Meier [70]. The fresh biomass was brought to the laboratory immediately and chopped with a commercial garden shredder. The shredded biomass was divided into three subsets, which then served as replicates. From each of these stocks about

500 g of fresh mass was drawn to determine the dry matter content and the biochemical analysis of the starting material. Subsets were taken from the bulk of the remaining plant material, both for biomethanization as fresh substrate (approx. 1500 g FM per sample, stored at −22 °C until the start of the batch trial) and as silage material (approx. 3000 g FM per sample, directly ensiled thereafter).

Collection II (stage collection) was carried out during the 2021 growing season with the aim of collecting common ragwort at its typical physiological development stages that, in practice, correspond to the usual cutting times of *J. vulgaris* - infested grasslands. On the following dates, 1 June (silage cutting), 8 June (early hay cutting), 16 June (late hay cutting), and 5 July (landscape conservation cutting), approximately 2000 g of ragwort plants were collected, and their growth height, as well as their development stages, were determined. Subsequently, the plant material was dried at 55 °C to a constant weight and ground to 1 mm sieve width using a hammer mill (Brabender, Duisburg, Germany).

4.3. Silage Preparation

The chopped *J. vulgaris* biomass was ensiled in three replicates in 3 L lab-scale silos (jars). The jars were cleaned and sterilized (180 °C, 8 h) before the substrates were filled in by hand. Fill weights of 1552–1563 g FM were realized, corresponding to a mean storage density of 0.52 g cm^{-3} FM. The filled jars were closed air-tight with a rubber-lined lid that was fixed by clips and stored for 92 days at 16 °C in a dark room. This was designed to allow gas to be released under pressure but prevented ambient air from flowing in. The jars were weighed after 30 days and again after the end of the ensiling period to determine the ensiling loss gravimetrically. After 92 days, the ensiled substrates were sealed air-tight in plastic bags and stored at −22 °C before analyzing their biochemical composition, including fermentation profiles.

4.4. Biochemical Analysis

4.4.1. Substrate Characteristics

DM content of the ragwort biomass was determined by oven drying at 55 °C to a constant weight, immediately after collection and additionally from the feedstock before and after ensiling. We analyzed the water-soluble carbohydrates (WSCH) and the enzyme-insoluble organic matter (EISOM) by near-infrared reflectance spectroscopy (NIRS, Bruker® MPA, Bruker, Ettlingen, Germany) with the enzymatic method, according to de Boever et al. [71], as the reference for EISOM. The dry combustion technique (Elementar® Analyzer Vario Max CNS, Elementar, Sprendlingen, Germany) has been used to determine crude protein contents (CP, N × 6.25). Neutral detergent fiber (NDF), acid detergent fiber (ADF), and crude fiber (CF) were determined by wet chemical analyses using a Fibretherm®, (Gerhardt, Königswinter, Germany). Hemicellulose contents have been estimated through the difference between NDF and ADF concentrations. Crude ash (CA) content was determined after incineration at 550 °C for 5 h in a muffle furnace. Volatile solids (VS) were calculated by the difference between total solids (TS) and the amount of CA.

4.4.2. Silage Fermentation Characteristics

The fermentation characteristics of the lab silages were determined according to the VDLUFA guidelines [72]. After thawing the frozen silage samples at room temperature, silage extracts were prepared from 50 g silage and 200 mL deionized water. The pH values of these extracts were measured potentiometrically by a calibrated pH analyzer (WTW pH3210, Xylem Analytics, Weilheim, Germany). Acetic and butyric acids were quantitatively detected by gas chromatography (6890 Series GC-FID, Agilent Scientific Instruments, Santa Clara, CA, USA).

4.4.3. Quantification of the Total PA/PANO Content

Plant material was homogenized (dried, mixed, and milled). Each sample of 300 mg was soaked in a 15 mL polypropylene conical centrifuge tube using 11 mL H_2SO_4 0.05 M and 40 μL of internal standard solution (2 μg/mL in methanol) and was added to each tube. The tubes were thoroughly mixed and extracted over night using a continuous tube rotator (Multi Bio RS-24, Biosan, Riga, Latvia) with the following settings: orbital: 21/01, reciprocal: 15/01, vibrio: 5/1, duration: 14 h. Sample preparation for analysis of the total PA content of the products was determined by using a modification of the method of Cramer et al. [64] and Letsyo et al. [73]. During the course of sample preparation all 1,2-unsaturated retronecine- and heliotridine-type ester PAs/PANOs were converted into the corresponding core structures, i.e., retronecine and/or heliotridine. After derivatization, these analytes were analyzed by HPLC-ESI-MS/MS, generating a single signal that could be quantified by the use of the added isotopically labeled IST, allowing calculation of the total sum of the PA/PANO content of the sample (Cramer et al., 2013). Data analysis and integration was achieved with Analyst 1.6.2 Software (Applied Biosystems MDS Sciex, Darmstadt, Germany).

4.4.4. Sample Preparation for GC-MS Analysis

Plant material of fresh *J. vulgaris* plant material was frozen (−20 °C), lyophilized, powdered, and homogenized with a coffee grinder. The homogenized powders were stored at room temperature in the dark until sample preparation and analysis. The powdered plant material was extracted with 0.05 M sulphuric acid (3 × 2.0 mL) after adding an appropriate volume of heliotrine as an internal standard solution (0.1 mg from MeOH stock 10 mg/mL stored at −20 °C). The extraction process was repeated three times and centrifuged (>3500× *g* for 5 min). The acid-aqueous supernatants were combined and reduced with zinc dust (~100 mg) for 3 h with continuous stirring at pH~2 to convert existing PANOs to the corresponding tertiary PAs. After 3 h, the mixture was adjusted to alkaline (pH~10) with ammonia solution (25%) and then eluted from 3 to 6 g Isolute® column (1 mL aqueous solution/g Isolute) with 25 mL dichloromethane. The organic phase, after the Isolute clean-up process, was dried at room temperature and reconstituted in exactly 1.0 mL of methanol for GC-MS analysis.

4.4.5. GC-EI-MS PA Pattern Analysis of Raw Plant Materials

An Agilent Technologies 6890 N (Agilent Technologies Deutschland GmbH, Böblingen, Germany) gas chromatograph was coupled to an Agilent Technologies 5975B mass spectrometer. Separation of the analytes was performed on a ZB-1MS capillary column (30 m × 0.25 mm; ft 0.25 μm; Phenomenex, Aschaffenburg, Germany). The following temperature program was used: 100 °C (3 min)–6 °C/min–310 °C (3 min). The following MS settings were applied: ionization voltage of 70 eV, ion source temperature of 230 °C, and interface temperature of 290 °C. A set of co-injected hydrocarbons (evenly numbered C14–C32) was used to calculate the retention indices by linear extrapolation, as described by Frölich et al. [74]. The PA compounds were identified by retention indices and an in-house mass spectra library of authentic reference compounds. PA patterns of the individual plant samples were compared using the Agilent ChemStation software (Version D.03.00.611). The respective ratio (relative abundance, RA) of each PA was calculated based on its TIC-area signal in relation to the sum of all TIC-area signals of all PAs of the sample.

4.5. Biogas Yield Determination

4.5.1. Potential Methane Yield Estimation

The potential for methane formation was calculated according to Weißbach [56] using the following estimation equations based on two main biochemical parameters of the substrates:

$$VS = 1000 - (CA) - 0.62\,(EULOS) - 0.000221\,(EULOS)^2 \quad (1)$$

The substrate's amount of fermentable organic substances (VS g kg^{-1} DM) was estimated using Equation (1). The substrate-specific biogas (BGY) and methane (CH_4Y) yield potentials of the ragwort biomass as feedstock substrates were derived from vs. according to Equations (2) and (3), accordingly.

$$BGY = 0.80\ (VS) \tag{2}$$

$$CH_4Y = 0.42\ (VS) \tag{3}$$

where BGY and CH4Y are given in norm liter per kg (Nl kg DM^{-1}) and are corrected for volatile fatty acids (VFA) [75].

4.5.2. Batch Wet Co-Fermentation Test

A batch test according to VDI 4630 [76] was carried out to determine the biogas and methane potentials. For this purpose, digesters of 1 L batch were used, which were placed in a water bath at 38 °C. Each sample was tested in triplicate. The ratio of organic matter of the sample to organic matter of inoculum was <0.5 and was used to calculate the initial weight. After the addition of the substrate inoculums mixture, the digesters were closed and flushed with N_2 atmosphere. The produced biogas from each replicate was stored in a gas bag. The methane and carbon dioxide contents of the biogas was analyzed every three days with a biogas analyzer (bm 2000, Ansyco®, Karlsruhe, Germany). The volume of biogas was measured via gas meter (Ritter Apparatebau GmbH, Bochum, Germany). The biogas and methane potentials were corrected as standard norm liters (273.15 k and 1013.25 mbar) and reported for each kilogram fresh matter or organic matter. The test was ended until the daily amount of biogas produced corresponded to 1% of the total amount of gas (here, after 30 days).

4.6. Data Analysis

Biochemical characteristics are presented as means and corresponding standard deviations of the mean (sd) in brackets. The test characteristics affected by the factors "substrate condition" (Collection I) and "development stage" (Collection II) were first tested for normal distribution using the Shapiro–Wilk test. For a given normal distribution, one-way analysis of variance (ANOVA) followed by an F-test was applied. When the trait values were not normally distributed and neither log nor sqrt transformations achieved a normal distribution, the Kruskal–Wallis test was preferred. In the case of significant factor effects ($p < 0.05$), post-hoc mean comparisons among the factor levels were calculated. After an ANOVA, the HSD Test was used for this purpose. After the non-parametric alternative Kruskal–Wallis test, the Dunn test was applied. The α-level was set to 0.05. All statistical analyses were performed by scripts using the R environment version 3.6.3. [77].

Criteria for the comparative analysis to place our own results on PA degradation by ensiling into the context of current knowledge (Figure 3) were as follows: (1) laboratory ensiling conditions similar to our study, (2) information on the proportions of PA-containing species in the substrate, (3) identification of the dominant biochemical PA type, (4) an ensiling period of 90–105 days, and (5) precise information on the dry matter content of the ensiled material.

5. Conclusions

The substrate of the biomass from common ragwort is characterized by a strong dependence of the utilization-determining composition in relation to the ontogenetic development stage. In the course of the phenological development of common ragwort until the end of flowering, younger side shoots and flower sets are obviously not able to compensate for the progressive lignification of the structural substances. At the recommended stage of cutting, the proportions of structural substances have reached such an extent that their biochemical conversion in wet fermentation processes is limited.

Moderate levels of crude protein and sufficiently high levels of water-soluble carbohydrates for lactic fermentation allow for ensiling of common ragwort-dominated feedstock without the addition of WSCH-rich substrates, even in the stage of full flowering. However, the level of ensiling technology has to be advanced, especially in terms of rapid silo filling, careful feedstock compaction, and prompt covering in order to ensure early onset of anaerobic conditions. The low dry matter content of a freshly harvested ragwort biomass is always below the threshold value of 28% dry matter. Hence, direct ensiling is not recommended. The harvested plant material must be wilted before chopping to avoid effluent runoffs from the feedstock.

For the senecionine-type PAs, ensiling proved to be a suitable method of PA degradation. The prerequisite for this is a rather low dry matter content. Hence, excessive high wilting degrees need to be avoided. As a result, the target corridor for significant PA reduction via ensiling is quite narrow (28–32% DM). Since ensiling alone cannot guarantee PA degradation to a level that reliably eliminates contamination risks, subsequent biomethanization is an advisable option. After biogas processing, the PA-containing feedstock is decontaminated to such an extent that the digestate can be applied as organic fertilizer to grain crops without further risks to the agri-food chain. This concerns not only the PA contamination risk, but also the risk of ragwort seed dispersal with the digestate, which is a very efficient method for preventing germination.

Specific methane yields of only 173.6 $LN \cdot kg^{-1}$ oDM (fresh) and 185.1 $LN \cdot kg^{-1}$ oDM (silage), as they occur at the time of full flowering (at this point, the recommended cutting scheme), are in the lower scale level of landscape conservation growth and thus in the border area of economic usability. Ragwort substrate utilization can be lucrative for a biogas plant operator under the condition of short transport distances from the feedstock to the plant and exemption from the prime costs of the biomass production. By factoring in the safe treatment and degradation of PA-contaminated biomasses of an otherwise critical plant material, this usage option is definitely advisable. Our recommendation for a utilization-friendly application and control of common ragwort mass infestations areas is linked to the average growth heights of *J. vulgaris* from about 45 to 55 cm, largely irrespective of its state of flowering. However, if the proportion of ragwort does not dominate the stand, it is possible to cut at full flowering, if the accompanying gramineous vegetation can compensate for the poorer microbial degradability of the fermentable content.

Author Contributions: Conceptualization, J.M., T.B. and D.W.; methodology plant analysis, J.M.; methodology toxin analysis, T.B. and M.S.C.; methodology ensiling, J.M. and D.W.; methodology biomethanization, D.W.; biometric analysis, J.M.; resources, D.W. and T.B.; data curation, J.M., T.B. and D.W.; writing—original draft preparation, J.M.; writing—review and editing, T.B., D.W. and M.S.C.; All authors have read and agreed to the published version of the manuscript.

Funding: This research received no external funding.

Institutional Review Board Statement: Not applicable.

Informed Consent Statement: Not applicable.

Data Availability Statement: Data are included in this article. In the event of justified inquiries in the interest of science, the authors will provide the original data.

Acknowledgments: M.S.C. thanks the Niedersächsisches Ministerium für Wissenschaft und Kultur for the scholarship within the framework of "Wissenschaft.Niedersachsen.Weltoffen". The authors would like to thank Rotraut Degner for her technical help with plant collection and sample preparation and Elvira Scherelis for her help with language polishing. The authors thank the laboratory staff for their analytical support.

Conflicts of Interest: The authors declare no conflict of interest.

Appendix A

Table A1. Summary of PA/PANO-degradation studies on PA-containing *Jacobaea/Senecio* species. The tabulated data matches the data shown in Figure 3 of the manuscript.

Authors/Study	Plant under Study	Main Types of Pyrrolizidine Alkaloids	Percentage of *Senecio* in the Feedstock (% FM)	Initial PA-Level before Ensiling (mg/kg DM)	Feedstock Dry Matter Content (% FM)	Extent of PA-Degradation (% DM)
Müller et al. (2022, this study) lab silo A	*Jacobaea vulgaris* Gaertn.	erucifoline	100	7714.6	22.9	80.3
lab silo B			100	7741.6	22	73.5
lab silo C			100	8429.1	21.4	76.2
Berendonk & Hünting (2011)	*Jacobaea vulgaris* Gaertn.	senecionine	100	487.5	36.4	0
			75	365.7	38.6	46.8
			50	243.8	39.8	34.5
			25	121.9	40.5	38.4
			10	48.8	39.9	9.4
			5	24.4	43.8	33.6
Candrian et al. (1984)	*Senecio alpinus* L.	seneciphylline	100	4066.7	19.1	95.5
			41	1637.4	26.2	90.9
			23	918.6	29.1	91.4
			7	280.6	30.9	80.4
			3.5	140	32.8	54.3
Gottschalk et al. (2015)	*Senecio vulgaris* L.	mixed type	4.5	59	25	91.8
Klevenhusen et al. (2019)	*Senecio vulgaris* L.	senkirkine	10	565	22.1	35
			10	635	22.9	40.9

References

1. Harper, J.L.; Wood, W.A. *Senecio jacobaea* L. *J. Ecol.* **1957**, *45*, 617. [CrossRef]
2. Jung, S.; Lauter, J.; Hartung, N.M.; These, A.; Hamscher, G.; Wissemann, V. Genetic and chemical diversity of the toxic herb Jacobaea vulgaris Gaertn. (syn. *Senecio jacobaea* L.) in Northern Germany. *Phytochemistry* **2020**, *172*, 112235. [CrossRef] [PubMed]
3. Buckland, S.; Thompson, K.; Hodgson, J.; Grime, J. Grassland invasions: Effects of manipulations of climate and management. *J. Appl. Ecol.* **2001**, *38*, 301–309. [CrossRef]
4. Hodkinson, D.J.; Thompson, K. Plant Dispersal: The Role of Man. *J. Appl. Ecol.* **1997**, *34*, 1484–1496. [CrossRef]
5. Wardle, D.A. The Ecology of ragwort (*Senecio jacobaea* L.)—A review. *N. Z. J. Ecol.* **1987**, *10*, 67–76.
6. Crawley, M.J.; Nachapong, M. The Establishment of Seedlings from Primary and Regrowth Seeds of Ragwort (*Senecio jacobaea*). *J. Ecol.* **1985**, *73*, 255. [CrossRef]
7. Bosshard, A.; Joshi, J.; Lüscher, A.; Schaffner, U. Tansy ragwort and other ragwort species in Switzerland: An overview: (Jakobs-und andere Kreuzkrautarten: Eine Standortbestimmung). *Agrarforschung* **2003**, *10*, 231–235.
8. Andersson, S. Floral display and pollination success in *Senecio jacobaea* (Asteraceae): Interactive effects of head and corymb size. *Am. J. Bot.* **1996**, *83*, 71–75. [CrossRef]
9. Kunin, W.E. Patterns of Herbivore Incidence on Experimental Arrays and Field Populations of Ragwort, *Senecio jacobaea. Oikos* **1999**, *84*, 515. [CrossRef]
10. Bradbury, R.B.; Culvenor, C.C. The Alkaloids of *Senecio jacobaea* L. *Aust. J. Chem.* **1954**, *7*, 378. [CrossRef]
11. Witte, L.; Ernst, L.; Adam, H.; Hartmannt, T. Chemotypes of two pyrrolizidine alkaloid-containing Senecio species. *Phytochemistry* **1992**, *31*, 559–565. [CrossRef]
12. Macel, M.; Vrieling, K.; Klinkhamer, P.G.L. Variation in pyrrolizidine alkaloid patterns of *Senecio jacobaea. Phytochemistry* **2004**, *65*, 865–873. [CrossRef] [PubMed]
13. Goeger, D.E.; Cheeke, P.R.; Schmitz, J.A.; Buhler, D.R. Toxicity of tansy ragwort (*Senecio jacobaea*) to goats. *Am. J. Vet. Res.* **1982**, *43*, 252–254. [PubMed]
14. Smith, L.W.; Culvenor, C.C. Plant sources of hepatotoxic pyrrolizidine alkaloids. *J. Nat. Prod.* **1981**, *44*, 129–152. [CrossRef]
15. Wiedenfeld, H.; Edgar, J. Toxicity of pyrrolizidine alkaloids to humans and ruminants. *Phytochem. Rev.* **2011**, *10*, 137–151. [CrossRef]

16. White, H.D.; Krumperman, P.H.; Cheeke, P.R.; Deinzer, M.L.; Buhler, D.R. Mutagenic responses of tansy ragwort (*Senecio jacobaea*) plant, pyrrolizidine alkaloids and metabolites in goat milk with the salmonella/mammalian-microsome mutagenic test. *J. Anim. Sci.* **1984**, *58*, 1245–1254. [CrossRef]
17. Xia, Q.; He, X.; Shi, Q.; Lin, G.; Fu, P.P. Quantitation of DNA reactive pyrrolic metabolites of senecionine—A carcinogenic pyrrolizidine alkaloid by LC/MS/MS analysis. *J. Food Drug Anal.* **2020**, *28*, 167–174. [CrossRef]
18. Edgar, J.A.; Molyneux, R.J.; Colegate, S.M. Pyrrolizidine Alkaloids: Potential Role in the Etiology of Cancers, Pulmonary Hypertension, Congenital Anomalies, and Liver Disease. *Chem. Res. Toxicol.* **2015**, *28*, 4–20. [CrossRef]
19. Cameron, E. A Study of the Natural Control of Ragwort (*Senecio jacobaea* L.). *J. Ecol.* **1935**, *23*, 265. [CrossRef]
20. Leiss, K.A. Management practices for control of ragwort species. *Phytochem. Rev.* **2011**, *10*, 153–163. [CrossRef]
21. Gottschalk, C.; Kaltner, F.; Zimmermann, M.; Korten, R.; Morris, O.; Schwaiger, K.; Gareis, M. Spread of Jacobaea vulgaris and Occurrence of Pyrrolizidine Alkaloids in Regionally Produced Honeys from Northern Germany: Inter- and Intra-Site Variations and Risk Assessment for Special Consumer Groups. *Toxins* **2020**, *12*, 441. [CrossRef] [PubMed]
22. Neumann, H.; Huckauf, A. Jakobs-Kreuzkraut (*Senecio jacobaea*): Eine Ursache für Pyrrolizidin-Alkaloide im Sommerhonig?: Tansy ragwort (*Senecio jacobaea*): A source of pyrrolizidine alkaloids in summer honey? *J. Verbr. Lebensm.* **2016**, *11*, 105–115. [CrossRef]
23. Schrenk, D.; Gao, L.; Lin, G.; Mahony, C.; Mulder, P.P.J.; Peijnenburg, A.; Pfuhler, S.; Rietjens, I.M.C.M.; Rutz, L.; Steinhoff, B.; et al. Pyrrolizidine alkaloids in food and phytomedicine: Occurrence, exposure, toxicity, mechanisms, and risk assessment—A review. *Food Chem. Toxicol.* **2020**, *136*, 111107. [CrossRef] [PubMed]
24. Kempf, M.; Wittig, M.; Reinhard, A.; von der Ohe, K.; Blacquière, T.; Raezke, K.-P.; Michel, R.; Schreier, P.; Beuerle, T. Pyrrolizidine alkaloids in honey: Comparison of analytical methods. *Food Addit. Contam. Part A Chem. Anal. Control Expo. Risk Assess.* **2011**, *28*, 332–347. [CrossRef]
25. Suter, M.; Siegrist-Maag, S.; Connolly, J.; Lüscher, A. Can the occurrence of *Senecio jacobaea* be influenced by management practice? *Weed Res.* **2007**, *47*, 262–269. [CrossRef]
26. Möhler, H.; Diekötter, T.; Herrmann, J.D.; Donath, T.W. Allelopathic vs. autotoxic potential of a grassland weed—evidence from a seed germination experiment. *Plant Ecol. Divers.* **2018**, *11*, 539–549. [CrossRef]
27. UK Department for Environment Food and Rural Affairs. The Weeds Act, Chapter 54. Available online: https://www.legislation.gov.uk/ukpga/2003/40/introduction/enacted (accessed on 6 January 2022).
28. Roberts, P.D.; Pullin, A.S. The effectiveness of management interventions used to control ragwort species. *Environ. Manag.* **2007**, *39*, 691–706. [CrossRef]
29. Augustin, B.; Oehme, S.; Mattern, G.; Roth, W. Influence of different measures against *Senecio jacobaea* concerning efficacy and biological diversity. *Julius-Kühn-Archiv* **2018**, *458*, 47–53. (In German) [CrossRef]
30. Möhler, H. Basic and Applied Studies to Evaluate Management Options for the Regulation of Tansy Ragwort (*Jacobaea vulgaris*). Inaugural Dissertation, Christian-Albrechts-University, Kiel, Germany, 2020.
31. Siegrist-Maag, S.; Luescher, A.; Suter, M. Reaktion von Jakobs-Kreuzkraut auf Schnitt: Sensitive reaction of ragwort (*Senecio jacobaea*) to cutting dates. *Agrarforschung* **2008**, *15*, 338–343.
32. Gottschalk, C.; Ronczka, S.; Preiß-Weigert, A.; Ostertag, J.; Klaffke, H.; Schafft, H.; Lahrssen-Wiederholt, M. Pyrrolizidine alkaloids in natural and experimental grass silages and implications for feed safety. *Anim. Feed Sci. Technol.* **2015**, *207*, 253–261. [CrossRef]
33. Driehuis, F.; Wilkinson, J.M.; Jiang, Y.; Ogunade, I.; Adesogan, A.T. Silage review: Animal and human health risks from silage. *J. Dairy Sci.* **2018**, *101*, 4093–4110. [CrossRef] [PubMed]
34. Kalač, P.; Kaltner, F. Pyrrolizidine alkaloids of European Senecio/Jacobaea species in forage and their carry-over to milk: A review. *Anim. Feed Sci. Technol.* **2021**, *280*, 115062. [CrossRef]
35. Hama, J.R.; Strobel, B.W. Occurrence of pyrrolizidine alkaloids in ragwort plants, soils and surface waters at the field scale in grassland. *Sci. Total Environ.* **2021**, *755*, 142822. [CrossRef] [PubMed]
36. Crews, C.; Driffield, M.; Berthiller, F.; Krska, R. Loss of pyrrolizidine alkaloids on decomposition of ragwort (*Senecio jacobaea*) as measured by LC-TOF-MS. *J. Agric. Food Chem.* **2009**, *57*, 3669–3673. [CrossRef]
37. Chmit, M.S.; Müller, J.; Wiedow, D.; Horn, G.; Beuerle, T. Biodegradation and utilization of crop residues contaminated with poisonous pyrrolizidine alkaloids. *J. Environ. Manag.* **2021**, *290*, 112629. [CrossRef]
38. Zhao, S.; Schmidt, S.; Gao, H.; Li, T.; Chen, X.; Hou, Y.; Chadwick, D.; Tian, J.; Dou, Z.; Zhang, W.; et al. *A Prominent Role for Precision Composting in Sustainable Agriculture*; Research Square Platform LLC: Durham, NC, USA, 2021.
39. Capodaglio, A.; Callegari, A.; Lopez, M. European Framework for the Diffusion of Biogas Uses: Emerging Technologies, Acceptance, Incentive Strategies, and Institutional-Regulatory Support. *Sustainability* **2016**, *8*, 298. [CrossRef]
40. Van Meerbeek, K.; Appels, L.; Dewil, R.; van Beek, J.; Bellings, L.; Liebert, K.; Muys, B.; Hermy, M. Energy potential for combustion and anaerobic digestion of biomass from low-input high-diversity systems in conservation areas. *GCB Bioenergy* **2015**, *7*, 888–898. [CrossRef]
41. Müller, J.; Hahn, J. Ensilability of Biomass From Effloresced Flower Strips as Co-substrate in Bioenergy Production. *Front. Bioeng. Biotechnol.* **2020**, *8*, 14. [CrossRef]
42. Berendonk, C.; Cerff, D.; Hünting, K.; Wiedenfeld, H.; Becerra, J.; Kuschak, M. Pyrrolizidine alkaloid level in *Senecio jacobaea* and Senecio erraticus-the effect of plant organ and forage conservation. *Grassl. Sci. Eur.* **2010**, *15*, 669–671.

43. Becerra-Jimenez, J.; Kuschak, M.; Röder, E.T.; Wiedenfeld, H. Toxic pyrrolizidinalkaloids as undesired contaminants in food and feed: Degradation of the PAs from *Senecio jacobaea* in silage. *Pharmazie* **2013**, *68*, 636–639. [CrossRef]
44. Ferrero, F.; Dinuccio, E.; Rollé, L.; Tabacco, E.; Borreani, G. Suitability of cardoon (Cynara cardunculus L.) harvested at two stages of maturity to ensiling and methane production. *Biomass Bioenergy* **2020**, *142*, 105776. [CrossRef]
45. Berendonk, C.; Hünting, K. Einfluss der Silierung auf den Gehalt an Pyrrolizidinalkaloiden von *Senecio jacobaea* in Abhängigkeit vom Senecio-Gehalt im Siliergut. *Mitteilungen AGGF* **2011**, *12*, 159–162.
46. Candrian, U.; Luethy, J.; Schmid, P.; Schlatter, C.; Gallasz, E. Stability of pyrrolizidine alkaloids in hay and silage. *J. Agric. Food Chem.* **1984**, *32*, 935–937. [CrossRef]
47. Klevenhusen, F.; Pieper, R.; Winter, J.; Ronczka, S.; Speer, K. Stability of pyrrolizidine alkaloids from Senecio vernalis in grass silage under different ensilage conditions. *J. Sci. Food Agric.* **2019**, *99*, 6649–6654. [CrossRef]
48. Meserszmit, M.; Chrabąszcz, M.; Chylińska, M.; Szymańska-Chargot, M.; Trojanowska-Olichwer, A.; Kącki, Z. The effect of harvest date and the chemical characteristics of biomass from Molinia meadows on methane yield. *Biomass Bioenergy* **2019**, *130*, 105391. [CrossRef]
49. Scherzinger, M.; Kulbeik, T.; Kaltschmitt, M. Autoclave pre-treatment of green wastes–Effects of temperature, residence time and rotation speed on fuel properties. *Fuel* **2020**, *273*, 117796. [CrossRef]
50. Herrmann, C.; Idler, C.; Heiermann, M. Biogas crops grown in energy crop rotations: Linking chemical composition and methane production characteristics. *Bioresour. Technol.* **2016**, *206*, 23–35. [CrossRef] [PubMed]
51. Sivamani, S.; Saikat, B.; Naveen Prasad, B.S.; Baalawy, A.A.S.; Al-Mashali, S.M.A. A Comprehensive Review on Microbial Technology for Biogas Production. In *Bioenergy Research: Revisiting Latest Development*, 1st ed.; Srivastava, M., Srivastava, N., Singh, R., Eds.; Springer: Singapore, 2021; pp. 53–78; ISBN 978-981-33-4614-7.
52. Triolo, J.M.; Pedersen, L.; Qu, H.; Sommer, S.G. Biochemical methane potential and anaerobic biodegradability of non-herbaceous and herbaceous phytomass in biogas production. *Bioresour. Technol.* **2012**, *125*, 226–232. [CrossRef]
53. Dandikas, V.; Heuwinkel, H.; Lichti, F.; Drewes, J.E.; Koch, K. Correlation between biogas yield and chemical composition of energy crops. *Bioresour. Technol.* **2014**, *174*, 316–320. [CrossRef]
54. Stachowiak–Wencek, A.; Bocianowski, J.; Waliszewska, H.; Borysiak, S.; Waliszewska, B.; Zborowska, M. Statistical prediction of biogas and methane yields during anaerobic digestion based on the composition of lignocellulosic biomass. *BioRes* **2021**, *16*, 7086–7100. [CrossRef]
55. Raffrenato, E.; Fievisohn, R.; Cotanch, K.W.; Grant, R.J.; Chase, L.E.; van Amburgh, M.E. Effect of lignin linkages with other plant cell wall components on in vitro and in vivo neutral detergent fiber digestibility and rate of digestion of grass forages. *J. Dairy Sci.* **2017**, *100*, 8119–8131. [CrossRef] [PubMed]
56. Weißbach, F. On Assessing the Gas Production Potential of Renewable Primary Products (Zur Bewertung des Gasbildungspotenzials von nachwachsenden Rohstoffen). *Landtechnik* **2008**, *63*, 356–358. [CrossRef]
57. Meak, P. Biochemische Charakterisierung von Pflanzen Unterschiedlicher Nutzungsintensität zur Ableitung von Parametern für die Ermittlung des Energetischen Futterwertes. Inaugural Dissertation, Humboldt University, Berlin, Germany, 2002.
58. Frost, R.A.; Wilson, L.M.; Launchbaugh, K.L.; Hovde, E.M. Seasonal Change in Forage Value of Rangeland Weeds in Northern Idaho. *Invasive Plant Sci. Manag.* **2008**, *1*, 343–351. [CrossRef]
59. Purcell, P.J.; Boland, T.M.; O'Brien, M.; O'Kiely, P. In vitro rumen methane output of forb species sampled in spring and summer. *AFSci* **2012**, *21*, 83–90. [CrossRef]
60. Millard, P. The accumulation and storage of nitrogen by herbaceous plants. *Plant Cell Environ.* **1988**, *11*, 1–8. [CrossRef]
61. Debessai, W.T. Evaluation of Endophyte-Infected Tall Fescue Products, Their Interaction with *Senecio jacobaea* in Ruminants, and Detoxification of Alkaloids by Ammoniation or Ensiling after Grinding. Ph.D. Thesis, Oregon State University, Corvallis, OR, USA, 1998.
62. McDonald, P.; Henderson, A.R.; Heron, S. *The Biochemistry of Silage*, 2nd ed.; Chalcombe Publications: Marlow, UK, 1991.
63. Pahlow, G.; Muck, R.E.; Driehuis, F.; Elferink, S.J.W.H.O.; Spoelstra, S.F. Microbiology of Ensiling. In *Silage Science and Technology*; Buxton, D.R., Muck, R.E., Harrison, J., Eds.; Soil Science Society of America: Madison, WI, USA, 2003; pp. 31–93; ISBN 9780891182344.
64. Cramer, L.; Schiebel, H.-M.; Ernst, L.; Beuerle, T. Pyrrolizidine alkaloids in the food chain: Development, validation, and application of a new HPLC-ESI-MS/MS sum parameter method. *J. Agric. Food Chem.* **2013**, *61*, 11382–11391. [CrossRef]
65. Murphy, J.D.; Thamsiriroj, T. Fundamental science and engineering of the anaerobic digestion process for biogas production. In *The Biogas Handbook: Science, Production and Applications*; Wellinger, A., Ed.; Woodhead Publishing: Cambridge, UK, 2013; pp. 104–130; ISBN 9780857094988.
66. Teixeira Franco, R.; Buffière, P.; Bayard, R. Ensiling for biogas production: Critical parameters. A review. *Biomass Bioenergy* **2016**, *94*, 94–104. [CrossRef]
67. Herrmann, C.; Heiermann, M.; Idler, C. Effects of ensiling, silage additives and storage period on methane formation of biogas crops. *Bioresour. Technol.* **2011**, *102*, 5153–5161. [CrossRef]
68. Mangold, A.; Lewandowski, I.; Hartung, J.; Kiesel, A. Miscanthus for biogas production: Influence of harvest date and ensiling on digestibility and methane hectare yield. *GCB Bioenergy* **2019**, *11*, 50–62. [CrossRef]
69. Wannasek, L.; Ortner, M.; Amon, B.; Amon, T. Sorghum, a sustainable feedstock for biogas production? Impact of climate, variety and harvesting time on maturity and biomass yield. *Biomass Bioenergy* **2017**, *106*, 137–145. [CrossRef]

70. Meier, U. *Growth Stages of Mono-and Dicotyledonous Plants: BBCH Monograph*, 2nd ed.; Federal Biological Research Centre for Agriculture and Forestry: Berlin/Braunschweig, Germany, 2001.
71. De Boever, J.L.; Cottyn, B.G.; Buysse, F.X.; Wainman, F.W.; Vanacker, J.M. The use of an enzymatic technique to predict digestibility, metabolizable and net energy of compound feedstuffs for ruminants. *Anim. Feed Sci. Technol.* **1986**, *14*, 203–214. [CrossRef]
72. Naumann, C.; Bassler, R. *Die Chemische Untersuchung von Futtermitteln, 3. Aufl.*; VDLUFA-Verl.: Darmstadt, Germany, 2012; ISBN 9783941273146.
73. Letsyo, E.; Jerz, G.; Winterhalter, P.; Lindigkeit, R.; Beuerle, T. Incidence of Pyrrolizidine Alkaloids in Herbal Medicines from German Retail Markets: Risk Assessments and Implications to Consumers. *Phytother. Res.* **2017**, *31*, 1903–1909. [CrossRef] [PubMed]
74. Frölich, C.; Ober, D.; Hartmann, T. Tissue distribution, core biosynthesis and diversification of pyrrolizidine alkaloids of the lycopsamine type in three Boraginaceae species. *Phytochemistry* **2007**, *68*, 1026–1037. [CrossRef] [PubMed]
75. Weißbach, F.; Strubelt, C. Correcting the dry matter content of grass silages as a substrate for biogas production. *Landtechnik* **2008**, *63*, 210–211. [CrossRef]
76. Verein Deutscher Ingenieure. *Fermentation of Organic Materials-Characterisation of the Substrate, Sampling, Collection of Material Data, Fermentation Tests: Vergärung Organischer Stoffe-Substratcharakterisierung, Probenahme, Stoffdatenerhebung, Gärversuche*; Beuth-Verlag: Berlin, Germany, 2016; ICS 13.030.30, 27.190 (VDI 4630). Available online: https://www.vdi.de/fileadmin/pages/vdi_de/redakteure/richtlinien/inhaltsverzeichnisse/2385990.pdf (accessed on 24 January 2022).
77. R Development Core Team. *R: A Language and Environment for Statistical Computing*; R Foundation for Statistical Computing: Vienna, Austria, 2020. Available online: https://www.eea.europa.eu/data-and-maps/indicators/oxygen-consuming-substances-in-rivers/r-development-core-team-2006 (accessed on 24 January 2022).

plants
MDPI

Article

LC–DAD–MS Phenolic Characterisation of Six Invasive Plant Species in Croatia and Determination of Their Antimicrobial and Cytotoxic Activity

Danijela Poljuha [1], Barbara Sladonja [1,*], Ivana Šola [2], Mateja Šenica [3], Mirela Uzelac [1], Robert Veberič [3], Metka Hudina [3], Ibukun Michael Famuyide [4], Jacobus N. Eloff [4] and Maja Mikulic-Petkovsek [3]

1 Department of Agriculture and Nutrition, Institute of Agriculture and Tourism, Karla Huguesa 8, 52440 Poreč, Croatia; danijela@iptpo.hr (D.P.); mirela@iptpo.hr (M.U.)
2 Department of Biology, Faculty of Science, University of Zagreb, Horvatovac 102a, 10000 Zagreb, Croatia; ivana.sola@biol.pmf.hr
3 Department of Agronomy, Biotechnical Faculty, University of Ljubljana, Jamnikarjeva 101, 1000 Ljubljana, Slovenia; mateja.senica@bf.uni-lj.si (M.Š.); robert.veberic@bf.uni-lj.si (R.V.); metka.hudina@bf.uni-lj.si (M.H.); maja.mikulic-petkovsek@bf.uni-lj.si (M.M.-P.)
4 Phytomedicine Programme, Paraclinical Sciences Department, University of Pretoria, P/Bag X04, Onderstepoort, Pretoria 0110, Gauteng, South Africa; adeyerimi@gmail.com (I.M.F.); kobus.eloff@up.ac.za (J.N.E.)
* Correspondence: barbara@iptpo.hr; Tel.: +385-395-52-308-304

Abstract: Invasive plants' phytochemicals are important for their invasiveness, enabling them to spread in new environments. However, these chemicals could offer many pharmaceutical compounds or active ingredients for herbal preparations. This study provides the first LC–MS phytochemical screening of six invasive alien plant species (IAPS) in the Istria region (Croatia): *Ailanthus altissima*, *Ambrosia artemisiifolia*, *Conyza canadensis*, *Dittrichia viscosa*, *Erigeron annuus*, and *Xanthium strumarium*. The study aims to identify and quantify the phenolic content of their leaf extracts and assess their antimicrobial and cytotoxic potential. A total of 32 species-specific compounds were recorded. Neochlorogenic, chlorogenic, and 5-*p*-coumaroylquinic acids, quercetin-3-glucoside, and kaempferol hexoside were detected in all the tested IAPS. Hydroxycinnamic acid derivatives were the main components in all the tested IAPS, except in *E. annuus*, where flavanones dominated with a share of 70%. *X. strumarium* extract had the best activity against the tested bacteria, with an average MIC value of 0.11 mg/mL, while *A. altissima* and *X. strumarium* extracts had the best activity against the tested fungi, with an average MIC value of 0.21 mg/mL in both cases. All the plant extracts studied, except *X. strumarium*, were less cytotoxic than the positive control. The results provided additional information on the phytochemical properties of IAPS and their potential for use as antimicrobial agents.

Keywords: antimicrobial; cytotoxicity; ecosystem services; invasive plant species; phenolics; phytopharmaceuticals

1. Introduction

Although invasive alien plant species (IAPS) impacts are unpredictable, they all have mechanisms that contribute to their invasiveness. They often include direct competition, indirect competition mediated by herbivores or changes in the soil community, and interference competition via allelopathy [1]. Successful alien species often owe their capability to repel native enemies to novel biochemistry [2]. Previous comparative metabolomics studies on invasive species showed that successful invasive alien plants have more total and more unique metabolites than native ones [3]. Phytochemical uniqueness appeared to be very important in the invasion ability of alien plants [4–6]. Therefore, phytochemicals from alien invasive species may be a leading cause of environmentally harmful effects.

On the other side, some invasive species have compounds potentially useful to humans and could provide valuable ecosystem services [7]. Many plant species possess antifungal and antibacterial activity [8–10]. Antimicrobial agents produced by plants' specialised metabolism serve as their natural protective mechanism, increasing their competitiveness.

Thus, plant extracts or plant specialised metabolites that are not toxic and are specific in their action are considered possible candidates for pharmaceuticals. They could also be active ingredients of natural preparations, such as pesticides and herbicides in organic crop production. Meela et al. [10] argued that fungal infections might control plant distribution, and the success of invasive plant species may be due to their antifungal activity, which they found in seven invasive species. This may help fend off fungal pathogens or disturb established symbiotic relationships in their new habitat. Eloff et al. [11] showed that an easy solvent–solvent fractionation process of acetone leaf extracts of the invasive *Melianthus comosus* Vahl led to a product with much higher activity than commercial fungicides. The activity of this product in field trials was also much better than that of the commercial product. They also stated that toxic antifungal extracts might be useful in the floricultural industry.

Up to date, there have been very little literature data on the phytochemical potential of IAPS or their exploration in Croatia [9,12–14]. For our study, we selected six IAPS, all included in the list of the most relevant invasive plant species of Croatia [15] and widely present in the Istria region (the peninsula in west Croatia). Previous phytochemical studies revealed the presence of different bioactive compounds in those species: *Ailanthus altissima* (Mill.) Swingle [16], *Ambrosia artemisiifolia* L. [17], *Conyza canadensis* (L.) Cronquist [18], *Dittrichia viscosa* (L.) Greuter [19], *Xanthium strumarium* L. [20], and *Erigeron annuus* (L.) Pers. [21]. The extracts of selected plant species are used in traditional medicine in China and India, and their antimicrobial properties have been reported [22–27]. Our previous research [9] showed that *A. altissima* from Croatia could be a valuable resource for antimicrobial activity since its leaf extracts were as effective against *Escherichia coli* and *Candida albicans* as standard antibiotics.

Phenolic compounds in plants are generally one of the most significant bioactive groups, playing an essential role in their development and defence against infection and injury [28]. Higher concentrations of phenolic compounds in invasive plants may provide a competitive advantage over native species [6]. Still, the role and function of high phenolic diversity remain to be elucidated.

The phytochemical characterisation presented in this study is one step forward in this task. This knowledge is also crucial to define the possible application areas for the rational use of these species.

In our study, we identified and quantified phenolics in leaves of six IAPS from the territory of Istria (Croatia) for the first time (except for *A. altissima*) and tested the antibacterial, antifungal, and cytotoxic activities of their extracts as a basis for new possible ecosystem services.

2. Results and Discussion

2.1. Extraction Yield and Phenolic Concentration

The extraction technique's efficacy is manifested in the yield of an extract with maintained functional properties. We prepared two types of extracts from six selected IAPS. We used methanol as an extractant to prepare phenolic extracts and acetone for the bioassays (cytotoxicity and antimicrobial testing) due to its lower toxicity to fungi [29] and based on the comparison of different extractants [30,31]. We obtained the highest phenolic concentration from *D. viscosa* (68 mg/g), *E. annuus* (46 mg/g), and *A. altissima* (42 mg/g), followed by *X. strumarium* (29 mg/g), *A. artemisiifolia* (15 mg/g), and *C. canadensis* (13 mg/g) (Table 1).

When we used acetone as an extractant, we obtained the highest yield from *A. artemisiifolia* (101 mg/g), followed by *E. annuus* (66 mg/g), *D. viscosa* (63 mg/g), *X. strumarium* (57 mg/g), *C. canadensis* (47 mg/g), and finally *A. altissima* with the lowest extraction yield of 42 mg/g (Tables 2 and 3).

Table 1. The concentrations of identified phenolics in six invasive plant species (mg/g DW ± standard error). Data are presented as a heat map in Supplementary Figure S1.

	Ailanthus altissima	*Ambrosia artemisiifolia*	*Conyza canadensis*	*Dittrichia viscosa*	*Erigeron annuus*	*Xanthium strumarium*	**Stat. Significance**
Dicaffeoylquinic acids	n.d.	3.78 ± 0.06	4.76 ± 0.52	22.36 ± 0.14	3.99 ± 0.33	11.46 ± 0.18	
3-caffeoylquinic acid	2.44 ± 0.20	0.37 ± 0.00	0.26 ± 0.02	0.24 ± 0.01	0.18 ± 0.01	0.61 ± 0.06	
4-caffeoylquinic acid	2.83 ± 0.28	0.28 ± 0.02	0.28 ± 0.03	n.d.	0.13 ± 0.02	0.20 ± 0.02	
5-caffeoylquinic acid	7.42 ± 1.11	5.04 ± 0.15	2.72 ± 0.10	8.42 ± 0.08	2.86 ± 0.21	11.25 ± 0.09	
triCQA	n.d.	n.d.	n.d.	n.d.	1.06 ± 0.04	0.83 ± 0.04	
Caffeic acid	n.d.	0.43 ± 0.00	n.d.	n.d.	n.d.	n.d.	
Other caffeic acid derivatives	n.d.	0.21 ± 0.01	n.d.	n.d.	0.15 ± 0.00	n.d.	
Caffeic acid derivatives total	**12.68 ± 1.62 [b]**	**10.11 ± 0.28 [b]**	**8.02 ± 0.77 [b]**	**31.02 ± 0.01 [a]**	**8.37 ± 0.72 [b]**	**24.35 ± 0.81 [a]**	***
Caftaric acids	**n.d.**	**3.72 ± 0.09**	**n.d.**	**n.d.**	**n.d.**	**n.d.**	
3-*p*-coumaroylquinic acid	0.09 ± 0.02	0.03 ± 0.01	0.04 ± 0.01	0.03 ± 0.01	0.05 ± 0.00	n.d.	
4-*p*-coumaroylquinic acid	0.24 ± 0.05	n.d.	n.d.	n.d.	n.d.	n.d.	
5-*p*-coumaroylquinic acid	0.68 ± 0.14	0.11 ± 0.01	0.04 ± 0.00	0.03 ± 0.01	n.d.	0.04 ± 0.00	
***p*-coumaric acid derivatives**	**1.02 ± 0.19 [a]**	**0.14 ± 0.01 [b]**	**0.09 ± 0.01 [b]**	**0.07 ± 0.01 [b]**	**0.05 ± 0.00 [b]**	**0.04 ± 0.00 [b]**	**
3-feruloylquinic acid	0.09 ± 0.01	0.03 ± 0.00	0.03 ± 0.01	n.d.	0.01 ± 0.00	0.01 ± 0.00	
5-feruloylquinic acid	n.d.	0.31 ± 0.02	n.d.	n.d.	n.d.	0.34 ± 0.04	
Ferulic acid derivatives	**0.09 ± 0.01 [b]**	**0.35 ± 0.01 [a]**	**0.03 ± 0.00 [c]**	**n.d.**	**0.01 ± 0.00 [c]**	**0.35 ± 0.04 [a]**	***
Hydroxycinnamic acid derivatives	**13.79 ± 1.82 [b]**	**10.60 ± 0.29 [b]**	**8.14 ± 0.78 [b]**	**31.08 ± 0.11 [a]**	**8.21 ± 0.73 [b]**	**24.39 ± 0.42 [a]**	***
Ellagic acid	n.d.	n.d.	3.24 ± 1.61	n.d.	n.d.	n.d.	
Ellagic acid pentoside	12.13 ± 2.31	n.d.	n.d.	n.d.	n.d.	n.d.	
Ellagic acid hexosides	n.d.	n.d.	n.d.	31.63 ± 0.43	1.07 ± 0.18	n.d.	
Ellagic acid rhamnoside	n.d.	n.d.	n.d.	n.d.	0.07 ± 0.01	n.d.	
Hydroxybenzoic acid derivatives	**12.13 ± 2.39 [b]**	n.d.	**3.24 ± 1.97 [bc]**	**31.63 ± 0.53 [a]**	**1.14 ± 0.23 [c]**	n.d.	***
Apigenin	n.d.	n.d.	n.d.	n.d.	n.d.	0.05 ± 0.01	
Apigenin hexoside	0.87 ± 0.07	n.d.	n.d.	n.d.	n.d.	n.d.	
Flavones	**0.87 ± 0.07 [a]**	n.d.	n.d.	n.d.	n.d.	**0.05 ± 0.01 [b]**	***
Procyanidin dimer	n.d.	n.d.	n.d.	n.d.	n.d.	1.19 ± 0.09	
Epicatechin	n.d.	n.d.	n.d.	n.d.	n.d.	0.02 ± 0.00	
Flavanols	n.d.	n.d.	n.d.	n.d.	n.d.	**1.21 ± 0.11**	
Quercetin-dihexoside	n.d.	1.19 ± 0.06	0.04 ± 0.01	n.d.	0.15 ± 0.02	n.d.	
Quercetin-galloyl-hexosides	0.69 ± 0.07	n.d.	n.d.	n.d.	n.d.	n.d.	
Quercetin-3-galactoside	2.81 ± 0.28	0.45 ± 0.01	n.d.	0.25 ± 0.10	n.d.	n.d.	
Quercetin-3-glucoside	0.56 ± 0.05	0.10 ± 0.00	0.05 ± 0.01	0.11 ± 0.01	0.01 ± 0.00	0.34 ± 0.00	***

Table 1. *Cont.*

	Ailanthus altissima	*Ambrosia artemisiifolia*	*Conyza canadensis*	*Dittrichia viscosa*	*Erigeron annuus*	*Xanthium strumarium*	Stat. Significance
Quercetin-3-rutinoside	n.d.	0.24 ± 0.00	n.d.	0.25 ± 0.01	n.d.	0.10 ± 0.02	
Quercetin-3-xyloside	n.d.	0.01 ± 0.00	0.06 ± 0.01	0.08 ± 0.00	n.d.	0.05 ± 0.01	
Quercetin-3-rhamnoside	n.d.	n.d.	n.d.	n.d.	0.27 ± 0.02	n.d.	
Quercetin-3-glucuronide	n.d.	0.43 ± 0.01	0.07 ± 0.02	0.24 ± 0.00	0.08 ± 0.01	0.75 ± 0.02	
Quercetin-3-arabinopyranoside	n.d.	0.10 ± 0.00	0.10 ± 0.01	1.32 ± 0.09	n.d.	n.d.	
Quercetin-acetyl hexoside	0.13 ± 0.02	n.d.	n.d.	n.d.	n.d.	n.d.	
Kaempferol-hexoside	1.04 ± 0.09	0.06 ± 0.00	0.07 ± 0.02	0.05 ± 0.00	0.01 ± 0.00	0.01 ± 0.00	
Kaempferol-acetyl-hexosides	0.43 ± 0.05	0.19 ± 0.00	n.d.	n.d.	n.d.	n.d.	
Kaempferol-3-rutinoside	n.d.	n.d.	n.d.	n.d.	0.31 ± 0.04	n.d.	
Kaempferol-3-glucuronide	n.d.	0.19 ± 0.01	1.05 ± 0.07	0.56 ± 0.02	2.95 ± 0.28	0.38 ± 0.01	
Kaempferol-rhamnoside-hexoside	0.10 ± 0.01	n.d.	n.d.	n.d.	0.09 ± 0.01	n.d.	
Kaempferol-galloyl-hexoside	0.22 ± 0.02	n.d.	n.d.	n.d.	0.02 ± 0.00	n.d.	
Kaempferol-glucuronyl-hexoside	n.d.	n.d.	n.d.	n.d.	0.02 ± 0.00	n.d.	***
Isorhamnetin hexoside	n.d.	0.19 ± 0.01	n.d.	1.34 ± 0.07	n.d.	n.d.	
Isorhamnetin acetyl hexoside	n.d.	0.31 ± 0.01	n.d.	n.d.	n.d.	n.d.	
Isorhamnetin-3-rutinoside	n.d.	0.03 ± 0.00	n.d.	n.d.	n.d.	n.d.	***
Myricetin-3-glucuronide	n.d.	n.d.	n.d.	n.d.	0.17 ± 0.03	n.d.	
Myricetin hexoside	n.d.	n.d.	n.d.	0.64 ± 0.02	n.d.	n.d.	
Laricitrin-3-glucuronide	n.d.	0.57 ± 0.01	n.d.	0.23 ± 0.01	n.d.	n.d.	
Laricitrin	n.d.	n.d.	n.d.	n.d.	n.d.	0.65 ± 0.03	
Syringetin	n.d.	n.d.	n.d.	n.d.	n.d.	0.44 ± 0.03	
Flavonols	**5.97 ± 0.58** ^{a}	**4.05 ± 0.14** ^{abc}	**1.45 ± 0.14** ^{c}	**5.07 ± 0.28** ^{ab}	**4.07 ± 0.49** ^{abc}	**3.23 ± 0.10** ^{bc}	***
Naringenin-hexosides	n.d.	n.d.	0.17 ± 0.05 ^{b}	n.d.	32.07 ± 2.05	n.d.	***
Flavanones	n.d.	n.d.	**0.17 ± 0.05** ^{b}	n.d.	**32.07 ± 2.05**	n.d.	***
Vescalagin	6.55 ± 1.27	n.d.	n.d.	n.d.	n.d.	n.d.	***
Ellagitannins	**6.55 ± 1.27**	**n.d.**	**n.d.**	**n.d.**	**n.d.**	**n.d.**	
Gallic acid	n.d.	0.35 ± 0.01 ^{b}	n.d.	n.d.	n.d.	n.d.	
Digalloyl-HHDP-hexoside isomer	2.39 ± 0.52 ^{a}	n.d.	n.d.	n.d.	n.d.	n.d.	***
Gallotannins	**2.39 ± 0.52** ^{a}	**0.35 ± 0.01** ^{b}	n.d.	n.d.	n.d.	n.d.	***
TOTAL	**41.70 ± 6.59** ^{ab}	**15.00 ± 0.42** ^{b}	**13.00 ± 2.90** ^{b}	**67.78 ± 0.57** ^{a}	**45.50 ± 4.32** ^{ab}	**29.02 ± 0.65** ^{ab}	

Different letters ($^{a–c}$) in the same row indicate significant differences, determined by LSD range test, in phenolic content between different invasive plant species; n.d.—not detected. (*–statistically significant differences at *p*-value < 0.05, **–statistically significant differences at *p*-value < 0.001, ***–statistically significant differences at *p*-value < 0.0001.)

Table 2. Yield of extract (mg/g), cytotoxicity (LC_{50}) against Vero African green monkey kidney cells (mg/mL), minimum inhibitory concentration (MIC; mg/mL), total antibacterial activity (TAA; yield/MIC) after 24 h incubation (mL/g), and selectivity indices (SI; LC_{50}/MIC) of the acetone extracts of different plant species and gentamicin (positive control) against test bacteria.

Plant Species	Yield	LC_{50}	*Escherichia coli*			*Enterococcus faecalis*			*Staphylococcus aureus*			*Pseudomonas aeruginosa*			Average		
			MIC	TAA	SI	MIC	TAA	SI	MIC	TAA	SI	MIC	TAA	SI	MIC	TAA	SI
A. altissima	42	0.17	0.23	183	0.74	0.04	1050	4.25	0.31	135	0.55	0.31	135	0.55	0.22	376	1.52
A. artemisiifolia	101	0.02	0.31	326	0.06	0.06	1683	0.33	0.31	326	0.06	0.63	160	0.03	0.32	624	0.12
C. canadensis	47	0.17	0.31	152	0.55	0.04	1175	4.25	1.88	25	0.09	1.25	38	0.14	0.87	348	1.29
D. viscosa	63	0.03	0.16	394	0.19	0.04	1575	0.75	0.31	203	0.10	0.31	203	0.10	0.21	594	0.33
E. annuus	66	0.15	0.31	213	0.48	0.04	1650	3.75	1.25	53	0.12	1.25	53	0.12	0.71	532	1.17
X. strumarium	57	0.001	0.23	248	0.004	0.04	1425	0.03	0.08	713	0.01	0.08	713	0.01	0.11	775	0.01
Average			0.26	253	0.34	0.04	1426	2.23	0.69	243	0.16	0.64	217	0.16			
Gentamicin			8×10^{-4}			2×10^{-4}			1.3×10^{-4}			3×10^{-4}			3.6×10^{-4}		
Doxorubicin		0.012															

Table 3. Yield of extract (mg/g), cytotoxicity (LC_{50}) against Vero African green monkey kidney cells (mg/mL), minimum inhibitory concentration (MIC; mg/mL), total antifungal activity (TAA; yield/MIC) after 24 h incubation (ml/g), and selectivity indices (SI; LC_{50}/MIC) of the acetone extracts of different plant species and amphotericin B (positive control) against test fungi.

Plant Species	Yield	LC_{50}	*Candida albicans*			*Cryptococcus neoformans*			*Aspergillus fumigatus*			Average		
			MIC	TAA	SI	MIC	TAA	SI	MIC	TAA	SI	MIC	TAA	SI
A. altissima	42	0.17	0.16	269	1.12	0.31	134	0.56	0.16	269	1.12	0.21	224	0.93
A. artemisiifolia	101	0.02	0.16	646	0.11	0.63	162	0.03	1.25	81	0.01	0.68	296	0.05
C. canadensis	47	0.17	0.16	301	1.09	0.31	150	0.55	0.63	75	0.27	0.37	175	0.64
D. viscosa	63	0.03	0.16	403	0.22	0.31	202	0.11	0.63	101	0.05	0.37	235	0.13
E. annuus	66	0.15	0.31	211	0.46	0.31	211	0.46	1.25	53	0.12	0.62	158	0.35
X. strumarium	57	0.001	0.23	243	0.003	0.16	365	0.01	0.23	243	0.003	0.21	284	0.003
Average			0.20	346	0.50	0.34	204	0.29	0.69	137	0.26			
Amphotericin B			8×10^{-3}			8×10^{-3}			8×10^{-3}					

2.2. LC–DAD–MS Identification and Quantification of Phenolic Compounds

The highest number of individual compounds was identified in *A. artemisiifolia* (30), followed by *E. annuus* (26), *X. strumarium* (22), *D. viscosa* (22), *A. altissima* (21), and *C. canadensis* (18) (Supplementary Table S1). The identified compounds belong to the groups of hydroxycinnamic and hydroxybenzoic acid derivatives, flavones, flavanols, flavonols, flavanones, ellagitannins, and gallotannins (Supplementary Table S1). As expected, the highest number of individual compounds were identified in the groups of flavonols and hydroxycinnamic acid derivatives because it is known that flavonols such as quercetin and kaempferol [32], as well as acids such as caffeic, *p*-coumaric, and ferulic, predominate in Asteraceae [33] and *A. altissima* [9,34]. The compounds recorded in each species were neochlorogenic, chlorogenic, and 5-*p*-coumaroylquinic acids, quercetin-3-glucoside, and kaempferol hexoside. This is in accordance with other studies. Neochlorogenic acid has been recognised as the main antioxidant component of one of the most invasive plant species, *Polygonum cuspidatum* Siebold & Zucc. [35]. Chlorogenic acid is one of the major phenolic components in *Lonicera maackii* (Rupr.) Maxim., an invasive shrub in North America [36]. In [37], the invasive species *Tithonia diversifolia* (Hemsl.) A.Gray also contained a significant amount of chlorogenic acids. Very recently, 5-*p*-coumaroylquinic acid was recorded in phytotoxic *Lolium multiflorum* Lam. [38], and quercetin glycosides were redundantly present in allelopathic *Trigonella foenum graecum* L. [39]. Further studies are needed to determine whether these compounds may have an effect on the invasive ability of invasive species. The proposed activities of these compounds range from antimicrobial activity, drought, or heat resistance to allelopathy or others. As shown in Supplementary Table S1, we also detected 32 compounds that were present in only one of the tested species, according to the following representation: 9 in *A. altissima* (4-*p*-coumaroylquinic acid, ellagic acid pentoside, apigenin hexoside, quercetin galloyl hexosides 1 and 2, quercetin acetyl hexoside, kaempferol acetyl hexoside 2, vescalagin isomer, and digalloyl-HHDP-hexoside isomer), 7 in *A. artemisiifolia* (caftaric acids 1 and 2, caffeic acid, caffeic acid derivative 2, isorhamnetin-acetyl-hexoside, isorhamnetin-3-rutinoside, and gallic acid), 6 in *E. annuus* (ellagic acid rhamnoside, quercetin-3-rhamnoside, kaempferol-3-rutinoside, kaempferol-glucuronyl-hexoside, myricetin-3-glucuronide, and naringenin hexoside 2), 5 in *X. strumarium* (apigenin, procyanidin dimer, epicatechin, laricitrin, and syringetin), 4 in *D. viscosa* (dicaffeoylquinic acids 4 and 5, ellagic hexoside 2, and myricetin hexoside), and 1 in *C. canadensis* (ellagic acid). The fact that 9 out of 21 identified compounds in *A. altissima* were present in this species only was expected since only this species belongs to the small family Simaroubaceae, while others are from the ubiquitously present Asteraceae family. While 5-*p*-coumaroylquinic acid was present in all the tested species except *E. annuus*, and 3-*p*-coumaroylquinic in all except *X. strumarium*, 4-*p*-coumaroylquinic acid was detected in *A. altissima* only, so we assume that the biosynthesis of this isomer of chlorogenic acid is more constrained. As well, 5-*p*-coumaroylquinic acid has been detected in another invasive species, for example, in *Ageratina adenophora* (Spreng.) R.M.King & H.Rob. [40]. All three *p*-coumaroylquinic acid isomers were recorded in *A. altissima* only (Supplementary Table S1). The combination of three isomers can be found in the leaves of *Vaccinium uliginosum* L. (Ericaceae) as well [41]. In *A. altissima* only, we also recorded digalloyl-HHDP (6,6′-dicarbonyl-2,2′,3,3′,4,4′-hexahydroxybiphenyl)-hexoside and vescalagin. These compounds are present in wild *Fragaria vesca* L. fruits [42] and *Alchemilla vulgaris* L. and *Alchemilla mollis* (Buser) Rothm. [43].

Only *A. altissima* and *A. artemisiifolia* contained gallotannins, namely, digalloyl-HHDP-hexoside isomer and gallic acid, respectively (Table 1). Gallic acid occurs widely in plants and has antioxidant activities. Plants contain many antioxidant compounds, especially in leaves, because photosynthesis generates free radicals that have to be neutralised. Gallic acid has previously been identified in the invasive shrub *Rhododendron ponticum* [44]. Moreover, it was recorded that it had a dose-dependent inhibitory effect on the seed germination and seedling growth of cucumber (*Cucumis sativus* L.) [45]. Furthermore, it has been detected as one of the compounds responsible for the allelopathy of *Polygonella myriophylla* [46]. However, some data show that this acid alone is not sufficient for the successful invasion of *Phragmites australis* [47]. Therefore, it might be that gallic acid contributes to plant allelopathy in combination with other compounds. Hydroxycinnamic and hydroxybenzoic acid derivatives were quantitatively the main components of the phenolic fraction in all the tested plant extracts, except *E. annuus*, where flavanones dominated with a share of 70% (Table 1, Supplementary Figure S1). The abundant representation of flavanones in *E. annuus* has already been documented [48]. Hydroxycinnamic acid derivatives were significantly more present in *D. viscosa* and *X. strumarium* than in other species (Table 1).

As shown in Table 1, the highest concentration of total identified phenolic compounds was recorded in the extracts of *D. viscosa* (67.78 ± 0.57 mg/g DW). Moreover, among the identified compounds, 93% belonged to the groups of hydroxycinnamic and hydroxybenzoic acid derivatives. The predominant ones were caffeic acid and quercetin derivatives, similar to the Tunisian ecotype [49]. At the same time, *A. artemisiifolia* and *C. canadensis* extracts contained significantly lower concentrations (15.00 ± 0.42 and 13.00 ± 2.90 mg/g DW, respectively) of total identified phenolic compounds. *C. canadensis* had the lowest concentration of total identified phenolic compounds (Table 1, Supplementary Table S1). According to Queiroz et al. [50], this species contains acetylenes, which could be responsible for its allelopathic activity. The highest concentration of a single group of compounds we recorded in *E. annuus*—naringenin hexosides, at a concentration of 32.07 ± 3.01 mg/g DW (Table 1). Moreover, it is only in this species that we have recorded these compounds. Naringenin-6,8-di-C-hexoside has already been detected in the wild plants *Lathyrus aureus* (Steven) D.Brandza and *Lathyrus pratensis* L., which have been recognised as new sources of biologically active compounds [51].

Regarding the flavonoids, all the tested species contained quercetin and kaempferol derivatives. *A. altissima*, *A. artemisiifolia*, *D. viscosa*, and *X. strumarium* had more quercetin derivatives, while in *C. canadensis* and *E. annuus*, kaempferol derivatives dominated. Isorhamnetin derivatives were detected in *A. artemisiifolia* and *D. viscosa*, and myricetin derivatives in *D. viscosa* and *E. annuus*. The two species that contained all the four flavonol derivatives were *D. viscosa* and *E. annuus*. Flavanols were detected in *X. strumarium* only, and ellagitannin vescalagin in *A. altissima* only. The flavanones naringenin-hexosides were present in *C. canadensis* and *E. annuus*, the flavones apigenin and apigenin-hexoside in *X. strumarium* and *A. altissima*, respectively. The gallotannins digalloyl-hexahydroxydiphenoyl (HHDP)-hexoside isomer and gallic acid were present in *A. altissima* and *A. artemisiifolia*, respectively. Digalloyl-HHDP-glucoside has already been identified in flowers of pomegranate (*Punica granatum* L.) [52].

As shown in Table 1, we recorded the highest concentration of a particular phenolic group in *E. annuus* for flavanones (32.07 ± 3.01 mg/g DW, in particular, naringenin hexosides 1 and 2) and in *D. viscosa* for hydroxybenzoic acid derivatives (31.63 ± 0.53 mg/g DW, ellagic acid hexosides 1 and 2) and hydroxycinnamic acid derivatives (31.08 ± 0.11 mg/g DW, nine different types) (Supplementary Table S1). At the same time, these two species also had the highest concentration of all the identified compounds. Among the hydroxybenzoic acid derivatives, in all the species, the predominant were caffeic acid derivatives with even 31.02 ± 0.01 mg/g DW in *D. viscosa* (Table 1). This species also contained the highest concentration of hydroxybenzoic acid derivatives, in particular, ellagic acid hexosides at a concentration of 31.63 ± 0.53 mg/g DW.

Statistical comparison of the phenolic groups' concentrations between the species revealed the following: *D. viscosa* and *X. strumarium* were dominant over the other species in hydroxycinnamic acid derivative concentration, *D. viscosa* had a higher concentration of hydroxybenzoic acid derivatives than other species, and *E. annuus* was predominant in flavanone concentration (Table 1). *A. altissima* had significantly higher flavones and tannin concentrations than the other species.

In all the species, hydroxycinnamic and hydroxybenzoic acid derivatives dominated, with the share in total identified phenolic compounds between 62% (*A. altissima*) and 93% (*D. viscosa*). The only exception was *E. annuus*, where the predominant compounds were flavanones with a share of 70% (Table 1, Supplementary Figure S1). Caftaric acid was recorded only in *A. artemisiifolia*. This type of acid can otherwise be found in the leaf and flower of *Lapsana communis* L. subsp. *communis* (Asteraceae), a plant used in folk medicine as a poultice against chapped nipples or hands [53]. The isomer of catechin, epicatechin, was detected in *X. strumarium*. It has already been shown previously that catechin contributes to the invasiveness of *Centaurea stoebe* Tausch [54] and *Rhododendron ponticum* L. [44]. Laricitrin, which we also recorded only in *X. strumarium*, was recently detected in the invasive alien species *Rhododendron ponticum* [44], which suggests the implication of this compound in the plants' invasion process.

2.3. Antimicrobial Activity and Bioautography

The ethyl acetate/methanol/water (40:5.4:5) (EMW) eluent system defines separated components in almost all plant species' crude extracts; however, it moves most of the components up close to the solvent front, making it difficult to notice good separation. Most of the compounds in the extracts were of intermediate polarity or nonpolar (Supplementary Figure S2a). The bioautography assay showed that at least one compound from each plant extract was active against the tested bacteria with a relatively clear zone on TLC bioautograms, indicating growth inhibition (Supplementary Figure S2b–d), except those from *A. artemisiifolia* and *C. canadensis* against *Pseudomonas aeruginosa* (Supplementary Figure S2d). This suggests a higher resistance of *P. aeruginosa* to *A. artemisiifolia* and *C. canadensis* bioactive compounds.

The antimicrobial activity of plant extracts was classified based on the MIC values as follows: outstanding (0.02 mg/mL and below), excellent (0.021–0.04 mg/mL), very good (0.041–0.08 mg/mL), good (0.08–0.16 mg/mL), average activity (0.161–0.32 mg/mL), and weak (>0.32 mg/mL) [55].

The extract of *X. strumarium* had good activity against three of the four tested bacteria, with an average MIC value of 0.11 mg/mL, which is a promising antibacterial activity. The extracts of *D. viscosa* and *A. altissima* had an average activity with an average MIC value of 0.21 and 0.22, respectively (Table 2). *D. viscosa* had the highest concentration of total identified compounds (67.78 $\pm$ 0.57 mg/g DW), and especially hydroxycinnamic and hydroxybenzoic acid derivatives (31.08 $\pm$ 0.11 mg/g DW and 31.63 $\pm$ 0.53 mg/g DW, respectively) (Table 1). Since this species also showed the lowest MIC value against *E. coli* (Table 2), we hypothesise that activity against this bacteria strain might have been mediated via the identified hydroxycinnamic and hydroxybenzoic acid derivatives. The highest activity of all plant extracts was obtained against *Enterococcus faecalis*, with each observed MIC value below 0.1 mg/mL. The lowest activity was obtained against *Staphylococcus aureus* and *P. aeruginosa*, with average MIC values of 0.69 and 0.64 mg/mL, respectively (Table 2). Many of these activities are below values considered interesting for developing new antibiotics (MIC <1 mg/mL) [56]. Still, they could be very useful when used in the organic production of plant products if they are not edible because a higher concentration can be applied.

In addition to the MIC values, we also determined the total antibacterial activity (TAA) value, which takes into account the mass extracted from each species to compare plant extracts. TAA value (obtained by dividing yield by the MIC and expressed in ml/g) shows the largest volume to which the biologically active compounds present in 1 g of extract can be diluted and still inhibit the growth of the test organism [56]. *X. strumarium* and *A. artemisiifolia* extracts had the highest average TAA of 775 and 624 mL/g, respectively (Table 2). The values varied between 248 and 1425 mL/g with *X. strumarium* and between 160 and 1683 mL/g with *A. artemisiifolia* extract against different bacteria. This is an important parameter to compare the efficacy of different plants. Thus, 1 g of *X. strumarium* extract can be diluted to 1425 mL, and 1 g of *A. artemisiifolia* extract to even 1683 mL, and will still inhibit the growth of *E. faecalis*. In this particular case, the extract of *A. artemisiifolia* with the highest TAA of 1683 mL/g was the most effective against *E. faecalis*.

The cytotoxicity was determined using an in vitro assay against Vero monkey kidney cells. The LC_{50} and the selectivity index (SI) values were calculated (Table 2). The LC_{50} values ranged from 0.002 to 0.13 mg/ ml. Extracts of *A. altissima* and *C. canadensis* had the highest LC_{50} (lowest toxicity) of 0.17 mg/mL, both, while the extract of *X. strumarium* was the most toxic (0.001 mg/mL) to the cells. In classifying the in vitro cytotoxicity of plant extracts, values below 20 μg/mL are considered highly cytotoxic [57]. However, caution must be taken in making conclusions based on in vitro cytotoxic data only, because in vitro toxicity data may not always agree with in vivo toxicity test owing to a number of factors, such as gut interactions, solubility, bioavailability, and other pharmacokinetic/dynamic parameters. Therefore, further toxicity testing in in vivo systems will be useful. The selectivity index (SI) is a ratio that measures the window between cytotoxicity and antimicrobial (antibacterial) activity by dividing the given LC_{50} by the MIC value. The higher the SI ratio, the theoretically more effective and safe an extract would be during in vivo treatment for a given bacterial (fungal) infection, and the more likely it is that the activity is not due to a general metabolic toxin. Extracts of *A. altissima, C. canadensis, E. annuus*, and *D. viscosa* had SI values higher than 1 (Table 2) against *E. faecalis.*

The extracts of *A. altissima* and *X. strumarium* had the highest average activity against two of the three tested fungi, with an average MIC value of 0.21 mg/mL (Table 3). *A. altissima* showed the greatest variety of compounds (most specific ones) and also the least MIC value toward *A. fumigatus*. Hence, we hypothesise that some or a combination of some compounds could be responsible for higher toxicity of *A. altissima* extract to *A. fumigatus*. The highest activity of all plant extracts was obtained against *C. albicans*, with an average MIC value of 0.20 mg/mL. *A. fumigatus* was the most resistant fungus against the plant extracts with an average MIC value of 0.69 mg/mL. In comparison with amphotericin B, used as a positive control, with an average MIC value of 8×10^{-3} mg/mL, tested extracts were much less active.

The *X. strumarium* and *A. artemisiifolia* extracts had the highest average TAA of 284 and 296 mL/g, respectively (Table 3). The highest value for *X. strumarium* extract was 365 mL/g, which means that the yield from 1 g of *X. strumarium* extract can be diluted to 365 mL and will still inhibit the growth of *C. neoformans*. *A. artemisiifolia* had the highest TAA value (646 mL/g) against *C. albicans*.

None of the tested extracts had an SI ratio greater than 1, implying that the antifungal activity of the plant extract is probably due to a general metabolic toxic effect (Table 3). It should be noted, however, that in vitro toxicity results may not always predict the toxicity behaviour of extracts when tested in in vivo systems due to pharmacokinetic and pharmacodynamic interactions within the body [58].

3. Materials and Methods

3.1. Plant Material

The leaves of six invasive species widespread in Croatia were used in this study: *Ailanthus altissima* (Mill.) Swingle (Simaroubaceae), *Ambrosia artemisiifolia* L. (Asteraceae), *Conyza canadensis* (L.) Cronquist (Asteraceae), *Dittrichia viscosa* (L.) Greuter (Asteraceae), *Erigeron annuus* (L.) Pers. (Asteraceae), and *Xanthium strumarium* L. (Asteraceae). The official blacklist of the invasive species of Croatia, according to the law on the introduction, prevention, and spread of invasive alien species and management [59], has not been delivered yet. However, Nikolić et al. [14] published a list of 70 most relevant invasive plant species of Croatia, which contains all the species involved in this research. Moreover, all of them, except *D. viscosa*, are included in the Global Register of Introduced and Invasive Species [60].

The plants were collected in the Istria region (Figure 1). Their identities were established by Istrian Botanic Society botanists and confirmed by comparison with the literature. For each species, plant material from different locations (latitude from 44°46′03″ N to 45°19′36″ N, longitude from 13°35′48″ E to 14°08′47″ E) was collected in the optimal maturity stage for determination, from June to August 2018. Approximately 250 g of plant material from every location was collected and pooled. For phenolic profiling, plant material was freeze-dried and stored in the dark at room temperature until analysis. For antimicrobial assays, plants were air-dried in the shade and stored at room temperature.

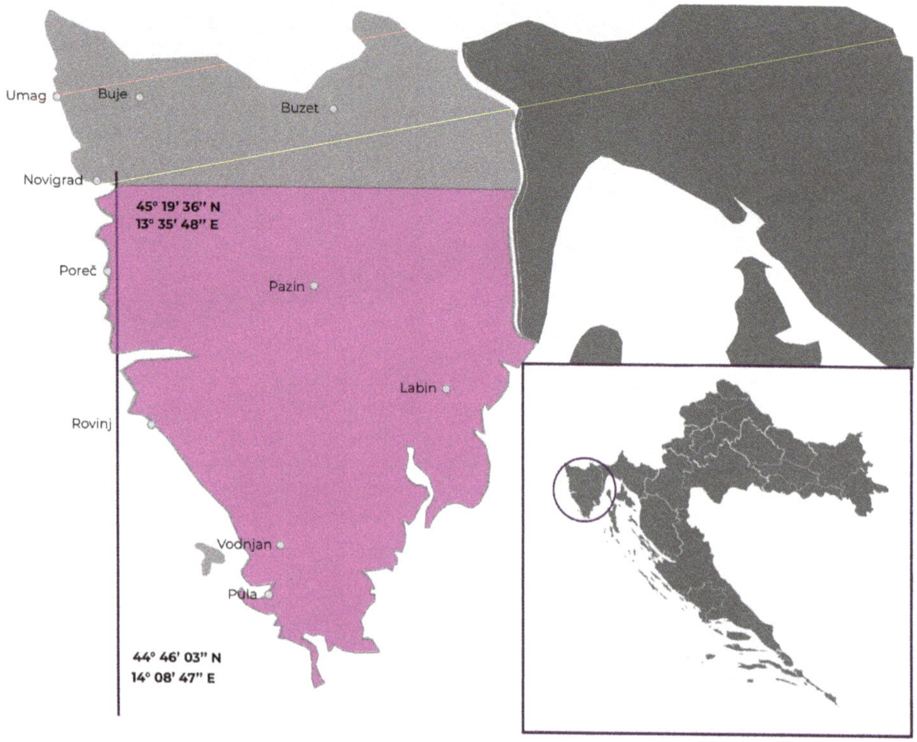

Figure 1. Location (geographic coordinates) in the Istria region where the six invasive alien plant species were collected. The location of the Istria Region in Croatia is given in the bottom right.

3.2. Extraction Procedures

The extraction of phenolic compounds for LC–MS identification and quantification was performed following the slightly modified procedure of Mikulic-Petkovsek et al. [61]. A hundred milligram of dried plant tissue was extracted with 6 mL of 70% methanol containing 3% (*v/v*) formic acid in a cooled ultrasonic bath for 60 min. Extracts were centrifuged for 10 min at 10,000× *g* and filtered through 20 µm polytetrafluoroethylene (PTFE) filters (Macherey-Nagel, Düren, Germany). Identification of individual phenolic compounds was qualitatively achieved using the method of external standards and quantitatively comparing peak area on chromatograms of samples with those of diluted standard solutions. The external standards used in this study were caffeic acid, apigenin-7-glucoside, ferulic acid, quercetin-3-*O*-rhamnoside, neochlorogenic (3-caffeoylquinic) acid, naringenin, ellagic acid, gallic acid, chlorogenic acid, and rutin (quercetin-3-*O*-rutinoside) from Sigma-Aldrich (St. Louis, MI, USA) Chemie; (-)epicatechin, quercetin-3-*O*-galactoside, quercetin-3-*O*-glucoside, *p*-coumaric acid, procyanidin B1, and kaempferol-*O*-glucoside from Fluka Chemie; quercetin-3-*O*-xyloside and quercetin-3-*O*-arabinopyranoside from Apin Chemicals (Abingdon, UK); and isorhamnetin-3-*O*-glucoside from Extrasynthese (Genay, France).

Acetone was used as an extractant in bioassays (cytotoxicity and antimicrobial activities). The choice of extractant was based on its efficacy in the quantity and diversity of compounds extracted, safety, low toxicity, and ease of removal [62]. The mixture of finely ground dry leaf samples (2 g) and acetone (20 mL) was sonicated for 20 min, vigorously shaken, and then centrifuged for 10 min at 4000× *g* (Hettich Centrifuge, Rotofix 32 A, Labotec, Johannesburg, South Africa). The supernatant was filtered through a filter paper

(Whatman No. 1) and concentrated by drying under a stream of cold air in preweighed glass vials. The yield was obtained by dividing the mass extracted by the initial mass. A stock solution (10 mg/mL) in acetone was prepared and used in the assays.

3.3. *Determination of Individual Phenolic Compounds Using LC–DAD–MS Analysis*

Phenolic compounds were analysed on an HPLC system (Thermo Finnigan Surveyor, Thermo Scientific, San Jose, CA, USA) equipped with a DAD detector set (280, 350, and 530 nm) following the procedure of Mikulic-Petkovsek et al. [61]. A Gemini C18 (Phenomenex, Torrance, CA, USA) column (150 × 4.6 mm i.d., 3 μm) was used at 25 °C. The samples were eluted with aqueous 0.1% formic acid and 3% acetonitrile in double-distilled water (A) and 0.1% formic acid/3% double-distilled water in acetonitrile (B) according to linear gradients reported by Wang et al. [63]. The Excalibur software was used for spectral data analyses (Thermo Fisher Scientific, Waltham, MA, USA). All phenolic compounds were identified using a mass spectrometer Thermo Finnigan LCQ Deca XP Mass (Thermo Fisher Scientific, Waltham, MA, USA) with an electrospray interface (ESI) operating in negative and positive ion modes, performing analyses under the same conditions as reported by Mikulic-Petkovsek et al. [64]. Identification of the phenolic compounds was established based on their retention times and their PDA spectra in comparison with standard phenolics and based on fragmentation patterns in different MSn modes compared with literature data. Contents of phenolics were expressed in mg/g dry weight (DW) of plant material. All analyses were performed in triplicate.

3.4. *Antimicrobial Testing*

3.4.1. Bioautography

Chemical constituents of the extracts were analysed by thin-layer chromatography (TLC) using aluminium-backed TLC plates (silica gel 60 F254, Merck, Kenilworth, NJ, USA). The TLC plates were developed under saturated conditions with ethyl acetate/methanol/water (40:5.4:5) (EMW) (polar/neutral) eluent systems developed in the laboratory of the Department of Paraclinical Sciences, University of Pretoria, which efficiently separate components of plant extracts [31]. To detect the separated compounds, vanillin–sulphuric acid (0.1 g vanillin (Sigma®, Pretoria, South Africa): 28 mL methanol: 1 mL sulphuric acid) was sprayed on the chromatograms and heated at 110 °C for optimal colour development.

A bioautographic method used by Kotze and Eloff [31] was used to determine the antibacterial activity of separated compounds. TLC plates (10 cm × 10 cm) were loaded with 10 μL of a 10 mg/mL solution of each of the extracts. The prepared plates were developed in an EMW mobile system. The chromatograms were dried at room temperature under a stream of air to remove the remaining solvent. The TLC plates were sprayed with a concentrated suspension of actively growing cells of four bacterial strains before incubating at 38 °C in a chamber at 100% relative humidity. Plates were sprayed with a 2 mg/mL solution of *p*-iodonitrotetrazolium violet (Sigma). Clear zones indicated inhibition of growth on the chromatogram after incubating for 1 h.

3.4.2. Antibacterial and Antifungal Activity

The minimal inhibitory concentrations (MICs) of extracts on the selected microbes were determined using the serial microdilution method in 96-well microplates previously developed by Eloff [62]. Antibacterial activity of plant extracts was tested against two Gram-positive (*Staphylococcus aureus* ATCC 29213 and *Enterococcus faecalis* ATCC 29212) and two Gram-negative (*Escherichia coli* ATCC 25922 and *Pseudomonas aeruginosa* ATCC 27853) bacterial strains. Bacteria strains were purchased from Anatech Instruments (PTY) LTD, Pretoria, South Africa. Frozen stocks of the bacteria were grown on Mueller Hinton agar (Sigma) and kept in the fridge. Prior to the assay, bacteria were grown in Mueller Hinton broth (Sigma) overnight and adjusted to a McFarland 1 standard. One hundred microlitres of each extract (10 mg/mL) dissolved in acetone was added to the first well of a round-bottomed sterile 96-well plate containing 100 μL of sterile distilled water and serially diluted (1:1). Then 100 μL of bacteria was added to the wells and incubated for 18–24 h. The bacteria were exposed to final extract concentrations of 2.5, 1.25, 0.63, 0.32, 0.16, 0.08, 0.04, and 0.02 mg/mL. Gentamicin and acetone served as positive and negative controls, respectively. Afterwards, 40 μL of *p*-iodonitrotetrazolium violet (INT; Sigma®) dissolved in hot sterile water was added to the wells with further incubation for 1–2 h. The MIC was determined visually as the lowest concentration inhibiting bacteria growth.

As regards antifungal activity, a slightly modified serial dilution method by Masoko et al. [65] was used to determine the MIC of extracts against fungi (*Candida albicans* ATCC 10231, *Cryptococcus neoformans* ATCC 32045, and *Aspergillus fumigatus* ATCC 204305). Fungi cultures were purchased from Anatech Instruments (PTY) LTD, Pretoria, South Africa. Fungi were maintained on Sabouraud agar (Sigma) prior to growth in Sabouraud broth (Sigma) overnight. The assay was performed as described above, except that amphotericin B served as a positive control, while INT (40 μL) was added to the wells prior to incubation for 24 h.

3.5. Cytotoxicity

Cytotoxicity (LC_{50}) assessment was performed using the 3-(4,5-dimethylthiazol-2-yl)-2,5-diphenyltetrazolium bromide (MTT) assay on Vero African green monkey kidney cells, according to Mosmann [66], with minor modifications [67]. Cells were kept in a minimal essential medium (Highveld Biological (PTY) LTD, Johannesburg, South Africa) supplemented with 5% foetal calf serum (Adcock Ingram, South Africa) and 0.1% gentamicin (Virbac, Johannesburg, South Africa) in a 5% CO_2 incubator. Cells were prepared from not more than 80% monolayers and seeded at a density of 1×10^4 cells/ well of sterile 96-well microtiter cell culture plates and incubated at 37 °C and 5% CO_2 in a humidified environment for 24 h. Following this, cells were exposed to extracts at varying concentrations in quadruplicate and incubated for 48 h. Doxorubicin (Pfizer Laboratories, Johannesburg, South Africa) served as a positive control, while untreated cells served as a negative control. Afterwards, the medium in each well was aspirated, and the wells were washed with PBS (Whitehead Scientific, Johannesburg, South Africa), and a fresh complete medium was added to each well. Then, 30 μL of 0.5% MTT (Sigma, Pretoria, South Africa) was added to the wells, and the plates were incubated at 37 °C for 4 h. Media in the wells were discarded, and DMSO was added to each well to solubilise the formazan crystals. The absorbance was measured using a BioTek Synergy microplate reader (Thermo Fisher Scientific, USA) at 570 nm, and LC_{50} values of the extracts were calculated. The selectivity index (SI) value for each sample was obtained by dividing the LC_{50} value by the respective MIC value (SI = LC_{50}/MIC).

3.6. Statistical Analyses

The data were analysed in the Statgraphics Plus 4.0 program (Manugistics Inc., Rockville, MD, USA) using one-way analysis of variance (ANOVA) and LSD multiple range test ($p \leq 0.05$).

4. Conclusions

The results of this study add valuable information to the existing knowledge on invasive plant species and might contribute to the use of these underutilised and undesirable species for the development of commercial natural products. All the extracts proved to be a rich source of phenolics, with very good to average antimicrobial activity against four bacterial strains and three fungi. The observed activity against fungi indicates that extracts of these plants could be useful to combat fungal infections of plants. In the cases where the selectivity index was too low, it could still be used on nonedible plants, such as flowers.

Supplementary Materials: The following are available online at https://www.mdpi.com/article/10.3390/plants11050596/s1: Figure S1: Heat map presenting the representations of phenolic groups in different invasive plants; Figure S2: Chromatogram of the acetone extracts of the plant species leaves developed in ethyl acetate/methanol/water (EMW) solvent system sprayed with vanillin-sulphuric acid and TLC bioautograms; Table S1: Spectrum, mass-to-charge ratio (m/z) values of the molecular masses, and main fragments (MS2—second-generation product ion, MS3—third-generation product ion) in negative ion mode ($(M\text{-}H)^-$) identified with ESI–MS and the distribution of individual compounds in different invasive plants.

Author Contributions: Conceptualisation, D.P., B.S., I.Š. and J.N.E.; methodology, D.P., B.S., M.M.-P., I.M.F. and J.N.E.; validation, M.M.-P. and J.N.E.; formal analysis, M.U., M.Š., I.M.F. and M.M.-P.; investigation, B.S. and D.P.; resources, B.S., M.M.-P., D.P., I.M.F. and J.N.E.; data curation, I.Š., M.U., D.P. and B.S.; writing—original draft preparation, D.P., B.S. and I.Š.; writing—review and editing, D.P., I.Š., B.S., M.U., M.M.-P., I.M.F., M.Š., R.V., M.H. and J.N.E.; visualisation, D.P.; supervision, J.N.E., R.V. and M.H.; project administration, B.S. and M.M.-P.; funding acquisition, D.P., B.S. and M.M.-P. All authors have read and agreed to the published version of the manuscript.

Funding: This research was supported by the Croatian Science Foundation under the project "NATURe as an ALLY: Alien Invasive Plants as Phytopharmaceuticals—NATURALLY" (IP-2020-02-6899) and the Slovenian Research Agency (ARRS) as a part of the program No. P4- 0013-0481. The work of Mirela Uzelac was supported by the Croatian Science Foundation "Young Researchers' Career Development Project–Training New Doctoral Students" (DOK-2021-02). The University of Pretoria, South Africa, provided postdoctoral support to Ibukun Michael Famuyide.

Institutional Review Board Statement: Not applicable.

Informed Consent Statement: Not applicable.

Data Availability Statement: Not applicable.

Acknowledgments: We especially thank Slavko Brana and the Istrian Botanical Society for the help in the fieldwork and availability. Special thanks to Tim Weber for editing the manuscript.

Conflicts of Interest: The authors declare no conflict of interest. The funders had no role in the design of the study; in the collection, analyses, or interpretation of data; in the writing of the manuscript; or in the decision to publish the results.

References

1. Bennett, A.E.; Thomsen, M.; Strauss, S.Y. Multiple mechanisms enable invasive species to suppress native species. *Am. J. Bot.* **2011**, *98*, 1086–1094. [CrossRef] [PubMed]
2. Lind, E.M.; Parker, J.D. Novel weapons testing: Are invasive pants more chemically defended than native plants? *PLoS ONE* **2010**, *5*, e10429. [CrossRef] [PubMed]
3. Macel, M.; de Vos, R.C.H.; Jansen, J.J.; van der Putten, W.H.; van Dam, N.M. Novel chemistry of invasive plants: Exotic species have more unique metabolomic profiles than native congeners. *Ecol. Evol.* **2014**, *4*, 2777–2786. [CrossRef] [PubMed]
4. Callaway, R.M.; Aschehoug, E.T. Invasive plants versus their new and old neighbors: A mechanism for exotic invasion. *Science* **2000**, *290*, 521–523. [CrossRef]
5. Cappuccino, N.; Arnason, J.T. Novel chemistry of invasive exotic plants. *Biol. Lett.* **2006**, *2*, 189–193. [CrossRef] [PubMed]
6. Kim, Y.O.; Lee, E.J. Comparison of phenolic compounds and the effects of invasive and native species in East Asia: Support for the novel weapons hypothesis. *Ecol. Res.* **2011**, *26*, 87–94. [CrossRef]
7. Sladonja, B.; Poljuha, D.; Uzelac, M. Non-Native Invasive Species as Ecosystem Service Providers. In *Ecosystem Services and Global Ecology*; Hufnagel, L., Ed.; IntechOpen: London, UK, 2018. [CrossRef]
8. Sahu, A.; Devkota, A. Antimicrobial activity of leaf extracts of some invasive alien plant species of Aster family against clinical bacteria. *Ecoprint Int. J. Ecol.* **2016**, *23*, 1–12. [CrossRef]
9. Poljuha, D.; Sladonja, B.; Šola, I.; Dudaš, S.; Bilić, J.; Rusak, G.; Motlhatlego, K.E.; Eloff, J.N. Phenolic composition of leaf extracts of *Ailanthus altissima* (Simaroubaceae) with antibacterial and antifungal activity equivalent to standard antibiotics. *Nat. Prod. Commun.* **2017**, *12*, 1609–1612. [CrossRef]
10. Meela, M.M.; Mdee, L.K.; Masoko, P.; Eloff, J.N. Acetone leaf extracts of seven invasive weeds have promising activity against eight important plant fungal pathogens. *S. Afr. J. Bot.* **2019**, *123*, 93–97. [CrossRef]
11. Eloff, J.; Angeh, I.; Mcgaw, L. Solvent-solvent fractionation can increase the antifungal activity of a *Melianthus comosus* (Melianthaceae) acetone extract to yield a potentially useful commercial antifungal product. *Ind. Crop. Prod.* **2018**, *111*, 69–77. [CrossRef]
12. Sladonja, B.; Poljuha, D.; Sušek, M.; Dudaš, S. Herbicidal effect of *Ailanthus altissima* leaves water extracts on *Medicago sativa* seeds germination. In Proceedings of the Conference VIVUS: Transmission of Innovations, Knowledge and Practical Experience into Everyday Practice, Naklo, Slovenia, 14–15 November 2014; Maček, M.A., Maček Jerala, M., Kolenc Artiček, M., Eds.; Biotehniški Center: Naklo, Slovenia, 2014; pp. 476–481.

13. Sladonja, B.; Poljuha, D.; Uzelac, M. Tree of Heaven leaf extract: A new generation of bioherbicides. In Proceedings of the International Conference on Biological Invasions: The Human Role in Biological Invasions—A case of Dr Jekyll and Mr Hyde? NEOBIOTA, Vodice, Hrvatsko ekološko društvo, Zagreb, Croatia, 15–18 September 2020; p. 107.
14. Peter, A.; Dujmović Purgar, D.; Voća, N.; Filipčić, P. Use of invasive plant species Japanese knotweed (*Reynoutria japonica* Houtt.) biomass in energy production. In Proceedings of the 3rd Croatian Symposium on Invasive Species with International Participation, Zagreb, Croatia, 26–27 November 2018; Jelaska, S.D., Ed.; Croatian Ecological Society: Zagreb, Croatia, 2018; p. 79.
15. Nikolić, T.; Mitić, B.; Boršić, I. *Flora Hrvatske—Invazivne biljke*; Izdavač ALFA dd.: Zagreb, Croatia, 2014.
16. Kožuharova, E.; Lebanova, H.; Getov, I.; Benbassat, N.; Kochmarov, V. *Ailanthus altissima* (Mill.) Swingle—A terrible invasive pest in Bulgaria or potential useful plant? *Bothalia J.* **2014**, *44*, 213–230.
17. Ding, W.; Huang, R.; Zhou, Z.; Li, Y. New Sesquiterpenoids from *Ambrosia artemisiifolia* L. *Molecules* **2015**, *20*, 4450–4459. [CrossRef] [PubMed]
18. Saluk-Juszczak, J.; Olas, B.; Nowak, P.; Wachowicz, B.; Bald, E.; Głowacki, R.; Pawlaczyk, I.; Gancarz, R. Extract from *Conyza canadensis* as a modulator of plasma protein oxidation induced by peroxynitrite in vitro. *Cent. Eur. J. Biol.* **2010**, *5*, 800–807. [CrossRef]
19. Brahmi-Chendouh, N.; Piccolella, S.; Crescente, G.; Pacifico, F.; Boulekbache-Makhlouf, L.; Hamri-Zeghichi, S.; Akkal, S.; Madani, K.; Pacifico, S. A nutraceutical extract from *Inula viscosa* leaves: UHPLC-HR-MS/MS based polyphenol profile, and antioxidant and cytotoxic activities. *J. Food Drug Anal.* **2019**, *27*, 692–702. [CrossRef] [PubMed]
20. Kamboj, A.; Saluja, A.K. Phytopharmacological review of *Xanthium strumarium* L. (Cocklebur). *Int. J. Green Pharm.* **2010**, *4*, 129–139. [CrossRef]
21. Jeong, C.H.; Nam, E.K.; Shim, K.H. Chemical Components in Different Parts of *Erigeron annuus*. *J. Korean Soc. Food Sci. Nutr.* **2005**, *34*, 857–861. [CrossRef]
22. Chalchat, J.C.; Maksimović, Z.A.; Petrović, S.D.; Gorunović, M.S.; Đordević, S.; Mraović, M. Chemical composition and antimicrobial activity of *Ambrosia artemisiifolia* L. essential oil. *J. Essent. Oil Res.* **2004**, *16*, 270–273. [CrossRef]
23. Kowarik, I.; Säumel, I. Biological flora of Central Europe: *Ailanthus altissima* (Mill.) Swingle. *Perspect. Plant Ecol. Evol. Syst.* **2007**, *8*, 207–237. [CrossRef]
24. Talib, W.; Abu Zarga, M.; Mahasneh, A. Antiproliferative, antimicrobial and apoptosis inducing effects of compounds isolated from *Inula viscosa*. *Molecules* **2012**, *17*, 3291–3303. [CrossRef]
25. Biswas, K.; Sinha, S.N. Evaluation of antibacterial activity of *Conyza canadensis* (L.) Cronquist collected from Nainital, Uttarakhand, India. *Int. J. Univ. Pharm. Biol. Sci.* **2014**, *3*, 46–53.
26. Sharma, R.K.; Verma, N.; Jha, K.K.; Singh, N.K.; Kumar, B. Phytochemistry, pharmacological activity, traditional and medicinal uses of *Erigeron* species: A review. *IJARI* **2014**, *2*, 379–383.
27. Al-Mekhlafi, F.A.; Abutaha, N.; Mashaly, A.M.A.; Nasr, F.A.; Ibrahim, K.E.; Wadaan, M.A. Biological activity of *Xanthium strumarium* seed extracts on different cancer cell lines and Aedes caspius, *Culex pipiens* (Diptera: Culicidae). *Saudi J. Biol. Sci.* **2017**, *24*, 817–821. [CrossRef] [PubMed]
28. Pietta, P.; Minoggio, M.; Bramati, L. Plant Polyphenols: Structure, Occurrence and Bioactivity. *Stud. Nat. Prod. Chem.* **2003**, *28*, 257–312. [CrossRef]
29. Eloff, J.N.; Picard, J.; Masoko, P. Resistance of animal fungal pathogens to solvents used in bioassays. *S. Afr. J. Bot.* **2007**, *73*, 667–669. [CrossRef]
30. Eloff, J.N. Which extractant should be used for the screening and isolation of antimicrobial components from plants? *J. Ethnopharmacol.* **1998**, *60*, 1–8. [CrossRef]
31. Kotze, M.; Eloff, J.N. Extraction of antibacterial compounds from *Combretum microphyllum* (Combretaceae). *S. Afr. J. Bot.* **2002**, *68*, 62–67. [CrossRef]
32. Panda, S.K.; Luyten, W. Antiparasitic activity in Asteraceae with special attention to ethnobotanical use by the tribes of Odisha, India. *Parasite* **2018**, *25*, 10. [CrossRef]
33. Sytar, O.; Hemmerich, I.; Zivcak, M.; Rauh, C.; Brestic, M. Comparative analysis of bioactive phenolic compounds composition from 26 medicinal plants. *Saudi J. Biol. Sci.* **2018**, *25*, 631–641. [CrossRef]
34. Sladonja, B.; Sušek, M.; Guillermic, J. Review on Invasive Tree of Heaven (*Ailanthus altissima* (Mill.) Swingle) Conflicting Values: Assessment of Its Ecosystem Services and Potential Biological Threat. *Environ. Manag.* **2015**, *56*, 1009–1034. [CrossRef]
35. Kurita, S.; Kashiwagi, T.; Ebisu, T.; Shimamura, T.; Ukeda, H. Identification of neochlorogenic acid as the predominant antioxidant in *Polygonum cuspidatum* leaves. *Ital. J. Food Sci.* **2016**, *28*, 25–31. [CrossRef]
36. Cipollini, D.; Stevenson, R.; Enright, S.; Eyles, A.; Bonello, P. Phenolic metabolites in leaves of the invasive shrub, *Lonicera maackii*, and their potential phytotoxic and anti-herbivore effects. *J. Chem. Ecol.* **2008**, *34*, 144–152. [CrossRef] [PubMed]
37. Omokhua, A.G.; Abdalla, M.A.; Van Staden, J.A. comprehensive study of the potential phytomedicinal use and toxicity of invasive *Tithonia species* in South Africa. *BMC Complement. Altern. Med.* **2018**, *18*, 272. [CrossRef] [PubMed]
38. Vitalini, S.; Orlando, F.; Palmioli, A.; Alali, S.; Airoldi, C.; De Noni, I.; Vaglia, V.; Bocchi, S.; Iriti, M. Different phytotoxic effect of *Lolium multiflorum* Lam. leaves against *Echinochloa oryzoides* (Ard.) Fritsch and *Oriza sativa* L. *Environ. Sci. Pollut. Res.* **2020**, *27*, 33204–33214. [CrossRef] [PubMed]
39. Omezzine, F.; Bouaziz, M.; Simmonds, M.S.J.; Haouala, R. Variation in chemical composition and allelopathic potential of mixoploid *Trigonella foenum graecum* L. with developmental stage. *Food Chem.* **2014**, *148*, 188–195. [CrossRef] [PubMed]

40. Zhang, M.; Liu, W.-X.; Zheng, M.-F.; Xu, Q.-L.; Wan, F.-H.; Wang, J.; Lei, T.; Zhou, Z.-Y.; Tan, J.-W. Bioactive quinic acid derivatives from *Ageratina adenophora*. *Molecules* **2013**, *18*, 14096–14104. [CrossRef] [PubMed]
41. Stefanescu, B.E.; Szabo, K.; Mocan, A.; Crisan, G. Phenolic compounds from five Ericaceae species leaves and their related bioavailability and health benefits. *Molecules* **2019**, *24*, 2046. [CrossRef]
42. Dias, M.I.; Barros, L.; Morales, P.; Camara, M.; Alves, M.J.; Oliveira, M.B.P.P.; Santos-Buelga, C.; Ferreira, I.C.F.R. Wild *Fragaria vesca* L. fruits: A rich source of bioactive phytochemicals. *Food Funct.* **2016**, *7*, 4523. [CrossRef]
43. Duckstein, S.M.; Lotter, E.M.; Meyer, U.; Lindequist, U.; Stintzing, F.C. Phenolic constituents from *Alchemilla vulgaris* L. and *Alchemilla mollis* (Buser) Rothm. at different dates of harvest. *Z. Für Nat. C* **2012**, *67*, 529–540. [CrossRef]
44. Jones, G.L.; Tomlinson, M.; Owen, R.; Scullion, J.; Winters, A.; Jenkins, T.; Ratcliffe, J.; Gwynn-Jones, D. Shrub establishment favoured and grass dominance reduced in acid heath grassland systems cleared of invasive *Rhododendron ponticum*. *Sci. Rep.* **2019**, *9*, 2239. [CrossRef]
45. Muzaffar, S.; Ali, B.; Wani, N.A. Effect of catechol, gallic acid and pyrogallic acid on the germination, seedling growth and the level of endogenous phenolics in cucumber (*Cucumis sativus* L.). *Inter. J. Life Sci. Biotechnol. Pharma. Res.* **2012**, *1*, 50–55.
46. Weidenhamer, J.D.; Romeo, J.T. Allelochemicals of *Polygonella myriophylla*: Chemistry and soil degradation. *J. Chem. Ecol.* **2004**, *30*, 1061–1078. [CrossRef] [PubMed]
47. Weidenhamer, J.D.; Li, M.; Allman, J.; Bergosh, R.G.; Posner, M. Evidence does not support a role for gallic acid in *Phragmites australis* invasion success. *J. Chem. Ecol.* **2013**, *39*, 323–332. [CrossRef] [PubMed]
48. Yoo, N.H.; Jang, D.S.; Yoo, J.L.; Lee, Y.M.; Kim, Y.S.; Cho, J.H.; Kim, J.S. Erigeroflavanone, a flavanone derivative from the flowers of *Erigeron annuus* with protein glycation and aldose reductase inhibitory activity. *J. Nat. Prod.* **2008**, *71*, 713–715. [CrossRef] [PubMed]
49. Rhimi, W.; Salem, I.B.; Immediato, D.; Saidi, M.; Boulila, A.; Cafarchia, C. Chemical composition, antibacterial and antifungal activities of crude *Dittrichia viscosa* (L.) Greuter leaf extracts. *Molecules* **2017**, *22*, 942. [CrossRef] [PubMed]
50. Queiroz, S.; Cantrell, C.; Duke, S.; Wedge, D.; Nandula, V.; Moraes, R.; Cerdeira, A. Bioassay-Directed Isolation and Identification of Phytotoxic and Fungitoxic Acetylenes from *Conyza canadensis*. *J. Agric. Food Chem.* **2012**, *60*, 5893–5898. [CrossRef] [PubMed]
51. Martinez, E.J.L.; Barrales, P.O.; Zengin, G.; Uysal, S.; Ceylan, R.; Guler, G.O.; Mocan, A.; Aktumsek, A. *Lathyrus aureus* and *Lathyrus pratensis*: Characterization of phytochemical profiles by liquid chromatography-mass spectrometry, and evaluation of their enzyme inhibitory and antioxidant activities. *RSC Adv.* **2016**, *6*, 88996–89006. [CrossRef]
52. Yisimayili, Z.; Abdulla, R.; Tian, Q.; Wang, Y.; Chen, M.; Sun, Z.; Li, Z.; Liu, F.; Aisa, H.A.; Huang, C. A comprehensive study of pomegranate flowers polyphenols and metabolites in rat biological samples by high-performance liquid chromatography quadrupole time-of-flight mass spectrometry. *J. Chromatogr. A* **2019**, *1604*, 460472. [CrossRef]
53. Fontanel, D.; Galtier, C.; Viel, C.; Gueiffier, A. Caffeoyl quinic and tartaric acids and flavonoids from *Lapsana communis* L. subsp. *communis* (Asteraceae). *Z. Für Nat. C* **1998**, *53*, 1090–1092. [CrossRef]
54. Duke, S.O.; Blair, A.C.; Dayan, F.E.; Johnson, R.D.; Meepagala, K.M.; Cook, D.; Bajsa, J. Is (−)-catechin a novel weapon of spotted knapweed (*Centaurea stoebe*)? *J. Chem. Ecol.* **2009**, *35*, 141–153. [CrossRef]
55. Eloff, J.N. A Proposal towards a Rational Classification of the Antimicrobial Activity of Acetone Tree Leaf Extracts in a Search for New Antimicrobials. *Planta Med.* **2021**, *87*, 836–840. [CrossRef]
56. Eloff, J.N. Quantifying the bioactivity of plant extracts during screening and bioassay-guided fractionation. *Phytomedicine* **2004**, *11*, 370–371. [CrossRef] [PubMed]
57. Kuete, V.; Krusche, B.; Youns, M.; Voukeng, I.; Fankam, A.G.; Tankeo, S.; Lacmata, S.; Efferth, T. Cytotoxicity of some Cameroonian spices and selected medicinal plant extracts. *J. Ethnopharmacol.* **2011**, *134*, 803–812. [CrossRef] [PubMed]
58. Ahmed, A.S.; Elgorashi, E.E.; Moodley, N.; McGaw, L.J.; Naidoo, V.; Eloff, J.N. The antimicrobial, antioxidative, anti-inflammatory activity and cytotoxicity of different fractions of four South African *Bauhinia* species used traditionally to treat diarrhoea. *J. Ethnopharmacol.* **2012**, *143*, 826–839. [CrossRef]
59. Invazivne Vrste. Available online: https://invazivnevrste.haop.hr/crna-i-bijela-lista/2 (accessed on 17 October 2021).
60. Global Register of Introduced and Invasive Species—GRIIS. Available online: http://www.griis.org (accessed on 20 December 2021).
61. Mikulic-Petkovsek, M.; Schmitzer, V.; Slatnar, A.; Weber, N.; Veberic, R.; Stampar, F.; Munda, A.; Koron, D. Alteration of the content of primary and secondary metabolites in strawberry fruit by *Colletotrichum nymphaeae* infection. *J. Agric. Food Chem.* **2013**, *61*, 5987–5995. [CrossRef] [PubMed]
62. Eloff, J.N. A sensitive and quick microplate method to determine the minimal inhibitory concentration of plant extracts for bacteria. *Planta Med.* **1998**, *64*, 711–714. [CrossRef]
63. Wang, S.Y.; Zheng, W.; Galletta, G.J. Cultural system affects fruit quality and antioxidant capacity in strawberries. *J. Agric. Food Chem.* **2002**, *50*, 6534–6542. [CrossRef]
64. Mikulic-Petkovsek, M.; Ivancic, A.; Todorovic, B.; Veberic, R.; Stampar, F. Fruit phenolic composition of different elderberry species and hybrids. *J. Food Sci.* **2015**, *80*, C2180–C2190. [CrossRef]
65. Masoko, P.; Picard, J.; Eloff, J.N. Antifungal activities of six South African *Terminalia* species (Combretaceae). *J. Ethnopharmacol.* **2005**, *99*, 301–308. [CrossRef]
66. Mosmann, T. Rapid colorimetric assay for cellular growth and survival: Application to proliferation and cytotoxicity assays. *J. Immunol. Method* **1983**, *65*, 55–63. [CrossRef]
67. McGaw, L.J.; Van Der Merwe, D.; Eloff, J.N. In vitro anthelmintic, antibacterial and cytotoxic effects of extracts from plants used in South African ethnoveterinary medicine. *Vet. J.* **2007**, *173*, 366–372. [CrossRef]

plants

MDPI

Review

Review of the Biology, Distribution, and Management of the Invasive Fireweed (*Senecio madagascariensis* Poir)

Kusinara Wijayabandara [1,*], Shane Campbell [1], Joseph Vitelli [2], Asad Shabbir [3] and Steve Adkins [1]

[1] School of Agriculture and Food Science, Faculty of Science, University of Queensland, Gatton 4343, Australia; shane.campbell@uq.edu.au (S.C.); s.adkins@uq.edu.au (S.A.)
[2] Department of Agriculture and Fisheries, Brisbane 40000, Australia; Joseph.Vitelli@daf.qld.gov.au
[3] Centre for Carbon, Water and Food, School of Life and Environmental Sciences, The University of Sydney, Camden 2006, Australia; asad.shabbir@sydney.edu.au
* Correspondence: k.wijayabandara@uq.edu.au; Tel.: +61-421509051

Abstract: Whilst exotic invasive species are a major threat to natural and modified ecosystems around the world, management programs to reduce their impacts often fail due to a lack of information about their biology and how best to control them in various situations. This paper reviews the currently available information on the biology, distribution, and management options for the invasive weed *Senecio madagascariensis* Poir. (fireweed). In addition, we developed a model to predict the climatic suitability of this weed around the world based on the current climate. *Senecio madagascariensis* originates from southern Africa but it has been introduced to several other countries including Australia. Climatic suitability suggests that there are large areas around the world suitable for the weed's growth where it is currently not present. The weed poses a major threat to livestock industries in these countries through its ability to reduce pasture production and poison animals. A range of control techniques have been used to try and manage *S. madagascariensis*. This paper highlights how a better understanding of the biology of *S. madagascariensis* can help determine the most effective treatments to impose and to further develop integrated management strategies. Besides using traditional approaches, the use of competitive pastures and more tolerant livestock (such as sheep and goats) are some of the other options recommended as part of an integrated approach. On-going research to identify host-specific biological control agents is also considered a priority.

Keywords: CLIMEX; ecology; herbicides; impact; integrated control; pasture management

1. Introduction

Senecio madagascariensis Poir. (fireweed), a native herbaceous plant from southern Africa [1] has been introduced to several countries including Australia, the United States of America (USA; Hawaii), Japan, Brazil, Argentina, Venezuela, Columbia, Uruguay, and Kenya [2]. *Senecio madagascariensis* plants contain pyrrolizidine alkaloids (PAs) and when eaten by some domestic livestock (particularly cattle and horses) it can lead to liver toxicity [3]. *Senecio madagascariensis* has an ability to spread into different habitats quickly through high seed production and multiple dispersal mechanisms, particularly wind. In some cases, effective *S. madagascariensis* management can be achieved using herbicides, but maintaining a healthy and competitive pasture and implementing multi-species grazing practices are also recommended for long-term control [4]. Nevertheless, the development of integrated management strategies that incorporate a range of options will be the most effective approach to reduce the impact of *S. madagascariensis* [5]. The aim of this paper is to evaluate the currently available information on the biology, distribution, and management of the invasive *S. madagascariensis* and to identify knowledge gaps that could be the catalyst for further research on this problematic weed.

2. Biology and Ecology of *S. madagascariensis*

2.1. Taxonomy

Botanical name *S. madagascariensis* Poir. *Senecio madagascariensis* was first described by JLM Poiret in Madagascar in 1817 [6]. It belongs to the Tracheophytes, in the class Magnoliopsida, order Asterales, and family Asteraceae with synonyms of *Senecio bakeri* S. Elliot, *Senecio incognitus* Cabrera, and *Senecio ruderalis* Hary [7]. In South Africa, it is called *S. ruderalis* Harv. or *S. junodianus* O.Hoffm, whereas cytological studies in Australia have confirmed it as *S. madagascariensis* [8]. Its name is derived from Latin, senex meaning 'old man' referring to its pappus appearing like a white beard and *madagascariensis* meaning 'from Madagascar' [9].

The taxonomic position of Australian *S. madagascariensis* is undetermined. Initially, it was assumed to be part of the native (Australia) *Senecio pinnatifolius; S. pinnatifolium* (coastal groundsel) group which was made up of four clearly recognised sub-species in Australia viz. *S. pinnatifolius* sub spp., *alpinus, S. dissectifolius, S. lanceolatus*, and *S. maritimus* [10]. This is consistent with results from genetic analysis undertaken on northern QLD populations in Australia, which showed a close relationship with the *S. madagascariensis* complex from South Africa and only a moderate similarity to *S. madagascariensis* coming from Madagascar [11]. Furthermore, the species can vary in chromosome number with the coastal groundsels having a diploid chromosome number of 40 while *S. madagascariensis* has 2n = 20 for Australian [12] and Argentinian populations [13,14].

2.2. Common Name

Senecio madagascariensis is known by various common names including fireweed (Australia), Madagascar ragwort (Australia), and South African ragwort (Australia). Due to its ability to spread like a wildfire, the common name 'fireweed' is most used. Another possible reason for this name might be its bright yellow flowers appearing like a wildfire across the landscape. Other common names include, in Argentina "golden button" and "yellow flower of Mar del Plata" [15].

2.3. Species Description

Senecio madagascariensis is an erect, herbaceous, short-lived perennial plant ca. 10 to 50 cm tall (occasionally up to 70 cm). It forms a single stem or occasionally multiple stems which arise from a central crown at its base [15]. Its stems are multi branched, especially towards the top of the plant. Leaves are typically lanceolate with tips that are acute with denticulate margins [16] and sessile or sub-sessile [15]. Individual plants possess a branched tap root which grows 10 to 20 cm deep, with numerous fibrous roots [16].

Senecio madagascariensis possess heterogamous, radiate daisy-like flower heads that are canary yellow in colour, having a total ca. 120 small tubular disk florets and petal-like ray florets [9]. Florets are enclosed in involucral bracts (21) which are greenish in colour with brown or blackish tips [15]. A single plant can produce up to 200 flower heads (capitula) and ca. 30,000 seeds per year [17]. Fruit is small, ca. 1.5 to 2.2 mm long and up to 0.5 mm wide. Each fruit contains a single seed and a short hair-like structure, the pappus [18].

In Australia, *S. madagascariensis* can be confused with the coastal groundsels (*S. pinnatifolius*) as they both produce showy radiate inflorescences and grow to a similar height [19]. Moreover, both types are predominantly insect-pollinated and form fruit that are wind-dispersed [19]. The main distinguishing characters that separate *S. madagascariensis* from the costal groundsels are the number of involucre bracts at the base of the capitulum and the morphology of the seed [19]. *Senecio madagascariensis* has 20 to 21 involucre bracts/phyllaries (Sindel, 1989) while the coastal groundsels have ca. 13 cup-shaped involucre bracts but rarely 20 (each 3.0 to 7.5 mm long) with an apex that is weakly to strongly pigmented brown, black, or purple [20].

3. Economic Impacts of *S. madagascariensis*

The major economic impacts of *S. madagascariensis* are associated with a decrease in pasture productivity due to competition and poisoning of livestock [9]. Due to the production of toxins in *S. madagascariensis*, the livestock productivity losses in Australia have been estimated at $2 million annually [21].

According to the results of a national survey of landholders undertaken in Armidale and Dorrigo, NSW in 1985, almost 50% of respondents said that they spent more than 50 h every year controlling *S. madagascariensis* on their property, while about 40% said they spent ≥ $1000 on control actions [4]. In Hawaii, *S. madagascariensis* is distributed over 162,000 hectares of pasture [22], and US$11 million is spent annually trying to control it [23].

4. Geographic Distribution

4.1. *Southern Africa—Native*

Senecio madagascariensis is native to the southern parts of Africa, where it is present from coastal regions up to high altitude areas (i.e., 1500 m above sea level). It is considered native to Madagascar and the Mascarene Islands, to coastal Mozambique and KwaZulu-Natal and the Eastern and Western Cape Provinces of South Africa and Eswatini. In Madagascar, *S. madagascariensis* occurs in small remote populations at lower elevations in the southeast and in the semi-arid southwest of the island [11]. Besides its native range in Africa, *S. madagascariensis* has also been recorded as an introduced species in the highlands of Kenya [24].

4.2. *Australia—Introduced*

Senecio madagascariensis was thought to have reached Australia in the ships trading between Europe and Australia that had passed through the Cape of South Africa [8]. In 1918, *S. madagascariensis* was first identified close to Raymond Terrace in the Hunter Region, of central NSW. By the late 1960s, it was found almost 650 km further south along the shoreline in the Bega Valley, of south-costal NSW, possibly arriving there as a fodder contaminant from the north-coast of NSW. *Senecio madagascariensis* had become a pasture threat by about 1983, despite being a very dry year with little fodder having been produced [25]. Large populations of *S. madagascariensis* continued to establish and spread rapidly in the mid-1980s throughout the pastures of coastal NSW as well as into Southeast Queensland (SEQ) [26]. According to Sindel [27] isolated plants were also encountered inland, especially at Dubbo on the Central Western Plains of NSW, inside the limits of the Western Plains Zoo.

Moreover, in southern Queensland, *S. madagascariensis* has spread northward from Brisbane, with isolated infestations occurring near Caboolture, Cooroy, Belli Park, Maleny, Yandina, Pelican Waters, and as far north as Malanda. There is also an herbarium record from near Longreach, although the specimen appears to have been collected near the roadside and it is not clear if it was an isolated plant or if an infestation was present. A prediction in view of climate change and land utilization, suggests that *S. madagascariensis* could become a serious problem as far north as Rockhampton [28].

In 2007, the weed was reported close to Rockhampton and on the Atherton Tablelands in far north Queensland (QLD), in accordance with the expectations of Sindel and Michael [24]. In SEQ, the most widespread incursions of *S. madagascariensis* have been identified in Logan City, Gold Coast City, and Scenic Rim Regional Council areas. Although locally abundant, it is generally less common in Brisbane City, Redland City, Ipswich City, and Moreton Bay Regional Council jurisdictions [29].

4.3. *Hawaii—Introduced*

Senecio madagascariensis was first discovered in the archipelago of Hawaii in the 1980s where it had established in pastures near Waimea and Kona on the Big Island [22]. Since then, it has spread to pastureland in the north eastern and western sides of the island and increased towards the southern areas [21]. The naturalised *S. madagascariensis* populations are found from sea level up to 1600 m in Maui [21] and Kaua'I [30].

Senecio madagascariensis plants in Hawaii, according to molecular investigation, are like those from South Africa more so than those from Madagascar or Eswatini [21]. These populations are thought to have arrived in Hawaii as contaminants of *Axonopus fissifolius* (Raddi) Kuhlm (carpet grass) seed lots sent from Australia. A detailed comparative study was undertaken on the alkaloid content of *S. madagascariensis* plant material collected from a single location in Australia (northern NSW) and multiple locations in Hawaii [3]. In total, 13 pyrrolizidine alkaloids (PAs) were identified and included: senecivernine, senecionine, integerrimine, senkirkine, mucronatinine, retrorsine, usaramine, otosenine, acetylsenkirkine, desacetyldoronine, florosenine, and doronine. Interestingly, while there was large variation in total PA content amongst individual plants within locations (217 to 1990 $\mu g\ g^{-1}$ on a dry weight basis); overall, the plant material collected from the Hawaiian Islands was found to be identical in pyrrolizidine alkaloids (Pas) content to the Australian collection. It was suggested that this was further evidence that *S. madagascariensis* in Hawaii may have originated from Australia [3].

4.4. *Japan—Introduced*

In 1976 *S. madagascariensis* was first identified in Naruto City in the Eastern Shikoku region of Japan. Gradually, it has spread widely in southern parts of the country. At present, *S. madagascariensis* covers a large portion of the southwestern part of Japan. Moreover, it has been observed on the Pacific coast and on the seacoast of Seto Island. According to these previous studies, *S. madagascariensis* has been confined to the warmer southwesten regions [2], including Tokyo where there is 1000 to 1700 mm of precipitation annually [8]. The pathway of its arrival into Japan is unknown.

4.5. *South America—Introduced*

Several countries in South America have recorded the presence of *S. madagascariensis*; Argentina, Brazil, Columbia, and Uruguay. In Argentina and Brazil, the first specimens of *S. madagascariensis* were collected in 1940 and 1995, respectively [31]. In Argentina it was found around cities and on roadsides in Buenos Aires province in the 1940s, but over about a 30-year period it spread into nearby provinces and is now widely distributed in northern and central Argentina [32]. In Brazil, it spread quickly throughout the Pampas region and is now widely distributed in southeastern Brazil [31,32]. In the 1980s, *S. madagascariensis* was recorded from Colombia for the first time [33]. It has invaded the cool moist Colombian highlands near Bogota [9]. In Uruguay, *S. madagascariensis* has been regarded as a serious threat to producers since its detection in late 1990s in grazing pastures near the western cities of La Concordia and Dolores in the region of Soriano and near the southwestern region of Colonia [34]. By 2010, farmers from the open eastern rangelands of Uruguay bordering Brazil were recording significant increases of seneciosis (intense acute or chronic necrosis of liver) in their grazing livestock [35,36].

5. Potential Spread and Future Distribution of *S. madagascariensis*

To assess the climatic suitability of *S. madagascariensis* around the world, a process-based model CLIMEX (version 3) has been used [37]. The CLIMEX model compares the response of a species to long term averages of climate for different locations. A series of growth and stress indices are inferred based on climates of known occurrences of a species. The annual growth index, GI, describes species response to temperature and soil moisture while stress indices (cold, wet, hot, dry, cold-wet, cold-dry, hot-wet, and hot-dry) exclude it from unfavourable locations. By combining these growth and stress indices, an Ecoclimatic Index (EI), ranging from 1–100 is calculated, which provide an overall measure of the suitability of a given location.

We used the 'Compare Locations' model in CLIMEX which was parametrised by using distribution data of *S. madagascariensis* from its native range in South Africa and

Madagascar, and its introduced range, such as Japan. Growth and stress parameters of *S. madagascariensis* were fitted using the methodology described in [37–39]. A modified CLIMEX temperate template was used as a starting point. The model fitting strategy started with fitting of stresses based on known occurrences in native range in Southern Africa (South Africa, Swaziland, Mozambique, Madagascar; n = 242) and introduced range in east Africa (Kenya; n = 13) and Southeast Asia (Japan; n = 66). The stress parameters were fitted iteratively by changing the values until a best fit is achieved. To exclude the *S. madagascariensis* suitability in tropical regions of the world, a hot-wet stress (HDS) was used. Similarly, the cold stress (THCS) value was increased to constrain weed's suitability from very cold regions of the world (Table 1).

Table 1. CLIMEX parameter values used for modelling the potential distribution of *S. madagascariensis*.

Parameter Group	Parameter	Value	Units
Temperature index	DV0 = limiting low temperature	5	°C
	DV1 = lower optimum temperature	18	°C
	DV2 = upper optimum temperature	24	°C
	DV3 = limiting high temperature	35	°C
Moisture index	SM0 = limiting low soil moisture	0.1	mm
	SM1 = lower optimum soil moisture	0.2	mm
	SM2 = upper optimum soil moisture	1	mm
	SM3 = limiting high soil moisture	1.5	mm
Cold stress	TTCS = temperature threshold	2.5	°C
	THCS = stress accumulation	−0.0001	week^{-1}
	DTCS = degree-day threshold	0	day °C
	DHCS = degree-day stress rate	0	week^{-1}
Heat stress	TTHS = temperature threshold	35	°C
	THHS = stress accumulation rate	0.001	week^{-1}
Dry stress	SMDS = wet stress threshold	0.1	
	HDS = stress accumulation rate	−0.0001	week^{-1}
Wet stress	SMWS = wet stress threshold	2	
	HWS = stress accumulation rate	0.001	week^{-1}
Hot-wet stress	PHW = stress accumulation rate	0.001	week^{-1}

The growth functions (temperature and moisture indices) of the model were fitted based on *S. madagascariensis* distribution in native range and introduced range, as well as reported experimental data [9]. To validate the model, the climatic suitability projection was matched with known occurrences of *S. madagascariensis* in Australia (n = 9283) and South America (n = 122). All occurrences in Australia and South America fall within projection, providing confidence that this model is reasonable representation of climatic suitability of *S. madagascariensis*.

The CLIMEX model, run under the current climate scenario, satisfies the present geographic range of *S. madagascariensis* around the world (Figure 1). All occurrences of *S. madagascariensis* matched well within the projection created by this model. For South America, the model suggests that most of the southern parts of Brazil, the whole of Uruguay and northeastern Argentina is highly suitable for *S. madagascariensis* (Figure 1). Most parts of Andes Mountain range in Bolivia, Peru, Ecuador, and Columbia are also suitable. In North America, southeastern States (Florida, Louisiana, Texas, Alabama, Georgia, Mississippi) and whole of east coast of USA, southern Mexico, and most of the Caribbean Islands are projected to be highly suitable for *S. madagascariensis*. One other highly suitable region, presently outside the range of *S. madagascariensis*, was in southeast Asia with southern and southwest parts of mainland China, southern Japan, and northern Vietnam being suitable with EI values of between 21 to 40 (Figure 1). Most of eastern and southern Africa, as well as northern most parts of African continent were predicted to carry a similar climate to that of the native range of the weed. According to the potential distribution model, most parts of western Europe including, Portugal, Spain, Italy Belgium, Hungary, France, Germany and the United Kingdom are all highly suitable with EI values of 30 to 60 (Figure 1). The model has predicted that southeastern Australia is climatically highly suitable for fireweed compared to northern Australia. This was further confirmed by concentration of most of the species' occurrences within this region. Inland QLD and NSW and northern parts of the Northern Territory (NT) and inland Western Australia (WA) have EI values in the range of 1 to 20. Furthermore, many parts of Victoria and Tasmania are suitable for *S. madagascariensis* to grow. Although *S. madagascariensis* has so far not been reported from New Zealand, our model predicts that most parts of both North and South Islands are climatically suitable for *S. madagascariensis*. However, CLIMEX projections are very coarse and do not include the regional microclimate variations, biotic factors (competition, natural enemies, allelopathy, etc.) and land use practices that can also affect a locations suitability for invasion.

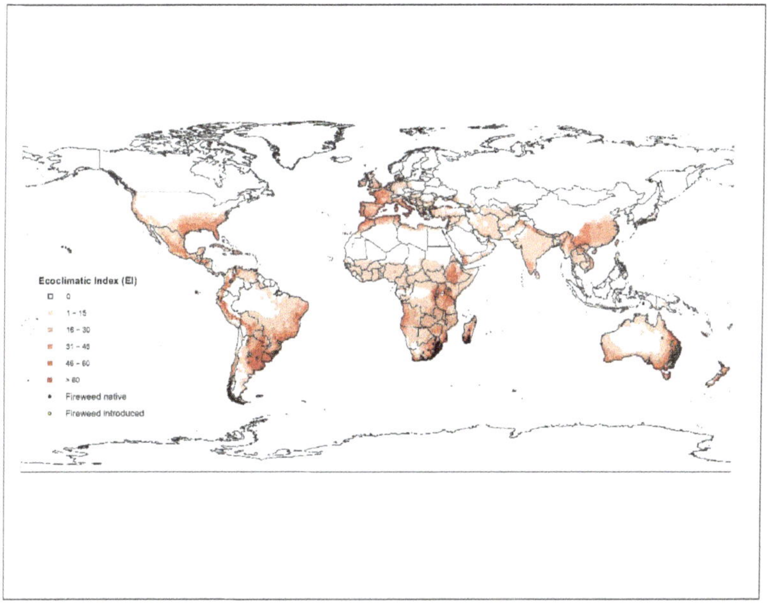

Figure 1. Current distribution (native—green dots; introduced—blue dots) and the climatic suitability for fireweed under the current climate modelled using CLIMEX.

Sindel and Michael [24] using BIOCLIM, suggested the potential range of *S. madagascariensis* in QLD was from the southern border, along the southern coast and north to about Gympie. However, Csurhes and Navie [29] using CLIMEX, suggested the potential range could go further, north into coastal central Queensland and into higher elevation areas including the Atherton Tableland in northern Queensland.

5.1. Preferred Habitat

Senecio madagascariensis can develop a different growth habit, with different leaf shapes on different soil types and in different habitats [11]. *Senecio madagascariensis* is an opportunistic weed and it can adapt and spread into new areas rapidly [8]. The preferred habitat of *S. madagascariensis* in the invaded range includes roadsides, livestock feeding areas, and areas around dams and other wet areas [18]. It can also be found in improved grasslands, in meadows and along riverbanks [40]. *Senecio madagascariensis* prefers well-drained, non-compact, high fertility soil, but it can grow on a wide range of soils including sands and limed soil [8,9].

5.2. Climatic Requirements

Senecio madagascariensis prefers a humid, maritime, and sub-tropical climate. It is found at similar latitudes on the eastern coast of the three continents of the southern hemisphere with an annual rainfall of between 500 to 1000 mm [9]. An annual mean temperature of 12.4 to 20.1 °C favours the establishment of *S. madagascariensis* in most locations where it has invaded [9]. However, young seedlings are sensitive to frost while older plants show tolerance [21], but frost can reduce their vigour [24]. This may lead to its absence in areas with a high frost rate [21]. *Senecio madagascariensis* can grow at altitudes of up to 2800 m above sea level in the tropical environments of Kenya and Columbia suggesting that the warmer altitudinal temperatures in these countries may allow it to grow at higher altitudes than would normally be seen in more temperate countries [24].

Based on an empirical modelling approach, Le Roux et al. determined the most climatically suitable areas for *S. madagascariensis* in Hawaii [41]. They concluded that they were low elevation and arid areas on the windward sides of all islands, with minimum and maximum annual temperatures between 12 to 18 °C and 21.2 to 26.5 °C respectively, elevations between sea level and 1000 m, solar radiation levels between 350 to 400 calories m^{-2} day^{-1}, and annual precipitation between 178 to 376 mm $year^{-1}$. This is consistent with other studies, although the estimation for precipitation is substantially lower. A potential limitation of this study is that the model developed is only based on climate data taken from positive locations on the five major islands of Hawaii. In addition, the LeRoux's projection (41) did not completely exclude the higher elevation and humid areas as possible invasion sites, as they were predicted to be reasonably suitable.

5.3. Growth and Development

Senecio madagascariensis is a short-lived perennial, however it often grows as an annual. Most plants finish their life cycle at the end of their first year [9]. However, a few plants will remain, continue to grow, and reproduce vigorously throughout their second year and therefore under these circumstances it is regarded to be a perennial. It can exhibit high plasticity with the ability to change its life cycle depending upon local conditions [9].

5.4. Reproduction and Seed Dynamics

Senecio madagascariensis reproduces primarily by seeds (i.e., achenes), although vegetative reproduction has been observed under certain conditions [8,42]. When its stems are trampled and contact moist soil, roots and shoots can sprout from the stem's nodes, resulting in new, self-supporting plants [42].

Flowering can take place throughout the year in some countries [23] however, in Australia, it is typically from late autumn (May) through to mid-summer (January). If conditions are favourable, some plants flower until late summer (February). Generally, plants flower 42 to 70 days after seedling emergence. Flowers are pollinated by insects such as European honeybees and hover flies [9]. Generally, the plants undergo senescence and die after flowering as this is an important part of the life cycle.

Shed seed can germinate as soon as it is dispersed [16]. Although, seed can germinate throughout the year, in Australia the highest peak in germination occurs from March to June [15]. Seeds germinate within a temperature range of 15 to 27 °C at the soil surface [18], and maximum germination occurs between 20 to 25 °C and germination ceases at 35 °C or above [9]. Germination is also highest in the presence of light compared to dark conditions [12]. Under normal conditions, *S. madagascariensis* seed dormancy is negligible and high temperatures induce seed dormancy [43]. Despite having relatively high germinability, viable seedbanks are thought to persist for between 3 to 5 years [18] and potentially up to 10 years under certain conditions [9].

5.5. Seed Dispersal

Human aided dispersal has been responsible for the introduction of *S. madagascariensis* into at least eight countries outside of its native range. Once in a new environment, the seeds can be dispersed from local populations through multiple mechanisms including wind, by attaching to animals and vehicles, or in contaminated agricultural produce [29]. Wind has been attributed to the rapid spread and expansion of *S. madagascariensis* in parts of several countries, including Australia, Argentina, Brazil, and the USA (i.e., Hawaii) [32,40,44]. It was hypothesised that the spread of *S. madagascariensis* by wind might have been enhanced through the rapid evolution of superior dispersal traits at the invasion front (i.e., range edge populations) [40], as has been reported for *S. inaequidens* in Europe [45]. However, after comparing key features of the propagules of *S. madagascariensis* plants (i.e., pappus size, achene size, and the ratio of pappus size to achene size) growing on the edge and from within the established range, no differences in dispersal potential were found, which is advantageous from a management perspective. Bartle et al. [40] recorded a similar response for South American populations. However, they did find that wind dispersal of *S. madagascariensis* was strongly affected by adaptation to prevailing geographic conditions. In low altitude areas of Argentina, plants had adapted to produce smaller seeds that could be dispersed further by wind. In higher altitude areas, larger seeds were produced, which suggested that seedling establishment was favoured over long-distance dispersal [44]. In-addition to wind dispersal, it can be dispersed by hay, grain, clothing, vehicles, livestock, birds, and other animals [9]. However, it is unknown whether *S. madagascariensis* seeds can go through the digestive tract of cattle, sheep, or birds and then germinate [9].

5.6. Toxicity

Among the 1200-worldwide species of *Senecio*, ca. 25 species (including *S. brasiliensis, S. heterotrichius, S. cisplatinus, S. selloi,* and *S. oxyphyllus*) are toxic to certain livestock, such as horses (*Equus caballus* L.) and cattle (*Bos taurus* L., *Bos indicus* Brahman) [3]. According to various studies, some *Senecio* species can be toxic to humans. For example, if milk products containing PAs from *S. madagascariensis* are consumed [46].

Due to the presence of pyrrolizidine alkaloids (PAs) in the leaves, *S. madagascariensis* is highly toxic to certain domesticated animals; that is, it can cause hepatopathy, decrease the development of young animals, and in some cases cause mortality when ingested [2]. According to Sheppard et al. the PAs from *S. madagascariensis* could cause chronic liver damage and fatality of horses [47]. Weeks or months after feeding on the plant has ended, the symptoms and death of animals can still occur. Although moving stock to areas free from *S. madagascariensis* can prevent further progress of the disease, ill health, and poor growth may continue [9].

In countries such as Brazil, Columbia, and Uruguay, where *S. madagascariensis* is spreading into new areas, there is growing concern that it is contributing to an increase in PA poisoning [33,35,48].

6. Management of *S. madagascariensis*

Several management techniques are used to control *S. madagascariensis*. These include cultural, physical, chemical, and biological methods, or a combination of these methods.

6.1. Legislation

In Australia, *S. madagascariensis* is a declared weed under the relevant legislation of all states and territories. However, the level of declaration and associated requirements varies greatly depending on the perceived level of risk. *Senecio madagascariensis* was added to the Hawaiian State Noxious Weed List by their Department of Agriculture in 1992 [30]. In Japan

S. madagascariensis has been declared an Invasive Alien Species under the Invasive Alien Species Act (which restricts importation, moving, or growing of species within Japan) [2].

6.2. Physical Control

Senecio madagascariensis seeds are mostly wind dispersed and therefore the physical hand pulling of plants must be completed before the seed is formed if it is to be effective. The pulled plants should then be burnt or deep buried to prevent plants re-growing and producing further seeds. This technique is only practical for isolated plants or small patches and not for large infestations, as it is time consuming and labour intensive [4,18,42].

In relatively flat and accessible areas, mowing and cutting style equipment (e.g., slashers, mulchers) can be used to weaken *S. madagascariensis* plants and help prevent them from reaching reproductive maturity [42]. To be effective, this technique needs to be repeated as few *S. madagascariensis* plants will be killed from a single operation [4,42]. It will also be most effective on smaller plants and if applied when the pasture is growing, enabling maximum competition to be imposed to restrict *S. madagascariensis* regrowth [18].

In the central coast region of NSW (Australia), slashing and mulching of paspalum (*Paspalum dilatatum* Poir.) and kikuyu (*Pennisetum clandestinum* Hochst. ex Chiov.) pastures infested with *S. madagascariensis* from mid-September onwards is an effective strategy for its management [16]. However, mulched *S. madagascariensis* can wilt and become more attractive to stock feeding on it and contains a greater concentration of PAs. Thus, after slashing or mulching, *S. madagascariensis* infested paddocks should not be grazed for at least 2 weeks [9]. Slashing can also have other downsides such as facilitating further spread of *S. madagascariensis* if plants are reproductive, and it may only delay flowering until later in the season and promote regrowth of plants in the following season [18].

6.3. Chemical Control

In terms of chemicals, 2,4-D formulations [16,22], dicamba, glyphosate, MCPA, tebuthiuron, triclopyr [22], bromoxynil, fluroxypyr/aminopyralid, metsulfuron-methyl, and triclopyr/picloram/aminopyralid [18,28] are some herbicides that have been found to be effective on one or more growth stages of *S. madagascariensis*. However, which chemicals and rates that can be legally applied, may vary between countries and even between jurisdictions within countries.

The selective herbicide bromoxynil can be very effective on young plants, but mature plants are more tolerant. Being a contact herbicide, only those parts of a plant that come directly in contact with the herbicide are killed and the plant will often regrow from unaffected parts. Significant seedling recruitment after spraying is also often observed. It has been reported [28,49] that bromoxynil at 3 L ha^{-1} was unsuccessful at controlling mature *S. madagascariensis*, with substantial regrowth recorded five months after spraying.

Glyphosate, a nonselective systemic herbicide proved equally as effective as bromoxynil in trials undertaken in Argentina [50]. However, as with bromoxynil, re-infestation will occur afterwards. Additionally, as glyphosate is nonselective it should only be applied where non-target damage can be tolerated or using target specific equipment such as wipe-on applicators [22].

In Australia, 2,4-D amine (3.2 kg ha^{-1}) and 2,4-D sodium salt (2 to 4 kg ha^{-1}) have been reported to give positive results after spraying, without harming the neighbouring pasture species, such as blue couch (*Digitaria didactyla* Wild), cogon grass (*Imperata cylindrica* (L.) Beauv), and white clover (*Trifolium repens* L.) [49]. Similarly, Ref. [22] suggested that in Hawaiian pastures where forage legumes are mixed with grasses, the amine salt formulation of 2,4-D would be preferable because of its mild effect on legumes. In contrast, metsulfuron-methyl at 40 to 80 g ha^{-1} provided effective control of *S. madagascariensis* in an Australian study, but it was severely damaging to any legumes (such as *T. repens*) present within the treated pasture [49].

Whilst the efficacy of several herbicides on plant mortality is known, there is minimal information on their effect on seeds located on plants at the time of their application. This is important, as landholders often spray mature plants, and it would be advantageous if the herbicides used not only killed the plants but also kill any seeds located on them at that time. It has been suggested that herbicides do not generally kill *S. madagascariensis* seeds if applied after flowering [18], but some success has been reported using certain herbicides on *C. odorata* (L.) R.M.King & H.Rob. (Siam weed) another Asteraceae species. According to one study metsulfuron-methyl was effective on immature and intermediate seed maturity stages but not mature seeds [51], whilst fluroxypyr was highly effective at causing mortality of immature and intermediate seeds and rendered a proportion of mature seeds non-viable after the plants were sprayed.

Even using the most effective herbicide, a single application will not suppress *S. madagascariensis* permanently [22] and follow up control will be necessary. Consequently, management strategies that rely solely on herbicide applications can become expensive if large areas must be controlled. Chemical control of *S. madagascariensis* in the Hawaiian archipelago, if undertaken, would cost USD $11 million year^{-1} [11], making it an uneconomical proposition [21].

6.4. Biological Control—Competitive Pastures

In an effective *S. madagascariensis* management program, maintaining a vigorous, competitive pasture through the autumn and winter months is an important step in providing best management practice [9]. The summer-growing and high fodder yielding perennial pasture species such as setaria (*Setaria sphacelata* Schum.), kikuyu (*Pennisetum clandestinum* Hochst. ex Chiov.), paspalum (*Paspalum dilatatum* Poir.), and Rhodes grass (*Chloris gayana* Kunth.) are also likely to be suppressive throughout the winter months. These predominantly summer-growing pasture species can be grown through the late summer and autumn months, to provide good groundcover, which then helps to prevent *S. madagascariensis* seedling establishment in the autumn and winter months [52]. However, according to a study undertaken in South Africa, *Eragrostis curvula* Schrad., *Cynodon dactylon* (L.) Pers., *Dactylis glomerata* L., *Festuca arundinacea* Schreb., *Pennisetum clandestinum* Hochst. ex Chiov., and *Themeda triandra* Forssk. grasses were all unable to suppress *S. madagascarinsis* due to its highly competitive ability [53].

6.5. Biological Control—Insects and Pathogens

Although a biocontrol program to control *S. madagascariensis* first commenced in Australia in 1987, only two insects were tested, and neither were released. *Aecidium* sp. is a rust fungus that was imported into quarantine in Australia for detailed studies. This rust was like a naturally occurring Australian isolate of the orange rust *Puccinia lagenophorae* Cooke [54] that is found on the family Asteraceae. The South African rust was less damaging than the Australian isolates of *P. lagenophorae.* These results imply that their introduction would be doubtful to Australia due to its low virulence, which would be expected to translate to poor control of *S. madagascariensis* [54].

In 2013, *Secusio extensa* (Erebidae: Arctiinae) was introduced to control *S. madagascariensis* in Hawaii. However, since release, this agent has undergone population outbreaks on a related alternative host, Cape ivy (*Delairea odorata*, Asteraceae: Senecioneae) which is another noxious weed in Hawaii. Given that this herbivore significantly preferred Cape ivy over *S. madagascariensis* for its oviposition and larval feeding, it has not proven to be an effective biological control agent in Hawaii [55]. Generalist herbivores (e.g., grasshoppers, weevils, thrips, aphids, scale insects, whiteflies, and vermin) that feed upon *S. madagascariensis* in Hawaii have been also shown to have an insignificant effect on the plant [11].

In the KwaZulu Natal Province of South Africa, 12 natural enemies have been recently identified in initial surveys feeding on five populations of *S. madagascariensis.* They comprise of three stem borers, four flower feeders, two sap suckers, and three plant pathogens [17]. Recently, four potential agents were prioritised for host-range assessments and these studies are currently in progress. The agents being tested are the capitulum-feeding *Homoeosoma stenotea* Hampson (Lepidoptera: Pyralidae), the stem-boring *Gaste-*

roclisus tricostalis (Thunberg) (Coleoptera: Curculionidae), and *Metamesia elegans* (Walsingham) (Lepidoptera: Tortricidae) and the root-feeding *Longitarsus basutoensis* Bechyné (Coleoptera: Chrysomelidae: Alticinae) [56].

6.6. Livestock Grazing

Using cattle as a control strategy for *S. madagascariensis* tends to be unsuccessful, as the reduced competition and improved light conditions that occur once the pasture is grazed down allows new seedlings to grow faster. Cattle also tend to avoid it in heavily grazed paddocks as it is more easily distinguished amongst the grass pasture [9]. In contrast, grazing with sheep or goats is considered an effective method to control *S. madagascariensis*. According to Watson et al. sheep and goats are about 20 times more tolerant to *S. madagascariensis* poisoning than horses and cattle [16]. Sheep might ingest and suppress *Senecio* spp. toxin due to their ability to undergo hepatic detoxification. This is related to their ruminal flora populations that can reduce the probability of poisoning [57].

In southern Brazil, high densities of *S. brasiliensis* and *S. madagascariensis* were controlled by using 16 sheep (3.0 stock units ha^{-1}) in a 5.5 ha experimental area [57]. A total of 28,629 plants of *S. brasiliensis* (10,122) and *S. madagascariensis* (18,507) were almost eliminated within a 2-year period. Before and after the experimental period, liver biopsies of sheep and cattle were examined but did not reveal any sign of seneciosis [57].

The effectiveness of using herbivores such as sheep and goats will depend a lot on the number of propagules available for recruitment to occur. In a Hawaiian study that investigated the effects of feral goats and sheep on *S. madagascariensis* in a natural environment (dry forest), Questad et al. found that in areas where there were high numbers of propagules, *S. madagascariensis* plants were smaller due to grazing, but there was largescale seedling recruitment [58]. In contrast, if there were few propagules at the start, recruitment and overall biomass of *S. madagascariensis* was greatly reduced.

Using livestock mediated herbivory, as part of a control strategy for *S. madagascariensis* is most applicable for modified environments such as pastures. In natural environments, careful consideration would need to be given as there is a higher risk of non-target damage occurring, but in some situations, it may still be applicable [58].

6.7. Integrated Management

Integrated weed management (IWM) is the use of a combination of methods, including cultural, physical, biological, and chemical approaches [59]. For *S. madagascariensis*, some trials to determine the effects of individual treatments have been undertaken, but testing of combinations of treatments is lacking. Despite this, Sindel and Coleman suggest that an effective IWM approach for *S. madagascariensis* control should include the use of perennial competitive pasture species, hand weeding (for isolated plants or small patches), the maintenance of good farm sanitation practices and the use of herbicides to control the weed [18]. Furthermore, for long-term control of this weed, reduced grazing within a competitive pasture may need to be used by landholders [18]. The integration of livestock that are more tolerant (e.g., sheep and goats) of *S. madagascariensis* should also be considered given the success obtained in reducing its density in some situations [57].

In Hawaii, Thorne et al. recommended that for the successful control of *S. madagascariensis*, federal, state, county, and private land managers need to develop IWM plans [42]. They further suggested that these plans be part of any management program for all land units, and that the key objectives should include preventive measures where the weed is not present, control measures where the weed is already established, and protocols for taking prompt action when *S. madagascariensis* first appears in an area.

To achieve this, an adaptive management approach was developed based on six key steps: establishing goals, setting management priorities, identifying appropriate methods, developing, and implementing an integrated weed management plan—monitoring results, modifying priorities, and improving the management plan. In terms of control options, they highlighted that they should, (a) not contribute to the spread of the weed, (b) be applied at the most effective time (i.e., point in the life cycle when it is most vulnerable), (c) minimise the risk to human health and potential damage to the general environment (e.g., non-target species), and (d) be cost effective. Key benefits and limitations associated with various options (i.e., pulling, mowing, cutting, cultural controls, livestock grazing, biological control agents, herbicides, burning) are also outlined to help landowners/managers to make informed decisions [42]. A working model is now proposed for the improved management of *S. madagascariensis* (Figure 2), however further refinement of the model is necessary.

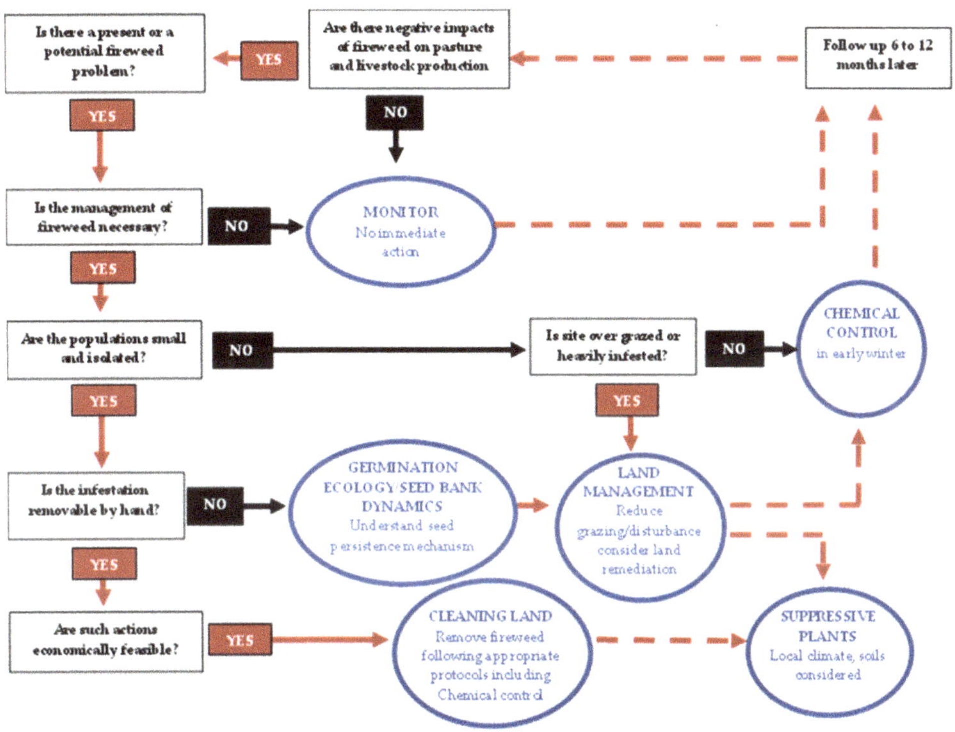

Figure 2. Conceptual decision diagram for proposing approaches to the integrated man-agement of fireweed incorporating managerial, chemical, and manual control.

7. Conclusions and Future Directions

Since *S. madagascariensis* contains PAs, that cause liver damage, certain domesticated animals (cattle and horses) can be poisoned after consuming this plant. It has been found to be a fast growing annual/short lived perennial plant that reaches reproductive maturity quickly and produces large quantities of viable seed for future recruitment events. The predicted distribution suggests that many regions across the world have a climate suitable for the growth of *S. madagascariensis*. This indicates that the weed is likely to expand its range in future, if not controlled. Although there are many management techniques available including legislation, cultural, physical, chemical, and biological (insects and suppressive plants), individually they have little effect on *S. madagascariensis* management. An IWM program incorporating a combination of techniques will generally be much more effective. However, for those landholders who do not have *S. madagascariensis* or are in the early stages of invasion, early detection is even more important to prevent its establishment in the field.

To build on the currently available information, further research is needed to improve management options for this problematic weed. In terms of ecology, seed dynamics appears to be a key factor influencing the spread and persistence of *S. madagascariensis*. Further research into the environmental conditions affecting germination, viability, and persistence would be advantageous, particularly if changing climate scenarios are included. From a control perspective, a range of herbicides are currently available to kill small to large *S. madagascariensis* plants. However, their effect on the germination and viability of seeds located on plants at the time of spraying has not been fully evaluated and warrants investigation. This is important as plants are often reproductive at the time of spraying and if a herbicide can not only kill the plant but also the seeds, subsequent seedling recruitment will be reduced. Control efforts will also benefit from the release of effective biological control agents that adversely affect key life stages of *S. madagascariensis*. Limited success has been achieved to date, but current efforts should continue in the search for host specific options. In the interim, improved pasture management and testing of suppressive pasture plants should be explored further as part of longer-term strategies to manage well established infestations. How these pastures are grazed, particularly in terms of the type of animals, is an area that could also be the focus of additional research. Animals such as sheep and goats are more tolerant of *S. madagascariensis* than cattle and utilising them in traditional cattle grazing operations could be beneficial. Finally, larger integrated management trials should be undertaken over different seasons and at several locations to encompass a range of pasture types, climatic conditions, and soil types. In doing so, detailed cost/benefit analysis data could be collected to enable landholders to make informed decisions about the most appropriate options for their situation.

Author Contributions: Conceptualization: K.W., S.C., J.V. and S.A. Writing—original draft: K.W. and A.S. Review and editing: K.W., S.C., J.V., A.S. and S.A. All authors have read and agreed to the published version of the manuscript.

Funding: The Queensland Department of Agriculture and Fisheries, Brisbane (RM 2018000131).

Institutional Review Board Statement: Not applicable.

Informed Consent Statement: Not applicable.

Data Availability Statement: Not applicable.

Acknowledgments: We thank the University of Queensland and the Queensland Department of Agriculture and Fisheries, Brisbane for providing funds for this study.

Conflicts of Interest: The authors declare no conflict of interest.

References

1. Egli, D.; Olckers, T. Establishment and impact of insect agents deployed for the biological control of invasive Asteraceae: Pro spects for the control of *Senecio madagascariensis*. *BioControl* **2017**, *62*, 681–692. [CrossRef]
2. Tsutsumi, M. Current and potential distribution of *Senecio madagascarensis* Poir. (fireweed), an invasive alien plant in Japan. *Grassl* **2011**, *57*, 150–157. [CrossRef]
3. Gardner, D.R.; Thorne, M.S.; Molyneux, R.J.; Pfister, J.A.; Seawright, A.A. Pyrrolizidine alkaloids in *Senecio madagascariensis* from Australia and Hawaii and assessment of possible livestock poisoning. *Biochem. Syst. Ecol.* **2006**, *34*, 736–744. [CrossRef]
4. Sindel, B.M.; Coleman, M.J.; Reeve, I.J. Putting fireweed on the front burner: Improvingmanagement and understanding impact. In Proceedings of the 18th Australian Weeds Conference, Eldershaw, VIC, Australia, 8–11 October 2012; pp. 166–169.
5. Mcfadyen, R.; Sparks, D. Biological control of fireweed. In Proceedings of the 11th Australian Weeds Conference, Melbourne, Australia, 30 September–3 October 1996; pp. 569–572.
6. Poiret, J.L.M. *Senecio madagascariensis, Encyclopédie Méthodique Botanique Supplément*; Chez Mme. veuve Agasse: Paris, France, 1817; p. 130.
7. Hassler, M. World Plants: Synonymic Checklists of the Vascular Plants of the World Species 2000 and It Is Catalogue of Life, Natu Ralis, Leiden, and The Netherlands. Available online: http://www.catalogueoflife.org (accessed on 17 November 2017).
8. Sindel, B.M. *Senecio madagascariensis* in Australia—Directions for Future Research. Available online: https://www.une.edu.au/__data/assets/pdf_file/0004/36868/fireweed-bvfa-report.pdf (accessed on 20 January 2018).
9. Sindel, B.M.; Radford, I.J.; Holtkamp, R.H.; Michael, P.W. The biology of Australian weeds: 33 *Senecio madagascariensis* Poir. *Plant Prot. Q.* **1998**, *13*, 2–15.

10. Scott, L.J.; Congdon, B.C.; Playford, J. Molecular evidence that fireweed (*Senecio madagascariensis*, Asteraceae is of South African origin. *Plant Syst. Evol.* **1998**, *213*, 251–257. [CrossRef]
11. Ramadan, M.M.; Murai, K.T.; Johnson, T. Host range of *Secusio extensa* (Lepidoptera: Arctiidae), and potential for biological control of *Senecio madagascariensis* (Asteraceae). *J. Appl. Entomol.* **2010**, *135*, 269–284. [CrossRef]
12. Radford, I.J.; Cousens, R.D. Invasiveness and comparative life-history traits of exotic and indigenous *Senecio* species in Australia. *Oecologia* **2000**, *125*, 531–542. [CrossRef]
13. Radford, I.J.; Liu, Q.; Michael, P.W. Chromosome counts for the Australian weed known as *Senecio madagascariensis* (Asteraceae). *Aust. Syst. Bot.* **1995**, *8*, 1029–1033. [CrossRef]
14. Lopez, M.G.; Wulff, A.F.; Poggio, L.; Xifreda, C.C. South African fireweed *Senecio madagascariensis* (Asteraceae) in Argentina: Relevance of chromosome studies to its systematics. *Bot. J. Linn* **2008**, *158*, 613–620. [CrossRef]
15. Sindel, B.M. The Ecology and Control of Fireweed (*Senecio madagascariensis* Poir.). Ph.D. Thesis, University of Sydney, New South Wales, Australia, 1 January 1989.
16. Watson, R.; Allan, H.; Munnich, D.; Launders, T.; Macadam, J. *Fireweed*; New South Wales Agriculture: Orange, Australia, 1997.
17. Egli, D.; Olckers, T. Abundance across seasons of insect herbivore taxa associated with the invasive *Senecio madagascariensis* (Asteraceae), in its native range in KwaZulu-Natal, South Africa. *Afr. Entomol.* **2015**, *23*, 147–156.
18. Sindel, B.; Coleman, M. *Senecio madagascariensis.* A Best Practice Management Guide for Australian Landholders. Available online: https://www.une.edu.au/__data/assets/pdf_file/0004/52366/2012.-Fireweed-A-Best-Practice-Management-Guide-for-Australian-Landholders.pdf (accessed on 24 January 2018).
19. Radford, I.J.; Muller, P.; Fiffer, S.; Michael, P.W. Genetic relationships between Australian fireweed and South African and Madagascan populations of *Senecio madagascariensis* Poir. And closely related *Senecio* species. *Aust. Syst. Bot.* **2000**, *13*, 409–423. [CrossRef]
20. Thompson, I.R. Taxonomic studies of Australian *Senecio* (Asteraceae): 5. The *S. pinnatifolius/S. pinnatifolius* complex. *Muelleria* **2005**, *21*, 23–76.
21. Le Roux, J.J.; Wieczorek, A.M.; Ramadan, M.M.; Tran, C.T. Resolving the native provenance of invasive fireweed (*Senecio madagascariensis* Poir.) in the Hawaiian Islands as inferred from phylogenetic analysis. *Divers. Distrib.* **2006**, *12*, 694–702. [CrossRef]
22. Motooka, P.; Nagai, G.; Onuma, K.; DuPonte, M.; Kawabata, A.; Fukumoto, G.; Powley, J. Control of Madagascar Ragwort (Aka Madagascar Fireweed, *Senecio madagascariensis*). Available online: https://scholarspace.manoa.hawaii.edu/bitstream/10125/12611/1/WC-2.pdf (accessed on 1 June 2020).
23. Egli, D. Native Range Studies on Insect Herbivores Associated with Fireweed (*Senecio madagascariensis*) in KwaZulu-Natal, South Africa, with Prospects for Biological Control in Invaded Countries. Ph.D. Thesis, University of KwaZulu-Natal, Pietermaritzburg, South Africa, 2017.
24. Sindel, B.M.; Michael, P.W. Spread and potential distribution of *Senecio madagaseariensis* Poir. (fireweed) in Australia. *Austral Ecol.* **1992**, *17*, 21–26. [CrossRef]
25. Meat & Livestock Australia, Scoping a Management Programme for Fireweed. Available online: https://www.mla.com.au/contentassets/af130e6381784cbba6a37d1e4aa37a2c/b.wee.0137_final_report.pdf (accessed on 10 December 2019).
26. Sindel, B.M. The ecology and control of fireweed (*Senecio madagascariensis* Poir.). *Plant Prot. Q.* **1986**, *1*, 163–172.
27. Sindel, B.M.; Michael, P.W.; McFadyen, R.E.; Carthew, J. The continuing spread of fireweed (*Senecio madagascariensis*)- the hottest of topics. In Proceedings of the Sixteenth Australian Weeds Conference, Brisbane, QLD, Australia, 18–22 May 2008; pp. 47–49.
28. Fireweed (*Senecio madagascariensis*). Available online: https://www.daf.qld.gov.au/__data/assets/pdf_file/0009/67167/fireweed.pdf (accessed on 13 August 2020).
29. Csurhes, A.; Navie, S. *Weed Risk Assessment: Fireweed (Senecio madagascariensis)*; The State of Queensland, Department of Employment, Economic Development and Innovation: Brisbane, QLD, Australia, 2010; pp. 3–19.
30. Starr, F.; Martz, K.; Loope, L. Fireweed (*Senecio madagascariensis*) An Alien Plant Report. Available online: http://www.hear.org/species/reports/senmad_fskm_awwa_report.pdf (accessed on 4 April 2020).
31. Mäder, G.; Castro, L.; Bonnato, S.L.; Freitas, L.B. Multiple introductions, and gene flow in subtropical South American populations of the fireweed, *Senecio madagascariensis* (Asteraceae). *Genet. Mol. Biol.* **2016**, *39*, 135–144. [CrossRef] [PubMed]
32. Dematteis, B.; Ferrucci, M.S.; Ortega-Baes, P.; Coulleri, J.P. Genome Size Variation between the Native and Invasive Ranges of *Senecio madagascariensis* (Asteraceae). *Syst. Bot.* **2020**, *45*, 212–218. [CrossRef]
33. Diaz, G.J. Toxicosis by Plant Alkaloids in Humans and Animals in Colombia. *Toxins* **2015**, *7*, 5408–5416. [CrossRef]
34. Villalba, J.; Ferna'ndez, G. Otra flor amarilla peligrosa: *Senecio madagascariensis*. *Rev. Plan Agropecu.* **2007**, *123*, 46–48. (In Spanish) Available online: https://www.planagropecuario.org.uy/publicaciones/revista/R123/R123_46.pdf (accessed on 17 June 2018).
35. García, J.A.; Santos, C.G.Y.; Rosas, J.; Dutra, F.; Gardner, D. A survey of *Senecio* spp. affecting livestock in Uruguay and their associated pyrrolizidine alkaloid content. *Cienc. Rural.* **2018**, *48*, e20170621. [CrossRef]
36. García, J.A.; Rosas, J.E.; Santos, C.G.Y.; Streitenberger, N.; Feijoo, M.; Dutra, F. *Senecio* spp. transboundary introduction and expansion affecting cattle in Uruguay: Clinico-pathological, epidemiological and genetic survey, and experimental intoxication with *Senecio oxyphyllus*. *Toxicon* **2020**, *173*, 68–74. [CrossRef]
37. Sutherst, R.W.; Maywald, G.F. A computerised system for matching climates in ecology. *Agric. Ecosyst. Environ.* **1985**, *13*, 281–299. [CrossRef]

38. Bourdôt, G.W.; Lamoureaux, S.I. *Abutilon theophrasti*—A comparison of two climate niche models. *N. Z. J. Agric. Res.* **2021**, *64*, 211–222. [CrossRef]
39. Kriticos, D.J.; Sutherst, R.W.; Brown, J.R.; Adkins, S.W.; Maywald, G.F. Climate change and the potential distribution of an invasive alien plant: *Acacia nilotica* ssp. indica in Australia. *J. Appl. Ecol.* **2003**, *40*, 111–124. [CrossRef]
40. Bartle, K.; Moles, A.T.; Bonser, S.P. No evidence for rapid evolution of seed dispersal ability in range edge populations of the invasive species *Senecio madagascariensis*. *Austral Ecol.* **2013**, *38*, 915–920. [CrossRef]
41. Le Roux, J.J.; Wieczorek, A.M.; Tran, C.T.; Vorsino, A.E. Disentangling the dynamics of invasive fireweed (*Senecio madagascariensis* Poir. species complex) in the Hawaiian Islands. *Biol. Invasions* **2010**, *12*, 2251–2264. [CrossRef]
42. Thorne, M.S.; Powley, J.S.; Fukumoto, G.K. Fireweed Control: An Adaptive Management Approach. Available online: https://www.ctahr.hawaii.edu/oc/freepubs/pdf/PRM-1.pdf (accessed on 1 June 2020).
43. Alonso, S.; Fernadez, O.; Langero, S.; Verona, C. Characteristics of seed germination of *Senecio madagascariensis* Poiret (Compositae). *Ecologia* **1982**, *7*, 95–116.
44. Dematteis, B.; Ferrucci, M.S.; Coulleri, J.P. The evolution of dispersal traits based on diaspore features in South American populations of *Senecio madagascariensis* (Asteraceae). *Aust. J. Bot.* **2019**, *67*, 358–366. [CrossRef]
45. Monty, A.; Mahy, G. Evolution of dispersal traits along an invasion route in the wind-dispersed *Senecio inaequidens* (Asteraceae). *Oikos* **2010**, *119*, 1563–1570. [CrossRef]
46. Johnson, A.E.; Molyneux, R.J.; Ralphs, M.H. Senecio: A dangerous plant for man and beast. *Rangelands* **1989**, *11*, 261–263.
47. Sheppard, A.; Olckers, T.; McFadyen, R.; Morin, L.; Ramadan, M.; Sindel, B. Biological control of *Senecio madagascariensis* (fireweed) in Australia—A long-shot target driven by community support and political will. In Proceedings of the XIII International Symposium on Biological Control of Weed, Waikoloa, HI, USA, 11–16 September 2011; pp. 123–127.
48. Stigger, A.L.; Estima-Silva, P.; Fiss, L.; Coelho, A.C.B.; Santos, B.L.; Gardner, D.R.; Marcolongo-Pereira, C.; Schild, A.L. *Senecio madagascariensis* Poir. (Asteraceae): A new cause of seneciosis in cattle in Southern Brazil. *Braz. J.* **2014**, *34*, 851–855.
49. Anderson, T.M.D.; Panetta, F.D. Fireweed response to boom spray application of different herbicides and adjuvants. *Plant Prot. Q.* **1995**, *10*, 152–153.
50. *Senecio madagascariensis* (Fireweed). Available online: https://www.cabi.org/isc/datasheet/49565 (accessed on 10 December 2018).
51. Patane, K.A.; Setter, S.; Graham, M. Effect of foliar herbicides on the germination and viability of Siam weed (*Chromolaena odorata*) seeds located on plants at the time of application. *Plant Prot. Q.* **2009**, *24*, 138–140.
52. Fireweed. Available online: https://www.dpi.nsw.gov.au/__data/assets/pdf_file/0007/49840/Fireweed.pdfW (accessed on 12 January 2019).
53. Fynn, O.G.; Harvey, K.J.; Tedder, M.; Olckers, T. Simulated herbivory counteracts the effects of grass competition on the invasive fireweed (*Senecio madagascariensis*). *Austral Ecol.* **2019**, *44*, 1410–1415. [CrossRef]
54. MaFadyen, R.; Morin, L. *Senecio madagascariensis* Poir.—Fireweed. In *Biological Control of Weeds in Australia*; Julien, M., McFadyen, R., Cullen, J., Eds.; CSIRO Publishing: Melbourne, Australia, 2012; pp. 526–536.
55. Krushelnycky, P.D.; Starr, F.; Starr, K.; Abran, M.; Thorne, M.; Leary, J.; Fukada, M.; Rubinoff, D. Performance of the control agent *Secusio extensa* (Lepidoptera: Erebidae) on its target host, *Senecio madagascariensis* (Madagascar fireweed), on an alternate host, *Delairea odorata* (Cape ivy), and on non-target plants, in Hawaii. *Biol. Control* **2018**, *121*, 234–246. [CrossRef]
56. Egli, D.; Olckers, T. Insect herbivores associated with the invasive herb *Senecio madagascariensis* (Asteraceae) in its native range in KwaZulu-Natal, South Africa and their potential as biological control agents in invaded countries. *Biocontrol Sci. Technol.* **2020**, *30*, 243–255. [CrossRef]
57. Bandarra, P.M.; De Oliveira, L.G.; Dalto, A.C.; Boabaid, F.M.; Juffo, G.; Riet-correa, F.; Driemeier, D.; Da Cruz, C.E.F. Sheep production as a *Senecio* spp. control tool. *Pesqui* **2012**, *32*, 1017–1022. [CrossRef]
58. Questad, E.J.; Uowolo, A.; Brooks, S.; Fitch, R.; Cordell, S. Resource Availability, Propagule Supply, and Effect of Non-native Ungulate Herbivores on *Senecio madagascariensis* Invasion. *Pac. Sci.* **2018**, *72*, 69–79. [CrossRef]
59. Swanton, C.; Weise, S. Integrated Weed Management: The Rationale and Approach. *Weed Technol.* **1991**, *5*, 657–663. [CrossRef]

plants

MDPI

Article

Exploring the Phytochemicals of *Acacia melanoxylon* R. Br.

Diana Alves [1], Sidónio Duarte [1], Pedro Arsénio [2], Joana Gonçalves [3], Cecília M. P. Rodrigues [3], Ana Lourenço [1,*] and Patrícia Máximo [1,*]

[1] LAQV-REQUIMTE, Department of Chemistry, NOVA School of Science and Technology, NOVA University, 2829-516 Caparica, Portugal; dfl.alves@campus.fct.unl.pt (D.A.); sidonioduarte88@gmail.com (S.D.)

[2] Linking Landscape, Environment, Agriculture and Food (LEAF)—Research Center, Associated Laboratory TERRA, Instituto Superior de Agronomia, Universidade de Lisboa, Tapada da Ajuda, 1349-017 Lisbon, Portugal; arseniop@isa.ulisboa.pt

[3] Faculty of Pharmacy, Research Institute for Medicines (iMed.ULisboa), Universidade de Lisboa, Av. Prof. Gama Pinto, 1649-003 Lisbon, Portugal; joanalmg@hotmail.com (J.G.); cmprodrigues@ff.ulisboa.pt (C.M.P.R.)

* Correspondence: ana.lourenco@fct.unl.pt (A.L.); psm@fct.unl.pt (P.M.)

Abstract: Invasive species are currently a world menace to the environment, although the study of their chemistry may provide a means for their future beneficial use. From a study of Portuguese *Acacia melanoxylon* R. Br. five known compounds were isolated: lupeol, 3β-*Z*-coumaroyl lupeol, 3β-*E*-coumaroyl lupeol (dioslupecin A), kolavic acid 15-methyl ester and vomifoliol (blumenol A). Their structures were elucidated by 1D and 2D NMR spectroscopy and mass spectrometry, and as a result some corrections are made to their previous ^{13}C NMR assignments. Cytotoxicity of 3β-*E*-coumaroyl lupeol (dioslupecin A) and kolavic acid 15-methyl ester was evaluated against HCT116 human colorectal cancer cells although biological activity was not evident.

Keywords: invasive species; *Acacia melanoxylon*; terpenoids; HCT116 human colorectal cancer cells; cytotoxicity

1. Introduction

Plant invasive species are one of the great threats to biodiversity since they establish and supersede native species, leading occasionally to the extinction of the latter, by disrupting the biotic and abiotic balance of the invaded ecosystem. Apart from this ecological impact, they also have a socio-economic impact by influencing human health, infrastructures and local economies [1,2]. These species have long been a concern in the Portuguese territory [3–5] and elsewhere, which resulted in the publication of legal regulations to prevent and manage the introduction and spread of invasive alien plants, both at the national [6] and European level [7,8]. The management of this problem in Europe costs millions of euros [9] and, since eradication is rarely achieved, actions end up frequently with periodic growth control and containment of these species [10]. The use of invasive species as a source of chemicals or pharmaceuticals allows a rational use of resources that can mitigate the cost of their control, turning a useless and abundant natural good into an added value resource [11]. Most likely, the prevalence of invasive species over endemic ones relies on the bioactivity of the metabolites they produce as invaders, that are surely responsible for their ease of expansion and dominance of the new habitat [11,12].

Bearing this in mind, we began our studies on the chemistry of invasive *Acacia* species. *Acacia* species, mainly from Australia, settled in the Mediterranean area and conquered significant lands. In Portugal they are distributed throughout the country, preferentially on acid substrates. Originally, they were introduced as a source for wood, as an erosion preventer, for reforestation purposes and for ornamental reasons and the perfume industry, among others [13,14]. Abiotic and biotic factors favored their establishment and with time some of their uses deteriorated and economic value lowered, leading to an increase in their abundance and to an invasion condition [14–16]. The management and control of Acacia invasions includes various steps, from risk assessment to containment, eradication and ecosystem restoration actions [13,15,17]. Some of the containment and control actions include mechanical removal (ring-barking, hand-pulling or cutting) followed by chemical control (with glyphosate), or biological control [18].

A. melanoxylon R. Br. (Australian blackwood) is a 15 m tree with evergreen leaves and pale-yellow flowers arranged in a globular head of 10–12 mm diameter. Flowering occurs in Portugal from February to June, and its fruits are brownish red pods. The seeds, encircled by an orange funicle, remain viable in the ground for more than 50 years, are dispersed by birds, wind, water, or rodents, and germinate after a space opening and/or fire occurrence. This species also propagates vegetatively, forming vigorous sprouts from the stump and roots [19].

Previous studies on the chemistry of *A. melanoxylon* include the isolation of hyperoside (quercetin-3-D-galactoside) from the flowers [20], dihydroflavonoids [21], hydroxyflavans [22], leucoanthocyanidins [23–25], a pyrogallol A-ring proanthocyanidin dimer [26], [4-*O*-4]-linked biflavanoids [27] from the heartwood, and acamelin (a furanoquinone) and a benzoquinone from undisclosed parts [28]. Studies from Portuguese invasive species include the identification of Δ^7 phytosterols, phytosteryl glucosides and long-chain *n*-alkyl caffeates by GC-EIMS from the dichloromethane extracts of wood and bark [29,30], as well as the antimicrobial activity of aqueous, ethanolic and methanolic leaf extracts, that showed no interesting results [31]. Other references to bioactivity of extracts of this species can be found in the recent review by Correia et al. on the biomass valorization of *Acacia* species [32].

Following our interest in the chemistry of invasive species, in this study we report on the isolation of five compounds: lupeol **1** [33], 3β-*Z*-coumaroyl lupeol **2** [34,35], 3β-*E*-coumaroyl lupeol (dioslupecin A) **3** [36,37], kolavic acid 15-methyl ester **4** [38–40] and vomifoliol (blumenol A) **5** [41–43] from a Portuguese invasive *A. melanoxylon* and demonstrate the utility of investigating its phytoconstituents, and as corollary of invasive species in general.

For compound **3**, previous biological activity studies regard the cytotoxicity studies on KB, COLO-205, HEPA-3B, and HELA cell lines and showed no activity [37].

For compound **4**, biological activity studies have been performed, namely inhibitory (*Trypanosoma brucei*) [38], and antimicrobial (*Escherichia coli, Proteus* sp., *Streptococcus aureus* and *Candida albicans*) [39] activities, as well as cytotoxicity (AGP01, HCT116, MCF07, NIHOVCAR, SKAMELL4 and SF295 cell lines) and anti-inflammatory activities [40]. Although antimicrobial activities and cytotoxicity were not observed, compound **4** showed high lipoxygenase inhibition activity when compared to standard quercertin and inhibited the production of IL-6 [40]. It also exhibited an inhibitory activity on the growth of *Trypanosoma brucei* with respect to the clinically used antitrypanosomal agents suramin and melarsoprol and showed a strong and selective inhibitory activity on the GAPDH enzyme of *T. brucei* [38].

For compound **5**, previous anticancer studies showed no significant activity (HIF-1 and NF-κB activities in reporter assays, and in cytotoxicity against A549, MDA-MB-231, MCF-7, KB, KB-VIN, HT29, A498, PC3 and PACA2 cell lines) [43–45]. Antimicrobial (*Micrococcus tetragenus, Escherichia coli, Staphylococcus albus, Bacillus cereus, Staphylococcus aureus, Micrococcus luteus, Bacillus subtilis, Pseudomonas aeruginosa*, methicillin-resistant *Staphylococcus aureus, Vibrio parahemolyticus* and *Candida albicans*), DPPH free radical scavenging, acetylcholinesterase inhibitory and brine shrimp larvicidal activities also showed no results [42,46].

2. Results and Discussion

2.1. Compound Identification

From the dichloromethane extract (17.63 g) of *A. melanoxylon* R. Br. (1250 g) collected at Peninha, Sintra, Portugal, lupeol **1** [33] (purified, <0.1 mg), 3β-*Z*-coumaroyl lupeol

2 [34,35] (purified, 15.6 mg), 3β-*E*-coumaroyl lupeol (dioslupecin A) **3** [36,37] (purified, 39.0 mg), and kolavic acid 15-methyl ester **4** [38–40] (purified 111.1 mg) were isolated (Figure 1). The study of the alkaloid content of this species led instead to the isolation of vomifoliol (blumenol A) **5** [41–43] (unpurified, 5.4 mg). Alkaloids were detected by TLC but their isolation was not achieved due to the low content in this species (this may be related to the time of harvest, since the accumulation of alkaloids is known to be seasonal, or due to other biotic or abiotic factors).

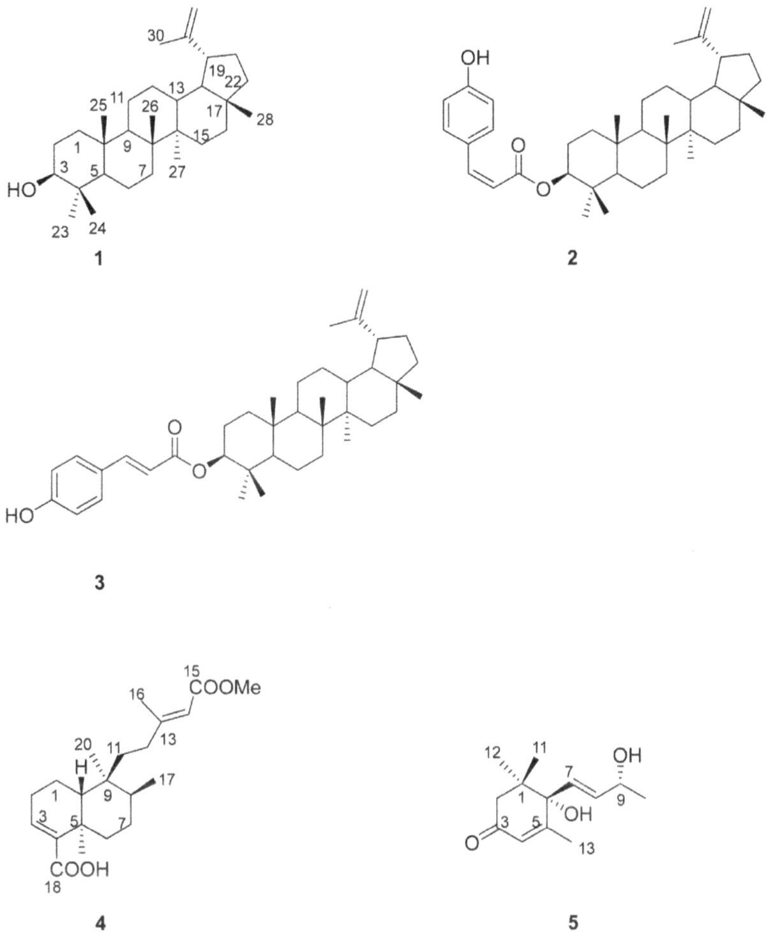

Figure 1. Lupeol **1**, 3β-*Z*-coumaroyl lupeol **2**, 3β-*E*-coumaroyl lupeol (dioslupecin A) **3**, kolavic acid 15-methyl ester **4** and vomifoliol (blumenol A) **5** isolated from *A. melanoxylon*.

The use of 1D and 2D NMR allowed the structure determination of all five compounds, further confirmed by mass spectrometry and comparison with literature data (Figure 1).

Compound **1** was identified by its characteristic H-3 (δ 3.19 ppm, *dd*, *J* = 5.0 Hz, *J* = 11.2 Hz), aliphatic methyl groups (δ 0.76 ppm, 0.79 ppm, 0.83 ppm, 0.94 ppm, 0.97 ppm and 1.03 ppm), CH_2-29 (δ 4.56 ppm, *sl* and δ 4.56 ppm, *sl*) and Me-30 (δ 1.68 ppm) resonances (Figure S1), all in agreement with literature values [33]. Comparison of the NMR spectra of **2** (Figures S2–S8) and **3** (Figures S9–S14) with that of **1** (Figure S1) allowed the recognition of a lupeol base skeleton substituted at C-3 with a coumaroyl unit—the structures of 3β-*Z*-coumaroyl lupeol **2** [34,35] and 3β-*E*-coumaroyl lupeol (dioslupecin A) **3** [36,37] were proposed, based on the *J* value couplings of the substituent's double bond (12.7 Hz and 15.9 Hz, respectively). Further confirmation came from comparison of the ^{1}H and ^{13}C resonance values with those of the literature [34–37].

For compound **3** corrections are made for the literature [36,37] resonances of the aliphatic methyl groups, based on HMBC and NOESY correlations (Table 1): HMBC between δ 0.89 and δ 0.92 with δ 81.2 (C-3) clearly indicates the presence of Me-23 and Me-24, distinguished by NOESY with H-3 (for δ 0.89); HMBC of δ 0.88 with δ 50.3 (C-9) and δ 37.1 (C-10) assigns Me-25; HMBC of δ 1.03 with δ 34.2 (C-7), δ 40.8 (C-8), δ 42.8 (C-14) and δ 50.3 (C-9) assigns Me-26; HMBC of δ 0.95 with δ 27.4 (C-15) and δ 42.8 (C-14) assigns Me-27; finally, HMBC of δ 0.79 with δ 35.5 (C-16), δ 43.0 (C-17) and δ 48.2 (C-18) assigns Me-28.

Table 1. ^{1}H and ^{13}C resonances of the aliphatic methyl groups of 3β-*E*-coumaroyl lupeol **3** ($CDCl_3$, 400 and 100 MHz).

	^{1}H (δ ppm)	^{13}C (δ ppm) *
Me-23	0.89	28.0
Me-24	0.92	16.6
Me-25	0.88	16.2
Me-26	1.03	15.9
Me-27	0.95	14.5
Me-28	0.79	18.0

* assigned by HSQC.

On purification and analysis, isomerization of the double bond was observed as described previously for coumaroyl esters [47]: the final ^{1}H NMR spectrum of the sample of **2** is composed of a 2.0:1.0 mixture of *E* and *Z* isomers (Figure S15), also present in the chromatogram of the GC-FID analysis (1.6:1.0, different ratios accountable by different isomerization times, Figure S16). Even the *E* isomer, compound **3**, obtained in pure form, seems to equilibrate in $CDCl_3$ to a 4.4:1.0 mixture of *E* and *Z* forms (Figure S17).

For compound **4**, a detailed analysis of the NMR data led to the proposed structure, confirmed by the $[M - H_2O + H]^+$ and $[M - H]^-$ ions present at *m/z* 331.3 u and *m/z* 347.2 u in the positive and negative ESI-MS spectra, respectively (Figures S18–S25). Analysis of the spectra and comparison of the NMR resonances with those of the literature [38–40] allowed us to make some corrections to previously assigned ^{13}C NMR values that should be interchanged: (a) C-2 and C-7: interchanged based on HMBC with H-3 and Me-17, respectively; (b) C-6 and C-11: interchanged based on COSY of H-12 with H-11 (C-11, assigned by HSQC); (c) Me-19 and Me-20: interchanged based on HMBC with C-11; (d) C-5 and C-9: interchanged based on HMBC correlations of Me-19 and Me-20, respectively; and (e) C-15 and C-18: interchanged based on HMBC of the OMe group (confirmed by HMBC of H-14). Table 2 lists the full ^{13}C NMR assignment and the ^{1}H NMR outstanding resonances (full ^{1}H NMR assignment can be found in the literature [38–40]).

Table 2. ^{13}C assignment and the ^{1}H outstanding resonances of kolavic acid 15-methyl ester **4** ($CDCl_3$, 125 and 500 MHz). *s* singlet, *d* duplet, *t* triplet.

	^{1}H	^{13}C		^{1}H	^{13}C
1		16.8	11		36.2
2		24.4	12		34.4
3	6.80 *t* 3.8	142.2	13		161.5
4		137.5	14	5.68 *d* 1.0	114.9
5		36.3	15		167.3
6		36.8	16	2.18 *d* 1.2	19.2
7		28.6	17	0.77 *d* 6.9	15.9
8		37.8	18		172.7
9		40.3	19	1.24 *s*	33.4
10		45.4	20	0.78 *s*	18.0
			COOMe	3.69 *s*	50.8

Finally, for compound **5**, the proposed structure based on NMR analysis was confirmed by the $[M - H_2O + H]^+$, $[M + H]^+$, $[M + Na]^+$ and $[M - H]^-$ ions present at *m/z* 207.1 u, *m/z* 225.1 u, *m/z* 247.1 u, and *m/z* 223.0 u in the positive and negative ESI-MS spectra, respectively (Figures S26–S32). Comparison of the ^{1}H and ^{13}C NMR resonances with those of the literature [41–43] confirm the structure, and allows us to correct the ^{13}C NMR δ values of the Δ^7 double bond that must be interchanged: C-7 δ 129.0 ppm and C-8 δ 135.7 ppm (ascertained by HSQC).

2.2. Cytotoxicity Evaluation

Cytotoxicity studies were performed for kolavic acid 15-methyl ester **4** and a sample of 3β-*E*-coumaroyl lupeol (dioslupecin A) **3**. The fact that compounds **2** and **3** equilibrate when in solution prevented us from testing both compounds separately. Since lupeol **1** and vomifoliol (blumenol A) **5** were isolated in small amounts their biological testing was also not performed.

To determine cellular toxicity, preliminary assays were performed with compounds **3** and **4** incubated in HCT116 cells for 72 h. Compound **3** showed to decrease cell viability to 64% only at the highest concentration tested (243 μM). However, at the same concentration, compound **4** decreased cell viability to 3%.

Therefore, only the IC50 of compound **4** was determined by the MTS metabolism assay, in order to evaluate its potential cytotoxic activity. In fact, compound **4** showed to have a relatively low cytotoxicity against the HCT116 colon cancer cell line, with an IC50 value of 176.3 μM (95% CI = 163.8 to 189.8 μM, Figure 2). This might suggest its use as an anti-inflammatory agent [40]. Nonetheless, additional studies using other cell lines are required to discard completely the cytotoxic effect of this compound. 5-Fluorouracil (5-FU), a cytotoxic agent in colon cancer treatment, was used as a positive control (IC50 value of 2.763 μM; 95% CI = 2.539 to 3.007 μM, Figure 3).

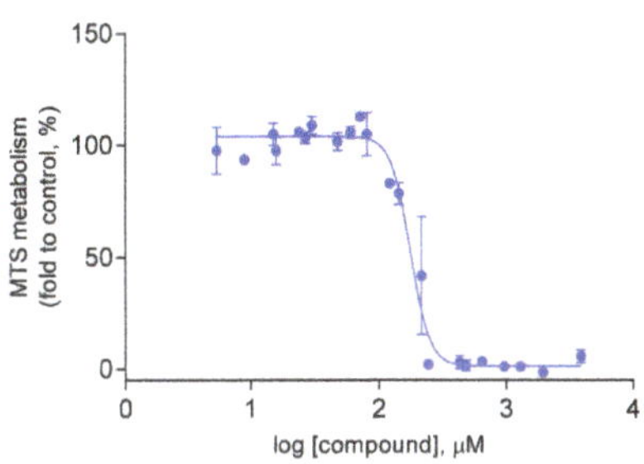

Figure 2. Dose-response curve in HCT116 cells upon incubation with compound **4** for 72 h. These results are representative of at least three independent experiments.

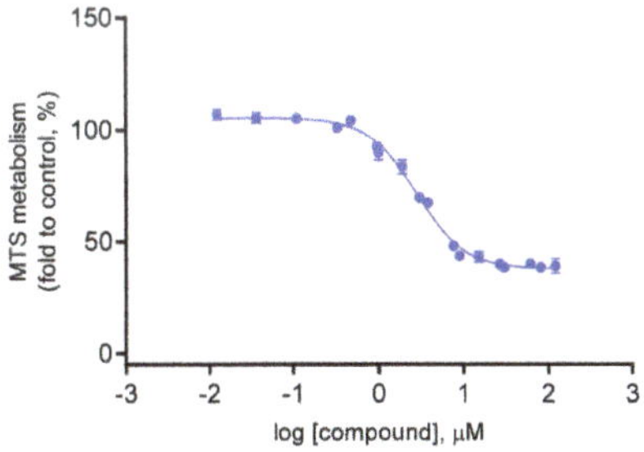

Figure 3. Dose-response curve in HCT116 cells upon incubation with 5-FU for 72 h. These results are representative of at least three independent experiments.

3. Materials and Methods

3.1. General Experimental Procedures

Silica gel 60 for chromatography (Merck 7734 or Merck 109385) was used for column chromatography (gravity and flash, respectively) and TLC was performed on Silica gel 60 Merck 5744 (0.5 mm) or 5554 TLC plates employing 254 nm and/or 366 nm UV-lamp for visualization. Molybdophosphoric acid or Munier Dragendorff reagent were used for additional spot development [48]. NMR spectra were recorded on a 400 MHz (100 MHz for ^{13}C) Brucker Avance III 400 or 500 MHz (125 MHz for ^{13}C) Bruker AVANCE Neo 500 spectrometer (Fällanden, Switzerland) at 25 °C using standard pulse programs. Residual solvent signals were used for calibration (^{1}H: $\delta(CHCl_3) = 7.26$ ppm, ^{13}C: $\delta(CDCl_3) = 77.00$ ppm). GC-FID was performed on GC Agilent 6890 operated at the following conditions: carrier gas Helium, split ratio 1:30; Column: ZB1HTinferno, L: 15 m, φ: 0.25 mm, df: 0.10 μm, injector temperature 280 °C; detector temperature 280 °C; temperature program 100 °C, hold for 0 min, increase 10 °C/min to 320 °C, hold for 20 min.

ESI-MS spectra were performed on a Thermo orbitrap Qexactive focus apparatus with direct inlet by a Thermo vanquish apparatus.

3.2. Plant Collection and Preparation

Branches and leaves of *A. melanoxylon* R. Br. were collected at Peninha, Sintra, Portugal (38°46′10″ N 9°27′33″ W), on 15 December 2019 and a voucher specimen (LISI032915, Patrícia Máximo and Ana Lourenço, *s.n.*, collected on 1 March 2020) was deposited at Herbário João de Carvalho e Vasconcellos (LISI), School of Agriculture (ISA), University of Lisbon. After air drying at room temperature in the dark, and milling, 1850 g were obtained.

3.3. Extraction and Isolation

3.3.1. Dichloromethane Extract

An amount of 1250 g of *A. melanoxylon* R. Br. was defatted with *n*-hexane and extracted with 2.5 L of dichloromethane (DCM), at room temperature for 5 h 30 min. After this time the plant material was separated and re-extracted with another 2.5 L of dichloromethane, at room temperature, for 18 h. The gathered filtrates were evaporated to yield 17.63 g of extract.

The extract was chromatographed on gravity column (diameter 15.0 cm, height 25 cm) with mixtures of *n*-hexane/ethyl acetate, ethyl acetate, two mixtures of ethyl acetate/methanol and methanol to yield 7 fractions (DCM1 (8/2), DCM2 (7/3), DCM3 (6/4), DCM4 (1/1), DCM5 (EtOAc), DCM6 (10% and 20% of MeOH), DCM7 (MeOH)).

The fraction DCM1 (4.13 g) was purified by flash chromatography with a mixture of *n*-hexane/ethyl acetate 9/1 to yield six fractions, two of them further purified: DCM1E and DCM1F. The fraction DCM1E (638.0 mg) was purified by flash column chromatography with mixtures of *n*-hexane/ethyl acetate 85/15 to yield the fraction DCM1E1 (586.0 mg). This fraction was further purified by flash column chromatography with mixtures of n-hexane/ethyl acetate 85/15, 8/2, and ethyl acetate to yield 4 fractions. One of them, DCM1E1B (385.0 mg), was purified twice by thin layer chromatography using mixtures of n-hexane/ethyl acetate 8/2 and 85/15 to yield fraction DCM1E1Bp1p, compound **1** (15.1 mg). Another, DCM1E1C (30.0 mg) was purified by thin layer chromatography using a mixture of *n*-hexane/ethyl acetate 8/2 to yield DCM1E1Cp1, compound **2** (23.1 mg). Another, DCM1E1D (64.0 mg) gave a mixture of compounds **2** and **3**. The fraction DCM1F (574.0 mg) was purified by flash column chromatography with a mixture of *n*-hexane/ethyl acetate 85/15 to yield fraction DCM1F2 (393.0 mg) that, after purification by thin layer chromatography using a mixture of *n*-hexane/ethyl acetate 8/2, yielded fraction DCM1F2p, compound **3** (57.0 mg).

The fraction DCM2 (2.91 g) was purified by flash column chromatography with mixtures of *n*-hexane/ethyl acetate 8/2 and 7/3 to yield three fractions; one of them, DCM2B (1.00 g), was further purified by flash column chromatography using mixtures of *n*-hexane/ethyl acetate 8/2 and 7/3, and ethyl acetate to yield six fractions; one of them, DCM2B5 yielded compound **4** (168.8 mg).

Final purification of the compounds for biological testing was achieved by flash chromatography for **4** (white solid, 111.1 mg, *n*-hexane/ethyl acetate 8/2) or thin layer chromatography for **1** (white solid, <1 mg, *n*-hexane/ethyl acetate 8/2), **2** (white solid, 3.5 mg, *n*-hexane/ethyl acetate 8/2), **3** (white solid, 27.8 mg, *n*-hexane/ethyl acetate 7/3), and for the mixture of **2** and **3** (white solids, 12.1 and 11.2 mg, respectively, *n*-hexane/ethyl acetate 8/2). On performing spectroscopic analysis isomerization of **2** and **3** was observed for some of the samples.

3.3.2. Vomifoliol 5 Extraction

Acidic extraction of the plant was performed for the isolation of alkaloids. However, a co-extraction compound, vomifoliol (blumenol A) **5**, was isolated instead. An amount of 500 g of *A. melanoxylon* R. Br. was extracted with 1.55 l of HCl 0.5 M, at room temperature, for 40 min. After centrifugation the supernatant was concentrated and basified with NH_4OH 1M. This solution was applied in an isolute®HM-N (Biotage, Uppsala, Sweden) column and after elution with dichloromethane and evaporation, the extract Amx (80.4 mg) was obtained.

This extract was purified by thin layer chromatography using a mixture of dichloromethane/methanol/NH_4OH 95/5/1 to yield 3 fractions. Of these, Amx-2 (11.8 mg) was purified by thin layer chromatography using a mixture of dichloromethane/methanol/NH_4OH 95/5/1 to yield fraction Amx-2.2B, compound **5** (5.4 mg).

3.4. Compound Characterization

Compound **1** (unpurified)—^{1}H NMR spectrum, see Supplementary Materials, Figure S1. Compound **2** (unpurified)—NMR spectra, see Supplementary Materials Figures S2–S8. Compound **3** (unpurified)—NMR spectra, see Supplementary Materials Figures S9–S14. Mixture 1.0:2.0 of compounds **2** and **3**—^{1}H NMR spectrum, see Supplementary Materials Figure S15. Mixture 1.0:1.6 of compounds **2** and **3**—GC-FID chromatogram, see Supplementary Materials Figure S16. Mixture 1.0:4.4 of compounds **2** and **3**—^{1}H NMR spectrum, see Supplementary Materials Figure S17. Compound **4**—NMR spectra and ESI-MS spectra, see Supplementary Materials Figures S18–S25. Compound **5** (unpurified)—NMR spectra and ESI-MS spectra, see Supplementary Materials Figures S26–S32.

3.5. Cell Culture and Treatments

HCT116 human colon carcinoma cells, commonly used in drug screens, were grown in McCoy's 5A modified medium supplemented with 10% heat-inactivated fetal bovine serum (FBS) and 1% antibiotic/antimycotic solution (Gibco, Life Technologies, Paisley, UK). Cells were cultured at 37 °C under a humidified atmosphere of 5% CO_2.

For cell viability experiments, HCT116 cells were seeded in 96-well plates, at a concentration of 5×10^3 cells/well and allowed to adhere for 24 h. Then, cells were exposed to the test compound, compound **4**, previously prepared in sterile DMSO. In order to plot a dose–response curve, cells were exposed to this compound in a range of concentrations between 0.04 µM and 4000 µM for 72 h. 5-Fluorouracil (5-FU), a cytotoxic agent used in colon cancer treatment, was used as a positive control and DMSO was used as vehicle control. Data are representative of three independent experiments.

3.6. Viability Assays

Cell viability of cells treated with compound **4** was evaluated using the CellTiter 96 AQueous Non-Radioactive Cell Proliferation Assay (Promega, Madison, WI, USA) according to the manufacturer's instructions. This colorimetric assay is based in the capacity of metabolic active cells to convert 3-(4,5-dimethylthiazo-2-yl)-5-(3-carboxymethoxyphenyl)-2-(4-sulfophenyl)-2H-tetrazolium inner salt (MTS) to formazan, a dye that is soluble in cell culture media. Formazan is quantified by measuring the absorbance at 490 nm and correlates with the amount of living cells in culture. Absorbance was measured using the GloMax-Multi+ microplate multimode reader (Promega, Madison, WI, USA) and the best-fit IC50 value, from at least three independent experiments, was calculated using the log (inhibitor) versus response (variable slope) function from GraphPad Prism software (version 8.0.2; San Diego, CA, USA).

4. Conclusions

In this study, five already known compounds were isolated and as a result some corrections are made to their previous ^{13}C NMR assignments. Cytotoxicity against HCT116 cells was evaluated for two of them, although no positive results were obtained.

As for the utility of the study of the chemistry of invasive species, here illustrated by two extracts of *A. melanoxylon*, we can propose that they can be used as a source of bioactive metabolites. Based on the literature and our own experimental results, lupeol derivatives **2** and **3** show no anticancer activity. Previous reports on vomifoliol (blumenol A) **5** for diverse bioactivities also showed no results. Nonetheless, kolavic acid 15-methyl ester **4**, the most abundant metabolite, is not cytotoxic and has previously been recognized as a bioactive naturally occurring trypanocide that may contribute to anti-inflammatory effects. *A. melanoxylon* can thus be considered a source for this metabolite.

We further add that it would be interesting to perform a detailed phytochemical study of bark samples of this species in general—ring-barking is presently one of the control measures for this species. Although Freire et al. [29,30] showed the presence of Δ^7 phytosterols and phytosteryl glucosides in the dichloromethane extracts of the bark by GC-EIMS, some of them with interesting reported bioactivities, many compounds may have escaped this screening; furthermore, although the antimicrobial bioactivity of the more polar ethanolic and aqueous extracts of the bark of this species is not noteworthy [49], again a detailed phytochemical approach could provide pure metabolites for which many more activities could be considered.

To conclude, although more studies are needed, this paper demonstrates that studying the chemistry of invasive species might provide a utility for this natural and abundant good that is currently underexplored.

Supplementary Materials: The following are available online at https://www.mdpi.com/article/10.3390/plants10122698/s1. Figure S1—^{1}H NMR spectrum of Lupeol **1** (unpurified) ($CDCl_3$, 400 MHz); Figure S2—^{1}H NMR spectrum of 3β-*Z*-coumaroyl lupeol **2** (unpurified) ($CDCl_3$, 400 MHz); Figure S3—^{13}C NMR spectrum of 3β-*Z*-coumaroyl lupeol **2** (unpurified) ($CDCl_3$, 100 MHz); Figure S4—DEPT135 spectrum of 3β-Z-coumaroyl lupeol **2** (unpurified) ($CDCl_3$); Figure S5—HSQC spectrum of 3β-Z-coumaroyl lupeol **2** (unpurified) ($CDCl_3$); Figure S6—HMBC spectrum of 3β-Z-coumaroyl lupeol **2** (unpurified) ($CDCl_3$); Figure S7—COSY spectrum of 3β-*Z*-coumaroyl lupeol **2** (unpurified) ($CDCl_3$); Figure S8—NOESY spectrum of 3β-*Z*-coumaroyl lupeol **2** (unpurified) ($CDCl_3$); Figure S9—^{1}H NMR spectrum of 3β-*E*-coumaroyl lupeol (dioslupecin A) **3** (unpurified) ($CDCl_3$, 400 MHz); Figure S10—^{13}C NMR spectrum of 3β-*E*-coumaroyl lupeol (dioslupecin A) **3** (unpurified) ($CDCl_3$, 100 MHz); Figure S11—DEPT135 spectrum of 3β-*E*-coumaroyl lupeol (dioslupecin A) **3** (unpurified) ($CDCl_3$); Figure S12—HSQC spectrum of 3β-*E*-coumaroyl lupeol (dioslupecin A) **3** (unpurified) ($CDCl_3$); Figure S13—HMBC spectrum of 3β-*E*-coumaroyl lupeol (dioslupecin A) **3** (unpurified) ($CDCl_3$); Figure S14—NOESY spectrum of 3β-*E*-coumaroyl lupeol (dioslupecin A) **3** (unpurified) ($CDCl_3$); Figure S15—^{1}H NMR spectrum of the mixture of 3β-*Z*-coumaroyl lupeol **2** and 3β-*E*-coumaroyl lupeol **3** (1.0:2.0) ($CDCl_3$, 500 MHz); Figure S16—GC-FID chromatogram of the mixture of 3β-*Z*-coumaroyl lupeol **2** and 3β-*E*-coumaroyl lupeol **3** (1.0:1.6); Figure S17—^{1}H NMR spectrum of the mixture of 3β-*Z*-coumaroyl lupeol **2** and 3β-*E*-coumaroyl lupeol **3** (1.0:4.4) ($CDCl_3$, 500 MHz); Figure S18—^{1}H NMR spectrum of kolavic acid 15-methyl ester **4** ($CDCl_3$, 500 MHz); Figure S19—^{13}C NMR spectrum of kolavic acid 15-methyl ester **4** ($CDCl_3$, 125 MHz); Figure S20—DEPT135 spectrum of kolavic acid 15-methyl ester **4** ($CDCl_3$); Figure S21—HSQC spectrum of kolavic acid 15-methyl ester **4** ($CDCl_3$); Figure S22—HMBC spectrum of kolavic acid 15-methyl ester **4** ($CDCl_3$); Figure S23—COSY spectrum of kolavic acid 15-methyl ester **4** ($CDCl_3$); Figure S24—NOESY spectrum of kolavic acid 15-methyl ester **4** ($CDCl_3$); Figure S25—ESI-MS spectra of kolavic acid 15-methyl ester **4**; Figure S26—^{1}H NMR spectrum of vomifoliol (blumenol A) **5** (unpurified) ($CDCl_3$, 500

MHz); Figure S27—^{13}C NMR spectrum of vomifoliol (blumenol A) **5** (unpurified) ($CDCl_3$, 125 MHz); Figure S28—DEPT135 spectrum of vomifoliol (blumenol A) **5** (unpurified) ($CDCl_3$); Figure S29—HSQC spectrum of vomifoliol (blumenol A) **5** (unpurified) ($CDCl_3$); Figure S30—HMBC spectrum of vomifoliol (blumenol A) **5** (unpurified) ($CDCl_3$); Figure S31—COSY spectrum of vomifoliol (blumenol A) **5** (unpurified) ($CDCl_3$); Figure S32—ESI-MS spectra of vomifoliol (blumenol A) **5** (unpurified).

Author Contributions: Conceptualization, P.M.; methodology, P.M., A.L., P.A. and C.M.P.R.; investigation, D.A., S.D. and J.G.; resources A.L., P.A. and C.M.P.R.; writing—original draft preparation, P.M.; writing—review and editing, P.M., A.L., P.A., J.G. and C.M.P.R.; supervision, P.M., A.L., P.A. and C.M.P.R.; funding acquisition, A.L., P.A. and C.M.P.R. All authors have read and agreed to the published version of the manuscript.

Funding: This work has received funding from FEDER through COMPETE 2020 under the Programme grant LISBOA-01-0145-FEDER-016405, and from National Funds through FCT under the Programme grant SAICTPAC/0019/2015 and PTDC/MED-FAR/29097/2017.

Institutional Review Board Statement: Not applicable.

Informed Consent Statement: Not applicable.

Data Availability Statement: Not applicable.

Acknowledgments: This work was supported by the Associate Laboratory for Green Chemistry-LAQV which is financed by national funds from FCT/MCTES (UIDB/50006/2020 and UIDP/50006/2020). GC-FID data obtained by the Analysis Laboratory, LAQV REQUIMTE—Chemistry department, FCT NOVA—PORTUGAL. FCT/MCTES is also acknowledged for supporting the National Portuguese NMR Network (ROTEIRO/0031/2013-PINFRA/22161/2016, co-financed by FEDER through COMPETE 2020, POCI, PORL, and FCT through PIDDAC).

Conflicts of Interest: The authors declare no conflict of interest.

References

1. Sharma, G.P.; Singh, J.S.; Raghubanshi, A.S. Plant invasions: Emerging trends and future implications. *Curr. Sci.* **2005**, *88*, 726–734.
2. Kumar Rai, P.; Singh, J.S. Invasive alien plant species: Their impact on environment, ecosystem services and human health. *Ecol. Indic.* **2020**, *111*, 106020. [CrossRef]
3. Almeida, J.D.D.; Freitas, H. The exotic and invasive flora of portugal. *Bot. Complut.* **2001**, *25*, 317–327.
4. Almeida, J.D.D.; Freitas, H. Exotic naturalized flora of continental portugal—A reassessment. *Bot. Complut.* **2006**, *30*, 117–130.
5. Almeida, J.D.D.; Freitas, H. Exotic flora of continental portugal—A new assessment. *Bocconea* **2012**, *24*, 231–237.
6. Decreto-Lei N.° 92/2019 de 10 de Julho da Presidência do Conselho de Ministros. Diário da República N.° 130/2019, Série I de 2019-07-10. Decreto-Lei N.° 92/2019 | DRE. Available online: https://files.dre.pt/1s/2019/10/19500/0000200010.pdf (accessed on 1 December 2021).
7. Tringali, C.; Oriente, G.; Piattelli, M.; Nicolosi, G. 2 minor dolabellane diterpenoid constituents from a dictyota species. *J. Nat. Prod.* **1985**, *48*, 484–485. [CrossRef]
8. Regulation (EU) No 1143/2014 of the European Parliament and of the Council of 22 October 2014 on the Prevention and Management of the Introduction and Spread of Invasive Alien Species. EUR-Lex—32014R1143—EN—EUR-Lex. Available online: https://eur-lex.europa.eu/legal-content/EN/ALL/?uri=CELEX:32014R1143 (accessed on 1 December 2021).
9. Brink, P.T. The Economic Costs of Invasive Alien Species (Ias). Available online: https://www.iucn.org/sites/dev/files/import/downloads/ten_brink_economic_impacts_of_ias__ptb_of_ieep_at_the_iucn_ep_event_21_feb_2013_final.pdf (accessed on 1 December 2021).
10. Weidlich, E.W.A.; Flórido, F.G.; Sorrini, T.B.; Brancalion, P.H.S.; Peralta, G. Controlling invasive plant species in ecological restoration: A global review. *J. Appl. Ecol.* **2020**, *57*, 1806–1817. [CrossRef]
11. Maximo, P.; Ferreira, L.M.; Branco, P.; Lourenço, A. Invasive plants: Turning enemies into value. *Molecules* **2020**, *25*, 3529. [CrossRef]
12. Fan, P.; Marston, A. How can phytochemists benefit from invasive plants? *Nat. Prod. Commun.* **2009**, *4*, 1407–1416. [CrossRef] [PubMed]
13. Souza-Alonso, P.; Rodríguez, J.; González, L.; Lorenzo, P. Here to stay. Recent advances and perspectives about *Acacia* invasion in mediterranean areas. *Ann. For. Sci.* **2017**, *74*, 55. [CrossRef]
14. Kull, C.A.; Shackleton, C.M.; Cunningham, P.J.; Ducatillon, C.; Dufour-Dror, J.-M.; Esler, K.J.; Friday, J.B.; Gouveia, A.C.; Griffin, A.R.; Marchante, E.; et al. Adoption, use and perception of australian *Acacias* around the world. *Divers. Distrib.* **2011**, *17*, 822–836. [CrossRef]

15. Vieites-Blanco, C.; González-Prieto, S.J. Invasiveness, ecological impacts and control of *Acacias* in southwestern europe—A review. *Web Ecol.* **2020**, *20*, 33–51. [CrossRef]
16. Le Maitre, D.C.; Gaertner, M.; Marchante, E.; Ens, E.-J.; Holmes, P.M.; Pauchard, A.; O'farrell, P.J.; Rogers, A.M.; Blanchard, R.; Blignaut, J.; et al. Impacts of invasive australian *Acacias*: Implications for management and restoration. *Divers. Distrib.* **2011**, *17*, 1015–1029. [CrossRef]
17. Wilson, J.R.U.; Gairifo, C.; Gibson, M.R.; Arianoutsou, M.; Bakar, B.B.; Baret, S.; Celesti-Grapow, L.; Ditomaso, J.M.; Dufour-Dror, J.-M.; Kueffer, C.; et al. Risk assessment, eradication, and biological control: Global efforts to limit australian *Acacia* invasions. *Divers. Distrib.* **2011**, *17*, 1030–1046. [CrossRef]
18. Lopez-Nunez, F.A.; Marchante, E.; Heleno, R.; Duarte, L.N.; Palhas, J.; Impson, F.; Freitas, H.; Marchante, H. Establishment, spread and early impacts of the first biocontrol agent against an invasive plant in continental europe. *J. Environ. Manag.* **2021**, *290*, 112545. [CrossRef]
19. Portugal, P.I.E. *Acacia melanoxylon*. Available online: https://invasoras.pt/en/invasive-plant/acacia-melanoxylon (accessed on 1 December 2021).
20. Falco, M.R.; Devries, J.X. Isolation of hyperoside (quercetin-3-d-galactoside) from flowers of *Acacia melanoxylon*. *Naturwissenschaften* **1964**, *51*, 462–463. [CrossRef]
21. Foo, L.Y. Configuration and conformation of dihydroflavonols from *Acacia melanoxylon*. *Phytochemistry* **1987**, *26*, 813–817. [CrossRef]
22. Hausen, B.M.; Bruhn, G.; Tilsley, D.A. Contact allergy to australian blackwood (*Acacia melanoxylon* R.Br.): Isolation and identification of new hydroxyflavan sensitizers. *Contact Dermat.* **1990**, *23*, 33–39. [CrossRef] [PubMed]
23. Foo, L.Y.; Wong, H. Diastereoisomeric leucoanthocyanidins from the heartwood of *Acacia melanoxylon*. *Phytochemistry* **1986**, *25*, 1961–1965. [CrossRef]
24. King, F.E.; Bottomley, W. The chemistry of extractives from hardwoods.17. The occurrence of a flavan-3-4-diol (melacacidin) in *Acacia melanoxylon*. *J. Chem. Soc.* **1954**, 1399–1403. [CrossRef]
25. Tindale, M.D.; Roux, D.G. A phytochemical survey of the australian species of *Acacia melanoxylon*. *Phytochemistry* **1969**, *8*, 1713–1727. [CrossRef]
26. Foo, L.Y. A novel pyrogallol a-ring proanthocyanidin dimer from *Acacia melanoxylon*. *J. Chem. Soc.-Chem. Commun.* **1986**, *3*, 236–237. [CrossRef]
27. Foo, L.Y. Isolation of 4-o-4 -linked biflavanoids from *Acacia melanoxylon*-1st examples of a new class of single ether-linked proanthocyanidin dimers. *J. Chem. Soc.-Chem. Commun.* **1989**, *20*, 1505–1506. [CrossRef]
28. Schmalle, H.W.; Hausen, B.M. Acamelin, a new sensitizing furano-quinone from *Acacia melanoxylon* R.Br. *Tetrahedron Lett.* **1980**, *21*, 149–152. [CrossRef]
29. Freire, C.S.R.; Coelho, D.S.C.; Santos, N.M.; Silvestre, A.J.D.; Neto, C.P. Identification of delta(7) phytosterols and phytosteryl glucosides in the wood and bark of several *Acacia* species. *Lipids* **2005**, *40*, 317–322. [CrossRef] [PubMed]
30. Freire, C.S.R.; Silvestre, A.J.D.; Neto, C.P. Demonstration of long-chainn-alkyl caffeates and δ7-steryl glucosides in the bark of *Acacia* species by gas chromatography–mass spectrometry. *Phytochem. Anal.* **2007**, *18*, 151–156. [CrossRef]
31. Silva, E.; Fernandes, S.; Bacelar, E.; Sampaio, A. Antimicrobial activity of aqueous, ethanolic and methanolic leaf extracts from *Acacia* spp. And eucalyptus nicholii. *Afr. J. Tradit. Complementary Altern. Med.* **2016**, *13*, 130–134. [CrossRef] [PubMed]
32. Correia, R.; Quintela, J.C.; Duarte, M.P.; Gonçalves, M. Insights for the valorization of biomass from portuguese invasive *Acacia* spp. In a biorefinery perspective. *Forests* **2020**, *11*, 1342. [CrossRef]
33. Pereira Beserra, F.; Xue, M.; Maia, G.; Leite Rozza, A.; Helena Pellizzon, C.; Jackson, C. Lupeol, a pentacyclic triterpene, promotes migration, wound closure, and contractile effect in vitro: Possible involvement of pi3k/akt and p38/erk/mapk pathways. *Molecules* **2018**, *23*, 2819. [CrossRef]
34. Chang, C.-I.; Kuo, Y.-H. Threee new lupane-type triterpenes from *Diospyros maritima*. *Chem. Pharm. Bull.* **1998**, *46*, 1627–1629. [CrossRef]
35. Karalai, C.; Laphookhieo, S. Triterpenoid esters from bruguiera cylindrica. *Aust. J. Chem.* **2005**, *58*, 556–559. [CrossRef]
36. Ali, M.; Heaton, A.; Leach, D. Triterpene esters from australian. *Acacia J. Nat. Prod.* **1997**, *60*, 1150–1151. [CrossRef]
37. Kuo, Y.H.; Chang, C.I.; Li, S.Y.; Chou, C.J.; Chen, C.F.; Kuo, Y.H.; Lee, K.H. Cytotoxic constituents from the stems of diospyros maritima. *Planta Med.* **1997**, *63*, 363–365. [CrossRef]
38. Nyasse, B.; Ngantchou, I.; Tchana, E.M.; Sonke, B.; Denier, C.; Fontaine, C. Inhibition of both trypanosoma brucei bloodstream form and related glycolytic enzymes by a new kolavic acid derivative isolated from entada abyssinica. *Pharmazie* **2004**, *59*, 873–875. [CrossRef] [PubMed]
39. Tchinda, A.T.; Fuendjiepa, V.; Mekonnenb, Y.; Ngoa, B.B.; Dagne, E. A bioactive diterpene from entada abyssinica. *Nat. Prod. Comm.* **2007**, *2*, 9–12. [CrossRef]
40. Vargas, F.D.S.; Almeida, P.D.O.D.; Aranha, E.S.P.; Boleti, A.P.D.A.; Newton, P.; Vasconcellos, M.C.D.; Junior, V.F.V.; Lima, E.S. Biological activities and cytotoxicity of diterpenes from copaifera spp. Oleoresins. *Molecules* **2015**, *20*, 6194–6210. [CrossRef]
41. González, A.G.; Guillermo, J.A.; Ravelo, A.G.; Jimenez, I.A.; Gupta, M.P. 4,5-dihydroblumenol a, a new nor-isoprenoid from perrottetia multiflora. *J. Nat. Prod.* **1994**, *57*, 400–402. [CrossRef]
42. Xiao, W.-L.; Chen, W.-H.; Zhang, J.-Y.; Song, X.-P.; Chen, G.-Y.; Han, C.-R. Ionone-type sesquiterpenoids from the stems of ficus pumila. *Chem. Nat. Compd.* **2016**, *52*, 531–533. [CrossRef]

43. Liu, X.; Tian, F.; Zhang, H.-B.; Pilarinou, E.; Mclaughlin, J.L. Biologically active blumenol a from the leaves ofannona glabra. *Nat. Prod. Lett.* **1999**, *14*, 77–81. [CrossRef]
44. Dat, N.T.; Jin, X.; Hong, Y.-S.; Lee, J.J. An isoaurone and other constituents from trichosanthes kirilowii seeds inhibit hypoxia-inducible factor-1 and nuclear factor-kb. *J. Nat. Prod.* **2010**, *73*, 1167–1169. [CrossRef] [PubMed]
45. Rahim, A.; Saito, Y.; Fukuyoshi, S.; Miyake, K.; Goto, M.; Chen, C.H.; Alam, G.; Lee, K.H.; Nakagawa-Goto, K. Paliasanines a-e, 3,4-methylenedioxyquinoline alkaloids fused with a phenyl-14-oxabicyclo[3.2.1]octane unit from melochia umbellata var. Deglabrata. *J. Nat. Prod.* **2020**, *83*, 2931–2939. [CrossRef]
46. Yang, J.M.; Liu, Y.Y.; Yang, W.C.; Ma, X.X.; Nie, Y.Y.; Glukhov, E.; Gerwick, L.; Gerwick, W.H.; Lei, X.L.; Zhang, Y. An anti-inflammatory isoflavone from soybean inoculated with a marine fungus aspergillus terreus c23-3. *Biosci. Biotechnol. Biochem.* **2020**, *84*, 1546–1553. [CrossRef] [PubMed]
47. Mclean, S.; Reynold, W.F.; Yang, J.-P.; Jacobs, H.; Jean-Pierre, L.L. Total assignment of the 1h and ^{13}C chemical shifts for a mixture of cis- and trans-p-hydroxycinnamoyl esters of taraxerol with the aid of high-resolution, ^{13}C-detected, ^{13}C-1h shift correlation spectra. *Magn. Reson. Chem.* **1994**, *32*, 422–428. [CrossRef]
48. Stahl, E. *Thin-Layer Chromatography a Laboratory Handbook*, 2nd ed.; Springer: Berlin/Heidelberg, Germany, 1969.
49. Neiva, D.M.; Luís, Â.; Gominho, J.; Domingues, F.; Duarte, A.P.; Pereira, H. Bark residues valorization potential regarding antioxidant and antimicrobial extracts. *Wood Sci. Technol.* **2020**, *54*, 559–585. [CrossRef]

plants MDPI

Review

New Approaches on Japanese Knotweed (*Fallopia japonica*) Bioactive Compounds and Their Potential of Pharmacological and Beekeeping Activities: Challenges and Future Directions

Alexandra-Antonia Cucu [1], Gabriela-Maria Baci [1], Ştefan Dezsi [2,*], Mircea-Emil Nap [3,4], Florin Ioan Beteg [5], Victoriţa Bonta [1], Otilia Bobiş [1,*], Emilio Caprio [6] and Daniel Severus Dezmirean [1]

1 Faculty of Animal Science and Biotechnology, University of Animal Sciences and Veterinary Medicine Cluj-Napoca, 400372 Cluj-Napoca, Romania; antonia.cucu@usamvcluj.ro (A.-A.C.); gabriela-maria.baci@usamvcluj.ro (G.-M.B.); victorita.bonta@usamvcluj.ro (V.B.); ddezmirean@usamvcluj.ro (D.S.D.)
2 Faculty of Geography, Babeş-Bolyai University, 400084 Cluj-Napoca, Romania
3 Faculty of Geodesy, Technical University of Civil Engineering Bucharest, 020396 Bucharest, Romania; mircea.nap@usamvcluj.ro
4 Faculty of Horticulture, University of Animal Sciences and Veterinary Medicine Cluj-Napoca, 400372 Cluj-Napoca, Romania
5 Faculty of Veterinary Medicine, University of Animal Sciences and Veterinary Medicine Cluj-Napoca, 400372 Cluj-Napoca, Romania; florin.beteg@usamvcluj.ro
6 Department of Agricultural Sciences, University of Naples "Federico II", Via Università, Portici, 100-80055 Naples, Italy; emilio.caprio@unina.it
* Correspondence: stefan.dezsi@ubbcluj.ro (Ş.D.); obobis@usamvcluj.ro (O.B.)

Abstract: Known especially for its negative ecological impact, *Fallopia japonica* (Japanese knotweed) is now considered one of the most invasive species. Nevertheless, its chemical composition has shown, beyond doubt, some high biological active compounds that can be a source of valuable pharmacological potential for the enhancement of human health. In this direction, resveratrol, emodin or polydatin, to name a few, have been extensively studied to demonstrate the beneficial effects on animals and humans. Thus, by taking into consideration the recent advances in the study of Japanese knotweed and its phytochemical constituents, the aim of this article is to provide an overview on the high therapeutic potential, underlining its antioxidant, antimicrobial, anti-inflammatory and anticancer effects, among the most important ones. Moreover, we describe some future directions for reducing the negative impact of *Fallopia japonica* by using the plant for its beekeeping properties in providing a distinct honey type that incorporates most of its bioactive compounds, with the same health-promoting properties.

Keywords: *Fallopia japonica*; Japanese knotweed; antimicrobial activity; antioxidant effect; bioactive compounds; honey; invasive species; phytopharmaceuticals

1. Introduction

Plants have always played an important role in human life because of their nutritional and health benefits and are therefore considered to be nutraceuticals [1–4]. Their use as alternative medicine continues to remain popular, despite the progress of the pharmaceutical industry. Moreover, in light of the impact that food can have on human health [5], consumers have become more sensitive towards balance-supporting food that can improve and enhance health [6,7].

In this light, Japanese knotweed, known also as: *Fallopia japonica*, *Reynoutria japonica (R. japonica)*, *Polygonum cuspidatum (P.cuspidatum)*, according to *Flora Europaea* [8], is well known to be used as part of alternative medicine, especially to prevent or to treat various disorders [9].

Native to Asia (China, Japan, Taiwan, North and South Korea) and North America, Japanese knotweed is an herbaceous perennial plant [10], from the *Polygonaceae* family [11]. It is found in sunny places, on the banks of rivers, or in pastures due to the presence of nitrogen from agricultural practices [12].

First introduced in Europe for ornamental purposes during the 19th century [13–15], Japanese knotweed has become one of the most invasive European alien species [16] because of its ability to expand in different types of habitats [17,18].

Possessing long-branches, thickened rhizomes and large ovate or elliptical blade leaves [10,19], this fast-growing invasive plant is distinguished by some specificities such as strong regeneration capacity [18], ample environmental resistance to various habitat conditions [20], high hybridization potential [21] and clonal and sexual reproduction [22]. Thus, it is tremendously difficult and costly to remove from invaded areas or to control their spread, as numerous studies have proven [23–29]. Moreover, it exhibits a menacing character towards native flora and ecosystems [30] because of the inhibition mechanism; its chemical composition manifests on local vegetation [31–33]. In this regard, alien plants in general, and the Japanese knotweed in particular, significantly affect the economy and ecosystem of the areas that they aggressively take over [15,34], and are therefore seen as a growing threat to global sustainability [35].

While Japanese knotweed can have a significant negative impact on its surrounding environment, including the endangerment and extinction of local species [33,36], effecting crop production [37], human health [38–40], forest regeneration [35,40] and urban and rural land management [41], this invasive plant also has several positive aspects. Reports in the literature show that Japanese knotweed can have a positive influence in terms of its use as food and fodder [39,42], medicine [10,43–45], fuel [46] and bioenergy [47,48], as well as an ecological indicator [49], insecticidal and fungicidal product [50–52], organic fertilizer [12] or carbon adsorbent [53]. These latest positive uses of *F. japonica* can act both as a preventive method for its invasiveness, as well as a pathfinder for further and innovative utilization of this troublesome plant species in the current circular economy.

It is well known that plants possess important sources of biologically active compounds that act as natural antioxidants, possessing nutritional, functional, as well as important health benefits [54]. It is also the case for Japanese knotweed and largely its rhizomes, whose phytochemical composition has indicated the presence of important amounts of anthraquinones (emodin, citreorosein, fallacinol, physcion, etc.), flavonoids (rutin, apigenin, quercetin, quercitrin, isoquercitrin, hyperoside, reynoutrin and kaempferol), and most important, stilbenes (resveratrol and polydatin), coumarins, lignans, essential oils along with other compounds [55–59].

Among these compounds, the most studied, because of their pharmacological values, are resveratrol, polydatin, emodin and physcion [60,61]. According to clinical trials, the four most studied bioactive compounds have revealed positive pharmacological effects, including antioxidant [61–63], antimicrobial [64–66], anti-inflammatory [67–69], neuroprotective [70–72] or anticancer activities [73–76], among others.

As current knowledge about the medicinal characteristics of Japanese knotweed are still incomplete, the aim of this review is to provide a deeper understanding of the pharmacological effects that the bioactive compounds of this plant possess and to draw attention to its possible use as an alternative medicine for specific diseases.

Furthermore, based on the therapeutic potential of the bioactive compounds found in different parts of Japanese knotweed, future directions can be taken into consideration, to both reduce its negative impact and offer new perspectives on the sustainable landscape management of knotweeds [77]. Thus, by creating new medical and economic value products, we can encourage the re-examination of the use of herbicides as the main eradication method for this invasive plant [78].

From this perspective, our last chapter will focus on the potential of Japanese knotweed flowers as a nectar source for bees and the valuable health-enhancing properties of the honey it can produce.

Consequently, to carry out this review, the authors researched literature using Japanese knotweed, *F. japonica*, *R. japonica* or *P. cuspidatum* as the main keywords, focusing especially on the papers that underlined its chemical composition and its potential use as a pharmaceutical, as well as the activities of its bioactive compounds. The denomination used by different authors was kept in our manuscript, due to the fact that we investigated the same plant, according to different botanical specialists. In an effort to obtain thorough information, the most accessed databases were Web of Science, ScienceDirect and Google Scholar, where resveratrol, emodin and polydatin have been carefully covered, as to achieve a comprehensive understanding of the holistic therapeutic effects that this invasive plant and its compounds can exert on an individual's health, highlighting in vitro and in vivo studies and results that demonstrate its antioxidant, anti-inflammatory, neuroprotective and anticancer properties. Moreover, in order to undertake a comprehensive research, the selection was unlimited and included articles up until the submission of the study.

2. Phyto-Chemical Constituents and Identification Methods

As stated before, Japanese knotweed (Figure 1) has been studied for its chemical composition and properties.

Figure 1. Japanese knotweed plant (Maris Marius Gabriel, beekeeper—personal collection).

Principal classes of compounds, considered as bioactive compounds, are flavonoids, stilbenes, anthraquinones, coumarins and lignans, found in roots, stems, leaves and flowers [10,57,58,79,80]. It is important to note that environmental factors, including climate or harvesting time, play a key role in the plant's chemical composition [81]. Moreover, as a balance, Japanese knotweed can have a great impact in modifying the physico-chemical parameters of the soils they invade [82], so that it is able to adapt to any kind of soil and even enrich it with a lot of nutrients [83]. As a consequence, it is challenging to compare the differences between the phyto-chemical constituents, depending on the region from where the samples were collected.

In this regard, Table 1 presents the most representative studies made on the chemical composition of Japanese knotweed, emphasizing the part of the plant analysed, the phyto-chemical constituents found, the identification method, and some results.

Table 1. Phyto-chemical constituents of Japanese knotweed and identification/quantification methods.

Root extract of *R. japonica* China (*R.j.C*); *R. japonica* Poland (*R.j.P*); *R.* x *bohemica* (*Rxb*) *R. sachalinensis* (*R.s.*)	HPLC-DAD-HR-MS gradient mode	water-formic acid (100:0.1 *v*/*v*) (solvent A) acetonitrile-formic acid (100:0.1 *v*/*v*) (solvent B)	- piceid (14.83 mg/g *R.j.C*; 7.45 mg/g *R.j.P*; 4.67 mg/g *Rxb*) - resveratrol (1.29 mg/g *R.j.C*; 0.65 mg/g *R.j.P*; 0.52 mg/g *Rxb*) - vanicoside B (0.64 mg/g *R.j.C*; 0.49 mg/g *R.j.P*; 0.77 mg/g *Rxb*; 2.25 mg/g *R.s.*) - vanicoside A (0.08 mg/g *R.j.C*; 0.04 mg/g *R.j.P*; 0.12 mg/g *Rxb*; 0.55 mg/g *R.s.*) - emodin (4.93 mg/g *R.j.C*; 4.01 mg/g *R.j.P*; 1.93 mg/g *Rxb*; 0.13 mg/g *R.s.*) - physcion (2.96 mg/g *R.j.C*; 1.23 mg/g *R.j.P*; 1.86 mg/g *Rxb*; 0.19 mg/g *R.s.*)	[57]
Root, stalk and leaves extract of *F. japonica* (Houtt) (*F.j.*) and *F. sachalinensis* (F. Schmidt)(*F.s.*)	LC-MS-Q/TOF isocratic mode UPLC-PDA gradient mode	methanol-acetonitrile (15:85 *v*/*v*) 0.25% aqueous acetic acid (solvent A) acetonitrile (solvent B)	- flavan-3-ols (leaves: 0.04–0.26 g/100 g *F.j.*; 0.04–0.23 g/100 g *F.s.*; stalk: 0.01–0.18 g/100 g *F.j.*; 0.02–0.24 g/100 g *F.s.*; roots: 0.02–1.48 g/100 g *F.j.*; 0.07–0.62 g/100 g *F.s.*) - phenolic acids (leaves: 0.01–0.49 g/100 g *F.j.*; 0.01–0.58 g/100 g *F.s.*; stalk: 0.01–0.08 g/100 g *F.j.*; 0.01–0.07 g/100 g *F.s.*; roots: 0.0–0.01 g/100 g *F.j.*; 0.0–0.03 g/100 g *F.s.*) - flavones/flavonols: (leaves: 0.01–1.71 g/100 g *F.j.*; 0.01–0.89 g/100 g *F.s.*; stalk: 0.01–0.28 g/100 g *F.j.*; 0.01–0.16 g/100 g *F.s.*; roots: 0.01 g/100 g *F.j.*; 0.01 g/100 g *F.s.*) - stilbenes: (leaves: 0.01–0.03 g/100 g *F.j.*; 0.01–0.02 g/100 g *F.s.*; stalk: 0.01–0.04 g/100 g *F.j.*; 0.01–0.06 g/100 g *F.s.*; roots: 0.02–0.50 g/100 g *F.j.*; 0.02–0.22 g/100 g *F.s.*)	[58]
Rhizome extracts of *R. japonica*, *R. sachalinensis* and *R.* x *bohemica*	LC-ESI-MS/MS gradient mode	water-formic acid (99.9:0.1) (solvent A) acetonitrile-formic acid (99.9:0.1) (solvent B)	- procyanidins with high degree of polymerization; dianthrone glycosides; phenylpropanoid disaccharide esters; hydroxycinnamic acid derivatives; lignin oligomers; izovitexin; izovitexin diglucoside	[63]
Leaves extract of *F. japonica* (*F.j.*) and *F.* x *bohemica* (*Fxb*)	HPTLC–MS/MS	developing agent: 0.1% TBHQ in methanol: acetone (1:1, *v/v*)	- violaxanthin: green leaves (4.9–53.3 mg/100 g *F.j.*; 3.9–39.9 mg/100 g *Fxb*), yellow leaves (<LOQ *F.j.*; 1.5 mg/100 g *Fxb*), green-yellowish leaves (4.2 mg/100 g *F.j.*; 7.1–96.8 mg/100 g *Fxb*) - neoxanthin: green leaves (38.2 mg/100 g *F.j.*; 24.4 mg/100 g *Fxb*), yellow leaves (<LOQ *F.j.* and *Fxb*), green-yellowish leaves (3.3 mg/100 g *F.j.*; 44.3 mg/100 g *Fxb*) - luteoxanthin: green leaves (2.9–6.3 mg/100 g *F.j.*; 2.2–5.4 mg/100 g *Fxb*), yellow leaves (<LOQ *F.j.* and *Fxb*), green-yellowish leaves (1.1 mg/100 g *F.j.*; <LOQ *Fxb*) - antheraxanthin: green leaves (10.3 mg/100 g *F.j.*; 12.8 mg/100 g *Fxb*), yellow leaves (1.0 mg/100 g *F.j.*; 2.0 mg/100 g *Fxb*), green-yellowish leaves (6.4 mg/100 g *F.j.*; 3.6 mg/100 g *Fxb*) - all-trans-lutein: green leaves (144.3 mg/100 g *F.j.*; 97.1 mg/100 g *Fxb*), yellow leaves (9.4 mg/100 g *F.j.*; 28.6 mg/100 g *Fxb*), green-yellowish leaves (55.8 mg/100 g *F.j.*; 127.9 mg/100 g *Fxb*) - all-trans-zeaxanthin: green leaves (3.4 mg/100 g *F.j.*; 2.7 mg/100 g *Fxb*), yellow leaves (1.8 mg/100 g *F.j.*; 6.1 mg/100 g *Fxb*), green-yellowish leaves (5.1 mg/100 g *F.j.*; <LOQ *Fxb*) - 13-cis-β-carotene: green leaves (1.4 mg/100 g *F.j.*; 0.9 mg/100 g *Fxb*), yellow leaves (<LOQ *F.j.* and *Fxb*), green-yellowish leaves (<LOQ *F.j.*; 1.9 mg/100 g *Fxb*) - all-trans-β-carotene: green leaves (97.3 mg/100 g *F.j.*; 68.7 mg/100 g *Fxb*), yellow leaves (8.0 mg/100 g *F.j.*; 12.7 mg/100 g *Fxb*), green-yellowish leaves (23.2 mg/100 g *F.j.*; 97.4 mg/100 g *Fxb*) - 9-cis-β-carotene: green leaves (8.6 mg/100 g *F.j.*; 6.1 mg/100 g *Fxb*), yellow leaves (<LOQ *F.j.* and *Fxb*), green-yellowish leaves (<LOQ *F.j.*; 9.8 mg/100 g *Fxb*) - other 8 carotenoid esters	[80]
Root, Stem, Leaf and Flower extract of *F. japonaica* (*F.j.*) and *F* x *bohemica* (*Fxb*)	UPLC gradient mode	methanol (solvent A) water (solvent B)	- polydatin (root: 5.72–13.38 mg/g *F.j.*; 0.43–13.68 mg/g *Fxb*; stem: 0.08–0.11 mg/g *F.j.*; 0.01–0.1 mg/g *Fxb*; stem and leaf: 0.16–0.3 mg/g *F.j.*; 0.2–0.28 mg/g *Fxb*; leaf: 0.13–0.25 mg/g *F.j.*; 0.05–0.41 mg/g *Fxb*; flowers: ND) - resveratrol (root: 0.83–12.07 mg/g *F.j.*; 0.05–2.74 mg/g *Fxb*; stem: ND; stem and leaf: 0.03–0.15 mg/g *F.j.*; 0.03–0.05 mg/g *Fxb*; leaf: ND; flowers: ND) - emodin (root: 0.55–13.38 mg/g *F.j.*; 0.0–5.42 mg/g *Fxb*; stem: 0.0–0.06 mg/g *F.j.*; 0.0–0.05 mg/g *Fxb*; stem and leaf: 0.06–0.41 mg/g *F.j.*; 0.11–0.12 mg/g *Fxb*; leaf: 0.0–0.05 mg/g *F.j.*; 0.0–0.05 mg/g *Fxb*; flowers: ND) - physcion (root: 3.97–15.72 mg/g *F.j.*; 0.0–9.71 mg/g *Fxb*; stem: 0.0–0.33 mg/g *F.j.*; 0.0–0.07 mg/g *Fxb*; stem and leaf: 0.16–0.77 mg/g *F.j.*; 0.24–0.39 mg/g *Fxb*; leaf: 0.0–0.89 mg/g *F.j.*; 0.0–0.06 mg/g *Fxb*; flowers: ND)	[81]

Table 1. *Cont.*

Root extract of *P. cuspidatum* Sieb. et Zucc.	HSCCC gradient mode	light petroleum-ethyl acetate-water (1:5:5, *v*/*v*) light petroleum-ethyl acetate-methanol-water (3:5:4:6, *v*/*v*) light petroleum-ethyl acetate-methanol-water (3:5:7:3, *v*/*v*)	- piceid (19.3 mg), anthraglycoside B (17.6 mg) from 200 mg sample - resveratrol (18.5 mg), emodin (35.3 mg) and physcion (8.2 mg) from 220 mg sample	[84]
Sprout extract of *R. japonica* (*R.j.*) *R. sachalinensis* (*R.s.*) and *Bohemian knotweed* (*B.k.*)	HPLC-DAD gradient mode	water-acetonitrile-orthophosphoric acid (94.9:5:0.1 *v/v*)(solvent A) water-acetonitrile-orthophosphoric acid (80:19.9:0.1 *v/v*)(solvent B)	- catechin (103 mg/kg *R.j.*; 167 mg/kg *R.s.*; 42 mg/kg *B.k.*) - epicatechin (568 mg/kg *R.j.*; 674 mg/kg *R.s.*; 230 mg/kg *B.k.*) - resveratroloside (48 mg/kg *R.j.*; 31 mg/kg *R.s.*; 11 mg/kg *B.k.*) - piceid (683 mg/kg *R.j.*; 502 mg/kg *R.s.*; 215 mg.kg *B.k.*) - resveratrol (64 mg/kg *R.j.*; 29 mg/kg *R.s.*; 23 mg/kg *B.k.*)	[85]
Root extract of *P. cuspidatum* Sieb. et Zucc.	HPLC-DAD and HPLC-ESI/MS gradient mode	water:acetic acid (95.5:0.5)(solvent A) acetonitrile (solvent B)	- piceid (1.75–5.03 mg/g) - resveratrol (0.378–1.15 mg/g) - emodin-8-β-D-glucoside (3.69–9.60 mg/g) - physcion-8-β-D-glucoside (0.299–0.854 mg/g) - aloe-emodin (0.032–0.109 mg/g) - emodin (1.05–2.50 mg/g) - physcion (0.180–0.456 mg/g)	[86]
Leaves extract of *F. japonica* Houtt (*F.j.*), *F. sachalinensis* F. Schmidt (*F.s.*) and *F. x bohemica* (*Fxb*)	HPTLC–MS/MS	developing agent: acetonitrile	- flavan-3-ols monomers: total content 84 mg/100 g *F.j.*; 236 mg/100 g *F.s.*; 139 mg/100 g *Fxb* - proanthocyanidin dimers: total content 99 mg/100 g *F.j.*; 206 mg/100 g *F.s.*; 140 mg/100 g *Fxb*	[87]
Roots extracts of *R. japonica*	UHPLC-DAD-ESI-MS^n gradient mode	water-formic acid (99.9:0.1) (solvent A) acetonitrile-formic acid (99.9:0.1) (solvent B)	- 4 stilbene (glycosides and aglycones) total amount of different extraction methods: 55.45 mg/g plant - 8 anthranoids (glycosides and aglycones) total amount of different extraction methods: 14.91 mg/g plant	[88]

HPLC-DAD-HR-MS-High-performance thin-layer chromatography hyphenated to high-performance liquid chromatography-diode array detection-mass spectrometry; LC-MS-Q/TOF-Liquid chromatography-photodiode detector-mass spectrometry/quadrupole time of flight; UPLC-PDA-Ultra-performance liquid chromatography with photodiode array detection; LC-ESI-MS/MS-Liquid chromatography electrospray ionization mass spectrometry/mass spectrometry; HPTLC–MS/MS-High performance thin layer chromatography and mass spectrometry; UPLC-Ultra Performance liquid chromatography; HSCCC-High-speed counter-current chromatography; HPLC-DAD-High-performance liquid chromatography with diode-array detection; HPLC-ESI/MS-High-performance liquid chromatography/electro-spray ionization tandem mass spectrometry; UHPLC-DAD-Highly sensitive ultra-high-performance liquid chromatographic-diode array detection; TBHQ- tertbuthyl hydroquinone; LOQ-limit of quantification.

Nawrot-Hadzik et al. (2018) revealed that the chemical composition of the roots of different species of *Fallopia (F. japonica, F. bohemica and F. sachalinensis)* contained a higher amount of piceid and resveratrol [57]. Comparing two *Fallopia* species, Lachowicz and Oszmianski (2019) reported that the leaves were a source of polymeric procyanidins, flavones, flavonols, phenolic acids, oleanolic and ursolic acids, while the roots contain flavan-3-ols and stilbenes, resveratrol being the most dominant [58]. Glavnik and Vovk (2020) extracted the anthraquinones from Japanese knotweed rhizomes using numerous extraction solvents and obtaining identical qualitative densitometric profiles in almost all the solvents with the exception of the sample test solution in dichloromethane, who allowed the extraction of physcion, emodin, and its equivalents (emodin-8-O-hexoside, emodin-O-acetyl-hexoside and emodin-O-malonyl-hexoside) [59]. Other compounds such as carotenoids were observed by Metličar et al. (2019) [80] when exploring the green and senescing leaves of two types of knotweeds, respectively, Japanese knotweed and Bohemian knotweed. Even if both plants indicated close pigment profiles, the total carotenoid content in green leaves exceeded the content revealed in senescing leaves for the two species, being comparable to one of the richest sources of carotenoids, specifically spinach [89,90]. Chen et al. (2013) have compared several Japanese knotweed samples from different parts of Canada and China, showing that the higher amount of resveratrol was found in the roots, compared to the stems and leaves. Moreover, it is suggested that the level of polydatin is similar in the samples from Prince Edward Island compared to those from China, while the level content of resveratrol turned up higher for the samples from China in contrast to those from Canada. In the case of emodin, Canadian samples contain a lower level than the Chinese samples, whereas physcion had higher levels in the Canadian samples compared to the Chinese ones [81]. However, Chu et al. (2005) founded in his study a higher amount of emodin, while physcion was found in the lowest quantity [84]. On the other hand, Vrchotová et al. (2007) evaluated the stilbene and catechin content of the spring sprouts and knotweed rhizomes of *Reynoutria species*, finding that the most predominant stilbene derivative in sprouts was piceid. By contrast, in the rhizome composition there was little resveratrol, while in the sprouts, the amount of resveratrol was much the same in all analysed knotweed samples [85]. Yi et al. (2020) managed to analyse five samples of the Japanese knotweed procured from different parts of China, reporting that from the seven compounds found, the greater proportion was represented by piceid [86]. It should be pointed out that according to the Chinese Pharmacopoeia, emodin and polydatin are considered to be standard indicators regarding the quality of dried material, thus a content of 1.0% emodin and 1.5% polydatin is required [19].

Studies made by Bensa et al. (2020) found that, concerning the degree of polymerization, leaves of the Japanese knotweed and other two knotweed species have similar chemical profiles of proanthocyanins [87]. This data is in accordance with the profiles of Japanese knotweed rhizomes examined by Glavnik et al. (2019) [91]. However, regarding the profile of gallates, leaves possess lower variety than rhizomes [87].

In his study, Alperth et al. (2021) detailed the presence of thirty-four substances, of which the greatest proportion was represented by anthranoids, while four new substances, namely demethylatedv torachrysonehexoside, sulfonyl torachryson, emodin methyl ether acetyl hexoside and malonylemodin, were declared new derivatives of the abovementioned compounds [88]. The aforementioned results are in agreement with the fact that the major chemical constituents found in the analysed parts of *F. japonica*, are a valuable source of polyphenolic compounds that exhibit a large spectrum of therapeutic properties and biological activities, which will be detailed as follows. As seen, the frequently used method for the identification of the biological constituents of this invasive alien species was chromatographic fingerprinting. This method is frequently employed for confirming the quality of plants used as alternative medical compounds or food supplements [87,92].

3. Biological Activities

In traditional medicine, *F. japonica* has been used for many years in Japan, China, Korea, and Taiwan. Recently, it became the spotlight of numerous studies, since its chemical composition has revealed promising therapeutic effects. Alcoholic, hydroalcoholic and water extracts from roots, rhizomes, and aerial parts (stems, leaves and flowers) have antibacterial activity [64,66,93–95], antioxidant effects [44,58,66,96–105], anticancer, antiproliferative and apoptotic properties [106–108], anti-inflammatory and antiviral activity [11,69,109–112] and many other bioactive properties.

The rich sources of biologically active substances found in *F. japonica* are essential for its specific therapeutic properties. Its applicability using in vitro and in vivo models has been largely investigated and is pointed out in the following section.

3.1. Antibacterial Activity

Recently, interest in antimicrobial substances originating from plants has extended to researchers investigating how different plants and their morphological extracts are associated with therapeutic activity and how they can be used as natural preservatives or ingredients for medicinal use. In this sense, *P. cuspidatum* exhibits significant antibacterial activities, mainly because of its bioactive compounds, especially stilbenes and anthraquinones. The most commonly used methods in analysing antimicrobial activity are disk diffusion, well diffusion and broth or agar dilution, together with the minimum inhibitory concentration (MIC) and minimum bactericidal value (MBC) [94]. Studies regarding antibacterial activity and bactericidal effects against different classes of microorganisms (Gram-positive and Gram-negative bacteria), including oral pathogens of *P. cuspidatum* extracts, were reported for several decades [93], without investigating the compounds, or classes of compounds responsible for this activity. Foodborne pathogens are represented by a large variety of microorganisms that lead to food spoilage and foodborne illnesses.

Lately, there has been an increase in interest, both from consumers and food producers, for safe food that is produced without using synthetic preservatives. Natural products with antimicrobial properties have been identified with plants as their main sources [64]. An excellent candidate for extracts with high antibacterial activity and, thus, important for food preservation is *P. cuspidatum*. A study by Shan et al. (2008), demonstrated that the major classes of constituents in the roots of *P. cuspidatum* are stilbenes and hydroxyanthraquinones [64]. The extract was tested on five foodborne pathogenic or faecal indicator bacteria, namely, *Bacillus cereus (B. cereus), Listeria monocytogenes (L. monocytogenes), Staphylococcus aureus (S. aureus), Escherichia coli* (*E. coli*) ATCC25922, and *Salmonella anatum (S. anatum)*. Results were reported using MIC, MBC, and observation of the bacterial morphology under scanning electron microscope of the incubated cells. Antibacterial results of the extract were verified using pure standards of the main compounds found in the root extract of *P. cuspidatum,* indicating high antibacterial activity for all of them. Except for *E. coli*, all other bacteria used in the study presented a high diameter of inhibition zone (DIZ) as average value compared to pure standards (6.4–20.2 mm compared to 15.2–35.4 mm). However, MIC and MBC were much better when compared to pure standards for Gram positive bacteria (*B. cereus, L. monocytogenes*), due to the synergism between compounds (156.3 μg/mL for *L. monocitogenes* compared to 312.5–625 μg/mL for pure standards and 312.5 μg/mL for *B. cereus* compared to 625 μg/mL for pure standards, excepting resveratrol with the same concentration as the extract). Another in vitro study demonstrated the potent inhibitory effect of *P. cuspidatum* and specifically of emodin against Gram-negative *Haemophilus parasuis* [113]. Therefore, the results indicated a MIC concentration of 32 g/mL and a MBC value of 64 g/mL for emodin.

Magacz et al. (2021) [95] proved the antibacterial effects of three *Reynoutria* species against *Streptococcus mutans (S. mutans)* and the positive impact on the prevention of caries formation. Due to the fact that extracts of the rhizomes of *Reynoutria species* are a rich source of polyphenols, they can balance the lactoperoxidase cycle, acting as activating/inhibiting agents for the antibacterial activity of the enzyme. The results showed that *R. japonica* extract exhibited the strongest inhibitory effect against the analysed microorganism using the acetone extracts (MIC 0.15 mg/mL), while for the bactericidal activity *R. Japonica* revealed a strong activity for the ethyl acetate fraction and acetone (MBC 0.60 mg/mL). Moreover, the authors observed that the content of polyphenols is a key factor for the presence of its antimicrobial activity, as the water fraction in the case of *R. japonica* revealed a very week inhibition against *S. mutans* (MIC 1.20 mg/mL) compared to the other studied fractions. On the same page, Nawrot-Hadzik et al. (2019a) [63] observed the dependence between the polyphenols and the antiseptic activity when testing the bacterial viability of the same three *Reynoutria* species (*R. Japonica, R. sachalinensis* and *R. bohemica)* against Gram positive bacteria (*Streptococcus mutans, Streptococcus salivarius, Streptococcus sanguinis, and Streptococcus pyogenes*). Compared to the other three species, *R. japonica* extract had the highest antibacterial activity against *S. mutans* (mean MIC 1000 μg/mL and MBC 2000 μg/mL), mainly because of the abundance of stilbene, aglycones and anthraquinone aglycone, respectively. The latest mentioned studies and their results indicate that *R. japonica* can be considered a useful antimicrobial agent for the profilaxy of dental pathogens.

3.2. Antioxidant Activity

The positive therapeutic effects of the major compounds found in *F. Japonica*, namely polyphenols, are associated with its antioxidant capacity and therefore can be related to the modulation of a range of receptors and metabolic pathways. [105]. Therefore, as inhibitors of free radicals, they may cause oxidative reactions in the human body and thus produce various disorders, especially cancer [96].

When evaluating antioxidant activity, several methods are used specific for different matrices. These in vitro methods have certain advantages, but on the other hand, they also exhibit several disadvantages and limitations when used on plants, foods, or different extracts. The most used methods are: 2,2′-azinobis-(3-ethylbenzothiazoline-6-sulphonate) scavenging activity, (ABTS+) 1,1-diphenyl-2-picrylhydrazyl (DPPH·) radical scavenging activity, $Fe^{3+}-Fe^{2+}$ transformation assay, ferric reducing antioxidant power (FRAP) assay, cupric ions (Cu^{2+}) reducing power assay (Cuprac), Folin–Ciocalteu reducing capacity (FCR assay), peroxyl radical (ROO·), superoxide radical anion (O_2·-), hydrogen peroxide (H_2O_2) scavenging assay, hydroxyl radical (OH·) scavenging assay, singlet oxygen ($1O_2$) quenching assay, nitric oxide radical (NO·), oxygen radical absorbance capacity assay (ORAC), to mention the most important.

Considerable studies regarding the antioxidant activity of *F. Japonica* have been carried out worldwide. For instance, Lachowicz and Oszmiańsk (2019) [58] compared the antioxidative activity between *F. japonica* and *F. sachalinensis* plants and their morphological parts (leaves, stalks, and roots), noting that the average values in *F. japonica* were higher than in *F. sachalinensis*. As for the morphological parts, the lowest level of antioxidative activity was found in the stalks of both *Fallopia* sp. samples, measuring an average of 31.79 for the ABTS assay and 14.11 mmol Trolox/100 g DW for the ORAC assay, respectively. Still, both the leaves of *F. japonica* and the roots of *F. sachalinensis* showed the highest antioxidative activity (81.12/71.22-ABTS assay, compared to 30.42/30.85—ORAC assay- mmol Trolox/100 g DW). This suggests that the antioxidative capacity of plants can be attributed to the polyphenolic content of its compounds. Another study showed strong antioxidant activity for 32 compounds from *P. cupidatum* samples collected in China [114], specifically 92.7% inhibition at a concentration of 64 $\mu g \cdot mL^{-1}$.

Studies have proven the high correlation between the phenolic content and the antioxidant capacity in *P. cupidatum* species. For instance, Lazurca et al. (2012) analysed the phenolic content and antioxidative activity of *P. cuspidatum* buds from Romania in different solvents. The analysis revealed that the total phenolic content (TPC) ranged from 1048 $\pm$ 13 mg GAE/100 g to 1426 $\pm$ 15 mg GAE/100 g fresh buds. The results are in agreement with the antioxidant capacity, which varied from 89.034 to 182.483 μM trolox/g fresh weight buds, using an extraction index of 96.3. This demonstrates a positive correlation between the antioxidant activities and the total polyphenol content, but not necessarily strong enough ($r = 0.5274$, $p < 0.01$) [98].

On the other hand, Ardelean et al. (2016) [44] observed by measuring the root extracts of this plant that the phenolic content was 146.34 ± 5.36, expressed as milligrams of gallic acid equivalents (GAE)/g extract. This result exhibited a notable correlation with its antioxidant effect, namely 92%, by using a concentration of 1 mg/mL in ethanol 50%. Other studies have also explored the phenolic and antioxidant activity of *P. cupidatum*, showing higher results, while Chen et al. (2010) identified for the ethanolic extract from *P. cuspidatum* roots a phenolic content of 276.78 ± 39.31 mg/mL and 231.73 ± 5.04 mg/mL expressed as gallic acid equivalents [102], Hsu et al. (2007) remarked that the phenolic and flavonoid total contents in *P. cuspidatum* have much higher inhibitory activity against the DPPH radical (641.1 ± 42.6 mg/g and 62.3 ± 6 mg/g). This latest study proved also an important antioxidative effect (the half maximal inhibitory concentration—IC_{50} value of *P. cuspidatum* extracts represents 110 µg/mL in free radical scavenging assays, 3.2 µg/mL in superoxide radical scavenging assays, and 8 µg/mL in lipid peroxidation assays, respectively) [100].

Among all polyphenols, the one most frequently associated with the antioxidative effect are stilbenes [115], mostly resveratrol [99,103].

Still, there are recent studies which demonstrate that stilbenes may not be the only contributing factors to the antioxidative activity of Japanese knotweed, and that different compounds may also play an important role. In this regard, Lee et al. (2011) has pointed out no correlation between the antioxidative property and resveratrol or emodin [116], as the smallest extracts of *R. japonica* showed the higher antioxidant activity.

In addition, Pan et al. (2007) revealed that resveratrol extract from *P. cuspidatum* roots had a lower antioxidant activity than ethanol extract [101]. Additionally, recent contributions confirmed the presence of polydatin and epicatechin in higher proportion than trans-resveratrol [117,118] showing at the same time a stronger antioxidant activity than the previously mentioned compound (higher IC_{50}; 7.08 g/mL or 31.02 M) [97].

Moreover, an investigation by Lachowicz et al. (2019) indicated that the greater radical scavenging capacity of the compounds found in two *Reynoutria* sp. extracts (*R. japonica* and *R. sachalinensis*) was given by catechins, procyanidin, trans-piceid and trans-resveratrolside (procyanidins derivatives) [58]. The same direction is confirmed by Nawrot-Hadzik et al. (2019a) who suggested that procyanidins can also be taken into consideration when speaking about the antioxidant potential of the the Japanese knotweed [63].

These data suggest that Japanese knotweed is a rich source of natural antioxidants that can be used in human diet as an ingredient in different foods or drugs in order to protect and prevent chronic diseases (cardiovascular diseases, cancer, diabetes, obesity, etc.).

3.3. *Anticancer, Antiproliferative and Apoptotic Activity*

The continuous research on cancer prevention and treatment has led to some notable studies that have revealed the chemopreventive role of the *Polygonum genus*, notably *P. cuspidatum*. As cancer is the result of some biochemical processes that produce oxidative damage and possible death to the cells, the phenolic and metabolite compounds found in the roots of this alien species plant have been screened in order to demonstrate their antiproliferative and apoptotic activities. The study carried out by Xie et al. (2014) [108] revealed that one of the active constituents of *P. cuspidatum*, namely resveratrol, had remarkable inhibition effects on three human hepatoma cell lines (HepG2, SMMC–7721 and BEL-7402), using concentrations ranging from 2.5 to 40 µM in a time-dependent manner. Among the tested cell lines, SMMC-7721 was the most sensitive to the resveratrol action. Moreover, the study also confirmed that resveratrol could prevent the apoptosis of caspase 3 and caspase 9, the main cells involved in this process. The antitumor effects of this compound were also proven in vivo, as doses of 10 and 30 mg/kg of resveratrol could significantly reduce tumour growth (41.7 ± 12.45% and 60.9 ± 9.9%, respectively). The extracts of the same plant, and once again resveratrol as the main compound, proved its efficiency in inhibiting the human oral cancer cells in a study carried out by Wang et al. (2019). Thus, the cisplatin-resistant cancer cells originated from a human tongue cancer cell line CAL 27 and were exposed to 150 μg/mL *P. cuspidatum* extract for different hours. The results have shown that for 48 h, the half maximal inhibitory concentration of the plant extract in the tested cancer cells was 110.34 ± 8.21 μg/mL [118]. Other antiproliferative effects of *P. cuspidatum* roots against human hepatocellular carcinoma cells of HepG2 and SMMC-7721 were investigated by Jiao et al. (2018) who observed that besides the cell apoptosis effects, polydatin could inhibit cell proliferation, invasion, and migration [106]. Furthermore, other research confirmed the chemopreventive properties of the juglone analogue 2-ethoxystypandrone isolated from the *P. cuspidatum* roots [107]. This compound is considered to be an inhibitor of the signal transducer and activator of transcription 3 (STAT3), an important constituent of hepatocellular carcinoma. Therefore, the study made by Li et al. (2019) demonstrated that 2-Ethoxystypandrone was able to block the STAT3 activation (7.75 ± 0.18 μM, using 50% concentration) to induce the death of hepatocellular carcinoma and to inhibit the proliferation of cancer cells (IC_{50} = 2.70 ± 0.28 μM) [107].

These dates are consistent with the idea that the extracts of the Japanese knotweed have shown promising healing effects on various types of cancer by inhibiting the growth and progression of cancer gene cells in different stages and in a time/dose-dependent manner.

3.4. Anti-Inflammatory and Antiviral Activity

Some bioactive compounds from the extracts of Japanese knotweed exhibit anti-inflammatory and antiviral activities. The anti-inflammatory and protective activities of *P. cuspidatum* on Dry Eye Disease (DED) was investigated by Park et al. (2018) with both in vitro and in vivo analysis. It was revealed that the active substances from the examined plant are represented by caftaric acid, rutin, quercitrin, resveratrol and polydatin, the latest being the crucial one. In vivo, these compounds had increased the tear volumes of the groups treated (using 100 and 250 mg/kg extract) to 5.95 ± 0.71 and 6.49 ± 0.89 mm ($p < 0.05$), respectively, in contrast to the exorbital lacrimal gland-excised group (4.33 ± 1.28 mm, $p < 0.0001$), whereas in vitro it had positive effects on inhibiting the hyperosmolar stress-induced inflammation by reducing the expression of COX-2,Bax, nuclear translocation of NF-B through the stimulation of PARP, NF-B and caspase 3 [11].

Furthermore, Yu et al. (2020) gave evidence that resveratrol, emodin-1-O-d-glucoside, and emodin-8-O-d-glucoside play crucial roles in the anti-inflammatory properties of *P. cuspidatum* [69]. In addition, the study revealed that different concentrations of resveratrol can have a significant contribution in modulating the inflammatory process, namely at a concentration of 100 M resveratrol mitigates the TNF- level with the inhibition rate of 51.55%, while half of the same concentration diminished the pro-inflammatory cytokines production levels of IL-6 and MCP-1 cells by 63.86% and 69.88% [69]. Besides, Hodgin et al. (2021) revealed that resveratrol extract from *P. cuspidatum* had a major contribution in attenuating the symptoms of the Gulf War Illness, associated in general with an inflammatory process. Results showed that by taking a dose of both 200 mg/day or 600 mg/day, the symptoms associated with this illness were diminished. On the contrary, the other bioactive compounds analysed from another plant (luteolin and fisetin) showed no significant effects [111].

As for the antiviral effects, several studies confirmed that the bioactive compounds extracted from the Japanese knotweed rhizomes possess high curative properties against viral microorganisms. For example, Liu et al. (2013) [109] underlined the positive effect that the roots and rhizomes of *P. cuspidatum* has on treating the Coxsackievirus B4 (CVB4) infection. Specifically, emodin was explored through in vivo and in vitro assays, suggesting that it can lessen both the replication of Hep-2 and CVB4 cells. Emodin together with ethyl acetate subfraction F3 from *P. cuspidatum* were also indicated to slow down the

Epstein–Barr virus (EBV) lytic cycle in a dose-dependent manner (5.94 μg/mL and 4.83 μg/mL, respectively, at a half maximal effective concentration [110].

Zhang et al. (2015) examined the cytotoxicity of both resveratrol and polydatin on Enterovirus 71 (EV71) infected cells, suggesting that while resveratrol has inhibition activity against the production of inflammatory cytokines of EV71 (IC50 of 21.36 μg/mL at a concentration of $\leq$40 μg/mL), polydatin revealed a lower antiviral activity (IC50 of 979.66 μg/mL, at a concentration of 300 μg/mL) [119].

The report made by Nawrot-Hadzik et al. (2021) showed significant effects of the butanol fractions from *R. japonica* and *R. sachalinensis* (especially vanicosides) on the inhibition of SARS-CoV-2 Mpro cells (7.877 g/mL, 4.031 g/mL, respectively at IC_{50}). However, these constituents may act in synergy with other compounds such as polymerized procyanidins that can also exhibit a strong inhibitory activity on the antiviral pathway, as suggested by previous authors [112].

3.5. Other Bioactive Properties

In addition to the aforementioned biological activities, Japanese knotweed possesses some other notable therapeutic activities. For instance, its cardioprotective effects were mentioned in studies focusing on hyperlipidaemia [120], ventricular remodelling [121], ischemia-reperfusion injury [122], myocardial infarction [123], where different compounds of *P. cuspidatum* (resveratrol and polydatin) were proven to improve the functions and mechanisms of some enzymes responsible for the good functioning of the heart. Moreover, promising results with regard to the neuroprotective effects of this plant were reported in animal models. Resveratrol exhibited positive effects on ameliorating anxiety [124] and alleviating depression activity by decreasing the malonaldehyde marker for oxidative stress, using 15 mg/kg/day extract [125]. Importantly, it has also been implicated in inhibiting the progression of Alzheimer Disease through the multiplication of neurons [126]. In addition, Liu et al. (2016) found out that resveratrol can reduce depression symptoms such as sucrose preference, lipid peroxidation, superoxide dismutase, immobility time and locomotor activity [127]. The same results have been confirmed by Wang et al. (2016) [114] and Moore et al. (2018) [128]. Polydatin [72], emodin-8-O-β-d-glucoside [70] and 2-methoxy-6-acetyl-7-methyljuglone [129], have also shown impressive results in improving the biological parameters for cognitive deficits or neuromuscular coordination.

The anti-diabetic activity of *F. Japonica* was demonstrated by its ability to scavenger the methylglyoxal reactions, responsible for the apparition of this chronic disease [130]. Several studies in this direction suggested its protective effects regarding diabetes by ensuring better glycaemic control [117,131–134]. Moreover, the extract of *P. cuspidatum* was described to be beneficial for gastrointestinal disorders, notably ulcers [135], as well as in respiratory dysfunctionalities [136] or obesity [121,137,138].

Subsequently, the positive effects on reproductive system disorders were outlined for some compounds found in the roots of Japanese knotweed. In particular, convincing evidence indicates that this invasive alien plant may be an ideal remedy for hormone replacement [139] through the enhancement of the oestrogen-sensitive cells.

All the described bioactive properties of *P. cuspidatum* confirm, once again, that this plant can be a valuable nutraceutical source for human health.

However, besides the positive activities of the biological active compounds of *F. japonica*, we cannot ignore the possible toxicity that may occur. In this regard, resveratrol has been revealed to exert undesirable effects depending on the dosage intake, the most common being abdominal discomfort, diarrhoea and nausea [140–143], while emodin inhibited human sperm functions [144]. Therefore, it is important to underline that further investigations and clinical trials are required in order to establish quantitative assessment of the standardized dose that can be used in order to maintain the health contribution of this compound and to avoid side-effects.

4. Future Directions

4.1. *Fallopia japonica Flowers: Important Nectar Source*

It is worth mentioning that *F. japonica* is one of the invasive plants that are considered to produce pollination services [145]. Even if the male flowers are very rare for pollen production, its female flowers are a great source of nectar rich fructose and glucose and, therefore, are intensively visited by pollination-insects, respectively bees [18,146–148].

To better appreciate its importance for beekeeping, but also to determine the massive unwanted invasion, different modern techniques may be used to detect the presence of Japanese knotweed. The newest and, at the same time, the most efficient of these methods, both in terms of actual working time in the field and of the surfaces covered in a certain time interval, is the technique that uses U.A.V. (Unmanned Aerial Vehicle) technology, which, depending on the chosen device model and the sensors mounted on it, can cover areas between 10 ha and 100 ha in less than two hours of flight [149]. During a photogrammetric operation, all data will be recorded in a series of three-dimensional coordinates, in a graphical form. From the existing varieties of this type of technology, drones that successfully combine all the good elements from each major category, in a single device, seem to be at present, the one best choice, for almost any kind of large surface work: this being about Fixed Wing Vertical Take-Off and Landing U.A.V. Some of the most relevant software for capturing or processing photogrammetric data are DroneDeploy, Agisoft Photoscan, and Pix4D. Referring to the correct choice of sensors to be used in this field, multispectral cameras are by far the most useful accessories. With their help, different types of vegetation can be identified and differentiated (either cultivated or wild), and it is also possible to study and establish their condition [149]. This method may be used successfully for the mentioned purposes.

It is known that the beekeeping value of *F. japonica* originates in the Far East, but as the plant became naturalized in Europe and America, this property was studied in the respective areas also. The study of Ferrazzi and Marletto (1990) is one of the few published investigating the beekeeping potential of the plant [146]. Conducted in the Piedmont region of Italy between 1988 and 1989, the study describes the feeding behaviour of the bees during the flowering period from July to October. The tiny greenish-white flowers, placed in axillary spikes, produce a high amount of nectar from non-differentiated nectar tissue located at the base of the stamens. The flower has eight stamens, presented regularly shaped anthers, in which however no pollen was ever observed. One of the few studies on nectar of *P. cuspidatum* was made in Poland in 2001 [150], where in three consecutive years (1998–2000), different parameters were measured, during the blooming period which was determined between 01.09 and 23.09. This period is also in accordance with different observations made by beekeepers in Romania. As stated by the previous study, *P. cuspidatum* (*F. japonica*) has a good potential for honey production, as the nectar of *Polygonum* flowers owns a concentration of sugars ranging from 30 to 56%, and a potential for honey production of 200–355 kg/ha, with a daily gain of honey production of 10–15 kg [150]. Still, the nutritional quality of the invasive *F. Japonica* nectar, compared with other invasive plant species, is unknown.

Foregoing descriptions of the plant demonstrate a very good source of secondary metabolites from different chemical classes, from root to the flowers. Nectary glands also contain high amounts of bioactive compounds from the class of polyphenols, which are transferred to honey when the bees are collecting the nectar.

On account of this, the management control and eradication of *F. japonica* has to be made using sustainable methods that will keep creating a healthy environment and protection for biodiversity [151]. Thereby, the effects that some methods may have on insect communities found in invaded locations need to be consciously evaluated [152].

4.2. *Knotweed Honey*

As known, honey is the only sweet natural product, made by the bees from flower nectar or from sweet secretions of plants or found on the plants, brought to the hive and left for ripening and maturation [153]. Depending on the nectar source, honey is classified as monofloral, multifloral or honeydew honey. Its aspect, taste, and chemical composition, depends also on the same parameters. Another important factor for honey colour is the harvest period. For instance, honey collected in spring and early summer is light in colour, whereas in mid-summer, the colour of honey transforms to dark yellow, orange or even brown. The darker the honey colour is, the higher its nutritional value, represented by vitamins, minerals and secondary metabolites coming from the plants which provide the sugar source. Japanese knotweed plants belong to the *Polygonaceae* family, from which dark coloured honeys are produced (e.g., buckwheat honey). Japanese knotweed honey makes no exception. It was known and consumed in Japan for many years, from where the plant is native, but its invasiveness in countless regions of the world made this honey known, analysed, and appreciated. Knotweed honey is dark in colour (Figure 2), very aromatic, with a particular flavour, and it is finely and uniformly crystallized during storage [146]. In addition, it has an extremely high content of minerals, but also possesses antioxidant and antibacterial properties [154].

Figure 2. Colour of Japanese knotweed honey (Maris Marius Gabriel, beekeeper—personal collection).

It is also known as "bamboo honey" due to the fact that the knotweed plant looks similar to a bamboo stick. Describing the chemical composition of Italian knotweed, Ferrazzi and Marletto (1990) stated that it has reduced humidity (17.9%) and medium electrical conductivity value, although it is very dark in colour (0.432 mS/cm), acidic pH (3.8) and has low acidity (free acidity 13 meq/kg, lactone acidity 1.25 meq/kg) and a small value of ash content (0.09%) [146].

A recent study [149] found similar characteristics for knotweed honey from western Romania. Due to the fact that this type of honey is harvested very late (generally October), there are few beekeepers in Romania producing it. Three samples were harvested in different locations from western Romania and analysed for their chemical composition, showing an average water content (15.6–20.0%) and electrical conductivity characteristic to multifloral honey (0.541–0.697 mS/cm), high diastazic index (11.52–18.74 DN) and low HMF content (0.29–9.93 mg/kg). Fructose and glucose are the main sugars found in knotweed honey, having similar values (38.9–39.8% fructose and 29.3–35.2% glucose). Therefore, the crystallization process installs very quickly from the time of harvesting. Other sugars identified were sucrose, turanose, maltose, trehalose and erlose (values from 0.13 to 3.18%). An important characteristic for this type of honey is total lipid and total nitrogen contents. High values (for honey) were quantified, ranging from 0.12 to 0.52% (total lipids) and 0.47–1.05% (total nitrogen). Further studies are necessary for the identification of amino acids and fatty acids, important compounds of this type of honey.

High amounts of minerals (K, Ca, Na and Mg) were also quantified in knotweed honey (6196.8 mg/kg for K). This amount is extremely high compared to other floral honeys, including buckwheat and manuka honey [155–158]. The polyphenolic content of honey came from the plants that ensure the nectar and pollen as raw materials for the bees. As stated before, Japanese knotweed is an important source of secondary metabolites from the classes of polyphenols, specifically stilbenes and anthraquinones. Total phenolic content of knotweed honey ranged between 100 and 195 mgGAE/100 g honey and 20–55 mgQE/100 g content of flavone/flavonols [154]. These values were higher compared to other dark coloured honeys, including buckwheat honey [159–162].

4.3. Comparison with Other Honey Types from the Same Plant Family

F. japonica (*R. japonica, P. cuspidatum*) belongs to the *Polygonaceae* family. According to Sanchez et al. (2009) [163] and Burke et al. (2010) [164], the phylogenetic tree for the *Fallopia, Reynoutria* and *Polygonum* is: *Polygonaceae, Polygonaceae, Polygonum, Fallopia* and *Reynoutria*. Searching for studies describing the chemical composition of Japanese knotweed honey, we could not find extensive literature, although this type of honey has been produced and consumed for many years, first in Japan, where the plant is native and later, all over the world where it has spread and maintained its common use. The few studies were compared to other dark coloured honeys, including buckwheat (from the same plant family) or manuka honey.

Manuka honey is known for its medicinal value more than for its taste, which is not appreciated by many consumers. It has a dark colour when fluid and freshly harvested, and its colour intensity decreases when crystallization occurs [155,165]. The smell is described as "pungent" most of the time and is not consumed for its aroma. Although, the health properties of dark honeys are by far superior to light colour honeys. From the chemical composition of honey, different classes of compounds found in very small amounts, determine their bioactive properties. The synergism between phenolics, organic acids, enzymes or peptides seems to be responsible for these properties. In the case of manuka honey, its bioactive properties are related to Unique Manuka Factor (UMF), representing the content of methylglyoxal and phenolic content. According to different studies [155,166–169], manuka honey contains high amounts of caffeic, p-coumaric, syringic acids, quercetin, luteolin, pinocembrin, isorhamnetin, kaempferol, chrysin, galangin, pinobanksin, methyl syringate, leptosin, glyoxal, and methyl glyoxal. The synergism between these classes of compounds present high antioxidant and antibacterial activity [170].

Buckwheat honey is produced in North America, but also in Europe, China and Russia. The main producers in Europe are Poland, Ukraine, the Netherlands, Germany and Moldova. This honey is characterised by a dark purple colour, almost black, with a malty aroma and strong animal odour [171]. Many scientific studies have shown the high antioxidant and antibacterial properties of this honey [172], especially its efficacy in respiratory tract infection. The study by Pasini et al. (2013) [171] highlighted the presence of p-hydroxybenzoic, p-coumaric, ferulic, syringic and caffeic acid, also abscisic acid, and quercetin, apigenin, pinobanksin, kaempferol, isorhamnetin, chrysin, pinocembrin and galangin in buckwheat honey. For this reason, the bioactive properties of this honey type are extremely high.

Furthermore, a study by Dzugan et al. (2020) [173], demonstrated that Polish buckwheat honey possesses the strongest antibacterial and antioxidant properties, because it is registered as the darkest Polish honey and, according to other scientific studies, the darker the honey, the higher the bioactive properties. Earlier in 2018, Deng et al. [157] comparatively evaluated the cellular antioxidant activity of buckwheat and manuka honey samples, using a cell-based model of HepG2 cells. The cellular antioxidant activity can determine the antioxidant activity more accurately than chemical-based methods, due to the fact that it takes into consideration the bioavailability, distribution of bioactive compounds in the interested biologic system and the cellular uptake. The EC_{50} value of buckwheat honey was lower than of manuka honey, indicating a potent cellular antioxidant activity for buckwheat honey.

Unpublished studies by Bobis et al. (2018) [174], presented at the Symposium Apimedica and Apiquality (Sibiu, Romania), highlighted the bioactive properties of knotweed honey (*F. japonica*), identified using HPLC, p-OH-hydroxybenzoic acid, syringic acid, p-coumaric acid, ferulic and t-cinnamic acid in honey samples from Romania. Additionally, a high percentage of inhibition of the DPPH radical was determined (61.69–71.75%), higher than registered for honeydew honey. Regarding the antibacterial activity, Japanese knotweed honey from Romania inhibits the growth of *S. aureus*, and is effective also on *Staphyloccocus pseudointermedius*, *Bacillus cereus* and *Salmonella enteritidis*.

Further studies of Japanese knotweed honey samples produced in different locations are needed to determine its chemical composition, biological properties and its best valorization.

5. Conclusions

Japanese knotweed, an all invasive plant species, represents a challenge because of its ecological and economic impact. Nevertheless, its chemical composition (from roots to leaves) has revealed the presence of unique active compounds that are reputable for their extensive variety of physicochemical features and biological activities. The aim of this review was to provide a comprehensive overview on the high therapeutic potential of Japanese knotweed by emphasizing its main compounds and their biological activities. Specifically, in vivo and in vitro studies on the chemical composition of *F. japonica* were covered, identifying resveratrol, emodin and polydatin as the main compounds that can exert therapeutic effects (antibacterial, antioxidant, anti-inflammatory and anticancer effects, among the most important ones). In addition, some future directions were proposed in order to valorise this troublesome plant for its nutraceutical and functional characteristics. Therefore, using *F. japonica* by exploiting its pharmacological properties through sustainable approaches such as beekeeping can lead to new and innovative opportunities, from which local communities and biodiversity can take advantage and, consequently, may reduce its negative impact. However, it should be noted that the management, control and eradication of *F. japonica* has to take into consideration the possible effects that the use of some insecticidal methods can have on the pollinators found in the invaded locations. Still, further research is needed to support the promising potential of this invasive plant as a honey source and to assess its health-supporting properties.

Author Contributions: Conceptualization, A.-A.C.; O.B.; M.-E.N. and D.S.D.; Data curation, G.-M.B.; Ş.D.; F.I.B. and V.B.; Resources, A.-A.C. and D.S.D.; Visualization, A.-A.C.; G.-M.B.; Ş.D.; M.-E.N.; V.B.; E.C. and D.S.D.; Supervision, E.C. and D.S.D.; Writing–original draft, A.-A.C.; G.-M.B.; M.-E.N. and O.B.; Writing–review & editing, A.-A.C.; G.-M.B. and O.B. All authors have read and agreed to the published version of the manuscript.

Funding: This research received no external funding.

Institutional Review Board Statement: Not applicable.

Informed Consent Statement: Not applicable.

Conflicts of Interest: The authors declare no conflict of interest.

References

1. Defelice, S. The Nutraceutical Revolution: Its Impact on Food Industry R & D. *Trends Food Sci. Technol.* **1995**, *6*, 59–61. [CrossRef]
2. Hardy, G. Nutraceuticals and Functional Foods: Introduction and Meaning A European Consensus of Scientific Concepts of Functional Foods. *Nutrients* **2002**, *16*, 688–689. [CrossRef]

3. Santini, A.; Novellino, E. Nutraceuticals: Beyond the Diet Before the Drugs. *Curr. Bioact. Compd.* **2014**, *10*, 1–12. [CrossRef]
4. Atazadegan, M.A.; Bagherniya, M.; Askari, G.; Tasbandi, A. The Effects of Medicinal Plants and Bioactive Natural Compounds on Homocysteine. *Molecules* **2021**, *26*, 3081. [CrossRef] [PubMed]
5. Mena, P.; Angelino, D. Plant Food, Nutrition, and Human Health. *Nutrients* **2020**, *12*, 2157. [CrossRef] [PubMed]
6. Küster-Boluda, I.; Vidal-Capilla, I. Consumer Attitudes in the Election of Functional Foods. *Span. J. Mark. ESIC* **2017**, *21*, 65–79. [CrossRef]
7. Augimeri, G.; Montalto, F.I.; Giordano, C.; Barone, I.; Lanzino, M.; Catalano, S.; Andò, S.; De Amicis, F.; Bonofiglio, D. Nutraceuticals in the Mediterranean Diet: Potential Avenues for Breast Cancer Treatment. *Nutrients* **2021**, *13*, 2557. [CrossRef]
8. Tutin, T.G.; Heywood, V.H.; Burges, N.A.; Moore, D.M.; Valentine, D.H.; Walters, S.M.; Webb, D.A. *Flora Europaea*; Cambridge University Press: Cambridge, UK, 2010; Volume 1.
9. Dong, X.; Fu, J.; Yin, X.; Li, X.; Wang, B.; Cao, S.; Zhang, H.; Zhao, Y.; Ni, J. Pharmacological and Other Bioactivities of the Genus Polygonum—A Review. *Trop. J. Pharm. Res.* **2014**, *13*, 1749–1759.
10. Patocka, J.; Navratilova, Z.; Ovando, M. Biologically Active Compounds of Knotweed (Reynoutria Spp.). *Mil. Med. Sci. Lett.* **2017**, *86*, 17–31. [CrossRef]
11. Park, B.; Lee, I.S.; Hyun, S.; Jo, K.; Lee, T.G.; Kim, J.S.; Kim, C. The Protective Effect of Polygonum Cuspidatum (PCE) Aqueous Extract in a Dry Eye Model. *Nutrients* **2018**, *10*, 1550. [CrossRef]
12. Cvejić, R.; Klages, S.; Pintar, M.; Resman, L.; Slatnar, A.; Mihelič, R. Invasive Plants in Support of Urban Farming: Fermentation-Based Organic Fertilizer from Japanese Knotweed. *Agronomy* **2021**, *11*, 1232. [CrossRef]
13. Bailey, P. Prize-Winners to Pariahs—A History of Japanese Knotweed s.l. (Polygonaceae) in the British Isles. *Watsonia* **2000**, *110*, 93–110.
14. Haeuser, E.; Dawson, W.; Thuiller, W.; Dullinger, S.; Block, S.; Bossdorf, O.; Carboni, M.; Conti, L.; Dullinger, I.; Essl, F.; et al. European Ornamental Garden Flora as an Invasion Debt under Climate Change. *J. Appl. Ecol.* **2018**, *55*, 2386–2395. [CrossRef]
15. Van Kleunen, M.; Essl, F.; Pergl, J.; Brundu, G.; Carboni, M.; Dullinger, S.; Early, R.; Gonz, P.; Groom, Q.J.; Hulme, E.; et al. The Changing Role of Ornamental Horticulture in Alien Plant Invasions. *Biol. Rev.* **2018**, *93*, 1421–1437. [CrossRef]
16. Nentwig, W.; Bacher, S.; Kumschick, S.; Pys, P. More than "100 Worst" Alien Species in Europe. *Biol. Invasions* **2017**, *20*, 1611–1621. [CrossRef]
17. Weston, L.A.; Barney, J.N.; Ditommaso, A. A Review of the Biology and Ecology of Three Invasive Perennials in New York State: Japanese Knotweed (*Polygonum Cuspidatum*), Mugwort (*Artemisia Vulgaris*) and Pale Swallow-Wort (*Vincetoxicum Rossicum*). *Plant Soil* **2005**, *277*, 53–69. [CrossRef]
18. Buhk, C.; Thielsch, A. Perspectives in Plant Ecology, Evolution and Systematics Hybridisation Boosts the Invasion of an Alien Species Complex: Insights into Future Invasiveness. *Plant Ecol. Evol.* **2015**, *17*, 274–283. [CrossRef]
19. Peng, W.; Qin, R.; Li, X.; Zhou, H. Botany, Phytochemistry, Pharmacology, and Potential Application of *Polygonum cuspidatum* Sieb. et Zucc.: A Review. *J. Ethnopharmacol.* **2013**, *148*, 729–745. [CrossRef]
20. Martin, F.; Müllerov, J.; Borgniet, L.; Dommanget, F.; Breton, V.; Evette, A. Using Single- and Multi-Date UAV and Satellite Imagery to Accurately Monitor Invasive Knotweed Species. *Remote Sens.* **2018**, *10*, 1662. [CrossRef]
21. Bailey, J. The Japanese knotweed invasion viewed as a vast unintentional hybridisation experiment. *Heredity* **2012**, *110*, 105–110. [CrossRef]
22. Bailey, J.P.; Bímová, K.; Mandák, B. Asexual Spread versus Sexual Reproduction and Evolution in Japanese Knotweed s.l. Sets the Stage for the "Battle of the Clones". *Biol. Invasions* **2009**, *11*, 1189–1203. [CrossRef]
23. Bashtanova, U.B.; Beckett, K.P.; Flowers, T.J.; Science, S.W. Review: Physiological Approaches to the Improvement of Chemical Control of Japanese Knotweed (Fallopia Japonica). *Weed Sci.* **2009**, *57*, 584–592. [CrossRef]
24. Kettunen, M.; Genovesi, P.; Gollasch, S.; Pagad, S.; Starfinger, U. *Technical Support to EU Strategy on Invasive Alien Species (IAS) Assessment of the Impacts of IAS in Europe and the EU*; No. 070307, 44; Institute for European Environmental Policy: Brussels, Belgium, 2009.
25. Lavoie, C. The Impact of Invasive Knotweed Species (*Reynoutria* spp.) on the Environment: Review and Research Perspectives. *Biol. Invasions* **2017**, *19*, 2319–2337. [CrossRef]
26. Cybill, S.; Soraya, R.; Hardion, B.J.; Nicolas, P.; Isabelle, C. Ecological Implications of the Replacement of Native Plant Species in Riparian Systems: Unexpected Effects of *Reynoutria japonica* Houtt. Leaf Litter. *Biol. Invasions* **2020**, *22*, 1917–1930. [CrossRef]
27. Diagne, C.; Leroy, B.; Vaissière, A.; Gozlan, R.E.; Roiz, D.; Jarić, I.; Salles, J.; Bradshaw, C.J.A.; Courchamp, F. High and Rising Economic Costs of Biological Invasions Worldwide. *Nature* **2021**, *592*, 571–576. [CrossRef] [PubMed]
28. Dusz, M.; Martin, F.; Dommanget, F.; Petit, A.; Dechaume-moncharmont, C.; Evette, A. Review of Existing Knowledge and Practices of Tarping for the Control of Invasive Knotweeds. *Plants* **2021**, *10*, 2152. [CrossRef] [PubMed]
29. Aguilera, A.G.; Alpert, P.; Dukes, J.S.; Harrington, R. Impacts of the Invasive Plant *Fallopia japonica* (Houtt.) on Plant Communities and Ecosystem Processes. *Biol. Invasions* **2010**, *12*, 1243–1252. [CrossRef]
30. Stoll, P.; Gatzsch, K.; Rusterholz, H.; Baur, B. Response of Plant and Gastropod Species to Knotweed Invasion. *Basic Appl. Ecol.* **2012**, *13*, 232–240. [CrossRef]
31. Smith, J.M.D.; Ward, J.P.; Child, L.E.; Owen, M.R.; Studies, E.; Road, A.; Le, L.; Sciences, M.; Ng, N. A Simulation Model of Rhizome Networks for *Fallopia japonica* (Japanese Knotweed) in the United Kingdom. *Ecol. Modell.* **2006**, *200*, 421–432. [CrossRef]

32. Fan, P.; Marston, A.; Hay, A. Short Communication Rapid Separation of Three Glucosylated Resveratrol Analogues from the Invasive Plant *Polygonum cuspidatum* by High-Speed Countercurrent Chromatography. *J. Sep. Sci.* **2009**, *32*, 2979–2984. [CrossRef]
33. Abgrall, C.; Forey, E.; Mignot, L.; Chauvat, M. Invasion by *Fallopia japonica* Alters Soil Food Webs through Secondary Metabolites. *Soil Biol. Biochem.* **2018**, *127*, 100–109. [CrossRef]
34. Bartz, R.; Kowarik, I. Assessing the Environmental Impacts of Invasive Alien Plants: A Review of Assessment Approaches. *NeoBiota* **2019**, *43*, 69. [CrossRef]
35. Rai, P.K.; Kim, K. Invasive Alien Plants and Environmental Remediation: A New Paradigm for Sustainable Restoration Ecology. *Restor. Ecol.* **2020**, *28*, 3–7. [CrossRef]
36. Colleran, B.; Lacy, S.N.; Retamal, M.R. Invasive Japanese Knotweed (*Reynoutria japonica* Houtt.) and Related Knotweeds as Catalysts for Streambank Erosion. *River Res. Appl.* **2020**, *36*, 1962–1969. [CrossRef]
37. Haines, A. Addressing Challenges to Human Health in the Anthropocene Epoch—An Overview of the Findings of the Rockefeller/Lancet Commission on Planetary Health. *Public Health Rev.* **2016**, *37*, 14. [CrossRef] [PubMed]
38. Schillberg, S.; Raven, N.; Spiegel, H.; Rasche, S.; Buntru, M. Critical Analysis of the Commercial Potential of Plants for the Production of Recombinant Proteins. *Front. Plant Sci.* **2019**, *10*, 720. [CrossRef]
39. Shackleton, R.T.; Shackleton, C.M.; Kull, C.A. The Role of Invasive Alien Species in Shaping Local Livelihoods and Human Well-Being: A Review. *J. Environ. Manag.* **2019**, *229*, 145–157. [CrossRef] [PubMed]
40. Jones, B.A.; Mcdermott, S.M. Health Impacts of Invasive Species Through an Altered Natural Environment: Assessing Air Pollution Sinks as a Causal Pathway. *Environ. Resour. Econ.* **2018**, *71*, 23–43. [CrossRef]
41. Rouifed, S. Landscape Perceptions and Social Representations of *Fallopia* spp. in France. *Sci. Nat.* **2018**, *105*, 67. [CrossRef] [PubMed]
42. Zhang, Y.; Zheng, L.; Zheng, Y.; Xue, S.; Zhang, J.; Huang, P.; Zhao, Y.; Hao, X.; He, Z.; Hu, Z.; et al. Insight into the Assembly of Root-Associated Microbiome in the Medicinal Plant *Polygonum cuspidatum*. *Ind. Crops Prod.* **2020**, *145*, 112163. [CrossRef]
43. Zhang, H.; Li, C.; Kwok, S.; Zhang, Q.; Chan, S. A Review of the Pharmacological Effects of the Dried Root of *Polygonum cuspidatum* (Hu Zhang) and Its Constituents. *Evid.-Based Complement. Altern. Med.* **2013**, *2013*, 208349. [CrossRef]
44. Ardelean, F.; Moacă, E.A.; Păcurariu, C.; Antal, D.S.; Dehelean, C.; Toma, C.C.; Drăgan, S. Invasive *Polygonum cuspidatum*: Physico-Chemical Analysis of a Plant Extract with Pharmaceutical Potential. *Stud. Univ. Vasile Goldis Arad Ser. Stiint. Vietii* **2016**, *26*, 415–421.
45. Ali, A.; Khalil, K.; Akter, K.; Kim, H.; Park, W.S.; Kang, D.; Koo, K.A.; Ahn, M. Comparative Inner Morphological and Chemical Studies on Reynoutria Species in Korea. *Plants* **2020**, *9*, 222. [CrossRef]
46. Pan, Z.; Qi, G.; Andriamanohiarisoamanana, F.J.; Yamashiro, T.; Iwasaki, M.; Umetsu, K. Anaerobic Co-Digestion of Dairy Manure and Japanese Knotweed (*Fallopia japonica*) under Thermophilic Condition: Optimal Ratio for Biochemical Methane Production. *Anim. Sci. J.* **2021**, *92*, e13523. [CrossRef] [PubMed]
47. Van Meerbeek, K.; Sciences, E.; Leuven, K.U. On the Map Biomass of Invasive Plant Species as a Potential Feedstock for Bioenergy Production. *Biofuels Bioprod. Biorefin.* **2015**, *9*, 273–282. [CrossRef]
48. Van Meerbeek, K.; Muys, B.; Hermy, M. Lignocellulosic Biomass for Bioenergy beyond Intensive Cropland and Forests. *Renew. Sustain. Energy Rev.* **2019**, *102*, 139–149. [CrossRef]
49. Pattison, Z.; Minderman, J.; Boon, P.J.; Willby, N. Twenty Years of Change in Riverside Vegetation: What Role Have Invasive Alien Plants Played? *Appl. Veg. Sci.* **2016**, *20*, 422–434. [CrossRef]
50. Bohinc, T.; Horvat, A.; Ocvirk, M.; Košir, I.; Rutnik, K.; Trdan, S. The First Evidence of the Insecticidal Potential of Plant Powders from Invasive Alien Plants against Rice Weevil under Laboratory Conditions. *Appl. Sci.* **2020**, *10*, 7828. [CrossRef]
51. Meela, M.; Mdee, L.; Masoko, P.; Eloff, J. Acetone leaf extracts of seven invasive weeds have promising activity against eight important plant fungal pathogens. *S. Afr. J. Bot.* **2019**, *121*, 442–446. [CrossRef]
52. Anžlovar, S.; Janeš, D.; Koce, J.D. The Effect of Extracts and Essential Oil from Invasive *Solidago* spp. and *Fallopia japonica* on Crop-Borne Fungi and Wheat Germination. *Food Technol. Biotechnol.* **2020**, *58*, 273–283. [CrossRef]
53. Koutník, I.; Vráblová, M.; Bednárek, J. Reynoutria japonica, an invasive herb as a source of activated carbon for the removal of xenobiotics from water. *Bioresour. Technol.* **2020**, *309*, 123315. [CrossRef]
54. Kasote, D.; Katyare, S.S.; Hegde, M.V.; Bae, H. Significance of Antioxidant Potential of Plants and its Relevance to Therapeutic Applications. *Int. J. Biol. Sci.* **2015**, *11*, 982–991. [CrossRef]
55. Park, J.H.; Moon, H.K.; Park, C.W. Flavonoid chemistry of Fallopia sect. Reynoutria (Polygonaceae) in Korea. *Korean J. Plant Taxon.* **2011**, *41*, 10–15. [CrossRef]
56. Xue, Y.; Liang, J. Screening of bioactive compounds in Rhizoma Polygoni Cuspidati with hepatocyte membranes by HPLC and LC-MS. *J. Sep. Sci.* **2013**, *37*, 250–256. [CrossRef]
57. Nawrot-Hadzik, I.; Granica, S.; Domaradzki, K.; Pecio, Ł.; Matkowski, A. Isolation and Determination of Phenolic Glycosides and Anthraquinones from Rhizomes of Various Reynoutria Species. *Planta Med.* **2018**, *84*, 1118–1126. [CrossRef]
58. Lachowicz, S.; Oszmiański, J. Profile of Bioactive Compounds in the Morphological Parts of Wild *Fallopia japonica* (Houtt) and *Fallopia sachalinensis* (F. Schmidt) and Their Antioxidative Activity. *Molecules* **2019**, *24*, 1436. [CrossRef]
59. Glavnik, V.; Vovk, I. Extraction of Anthraquinones from Japanese Knotweed Rhizomes and Their Analyses by High Performance Thin-Layer Chromatography and Mass Spectrometry. *Plants* **2020**, *9*, 1753. [CrossRef]

60. Zahedi, H.S.; Jazayeri, S.; Ghiasvand, R.; Djalali, M.; Eshraghian, M.R. Effects of *Polygonum cuspidatum* Containing Resveratrol on Inflammation in Male Professional Basketball Players. *Int. J. Prev. Med.* **2013**, *4*, S1–S4. [PubMed]
61. Chan, C.-L.; Gan, R.-Y.; Corke, H. The phenolic composition and antioxidant capacity of soluble and bound extracts in selected dietary spices and medicinal herbs. *Int. J. Food Sci. Technol.* **2016**, *51*, 565–573. [CrossRef]
62. Colica, C.; Milanović, M.; Milić, N.; Aiello, V.; De Lorenzo, A.; Abenavoli, L. A Systematic Review on Natural Antioxidant Properties of Resveratrol. *Nat. Prod. Commun.* **2018**, *13*, 1934578X1801300923. [CrossRef]
63. Nawrot-Hadzik, I.; Ślusarczyk, S.; Granica, S.; Hadzik, J.; Matkowski, A. Phytochemical Diversity in Rhizomes of Three Reynoutria Species and their Antioxidant Activity Correlations Elucidated by LC-ESI-MS/MS Analysis. *Molecules* **2019**, *24*, 1136. [CrossRef]
64. Shan, B.; Cai, Y.-Z.; Brooks, J.D.; Corke, H. Antibacterial properties of *Polygonum cuspidatum* roots and their major bioactive constituents. *Food Chem.* **2008**, *109*, 530–537. [CrossRef]
65. Mattio, L.M.; Dallavalle, S.; Musso, L.; Filardi, R.; Franzetti, L.; Pellegrino, L.; D'Incecco, P.; Mora, D.; Pinto, A.; Arioli, S. Antimicrobial activity of resveratrol-derived monomers and dimers against foodborne pathogens. *Sci. Rep.* **2019**, *9*, 19525. [CrossRef]
66. Nawrot-Hadzik, I.; Hadzik, J.; Fleischer, M.; Choromańska, A.; Sterczała, B.; Kubasiewicz-Ross, P.; Saczko, J.; Gałczyńska-Rusin, M.; Gedrange, T.; Matkowski, A. Chemical Composition of East Asian Invasive Knotweeds, their Cytotoxicity and Antimicrobial Efficacy Against Cariogenic Pathogens: An In-Vitro Study. *Med. Sci. Monit.* **2019**, *25*, 3279–3287. [CrossRef]
67. Bralley, E.E.; Greenspan, P.; Hargrove, J.L.; Wicker, L.; Hartle, D.K. Topical anti-inflammatory activity of *Polygonum cuspidatum* extract in the TPA model of mouse ear inflammation. *J. Inflamm.* **2008**, *5*, 1. [CrossRef] [PubMed]
68. Park, B.; Jo, K.; Lee, T.G.; Hyun, S.-W.; Kim, J.S.; Kim, C.-S. Polydatin Inhibits NLRP3 Inflammasome in Dry Eye Disease by Attenuating Oxidative Stress and Inhibiting the NF-κB Pathway. *Nutrients* **2019**, *11*, 2792. [CrossRef]
69. Yu, M.; Chen, T.-T.; Zhang, T.; Jia, H.-M.; Li, J.-J.; Zhang, H.-W.; Zou, Z.-M. Anti-inflammatory constituents in the root and rhizome of *Polygonum cuspidatum* by UPLC-PDA-QTOF/MS and lipopolysaccharide-activated RAW264.7 macrophages. *J. Pharm. Biomed. Anal.* **2020**, *195*, 113839. [CrossRef]
70. Wang, C.; Zhang, D.; Ma, H.; Liu, J. Neuroprotective effects of emodin-8-O-β-d-glucoside in vivo and in vitro. *Eur. J. Pharmacol.* **2007**, *577*, 58–63. [CrossRef]
71. Gonçalinho, G.; Roggerio, A.; Goes, M.; Avakian, S.; Leal, D.; Strunz, C.; Mansur, A. Comparison of Resveratrol Supplementation and Energy Restriction Effects on Sympathetic Nervous System Activity and Vascular Reactivity: A Randomized Clinical Trial. *Molecules* **2021**, *26*, 3168. [CrossRef] [PubMed]
72. Li, R.-P.; Wang, Z.-Z.; Sun, M.-X.; Hou, X.-L.; Sun, Y.; Deng, Z.-F.; Xiao, K. Polydatin protects learning and memory impairments in a rat model of vascular dementia. *Phytomedicine* **2012**, *19*, 677–681. [CrossRef] [PubMed]
73. Hsu, S.-C.; Chung, J.-G. Anticancer potential of emodin. *BioMedicine* **2012**, *2*, 108–116. [CrossRef]
74. De Maria, S.; Scognamiglio, I.; Lombardi, A.; Amodio, N.; Caraglia, M.; Cartenì, M.; Ravagnan, G.; Stiuso, P. Polydatin, a natural precursor of resveratrol, induces cell cycle arrest and differentiation of human colorectal Caco-2 cell. *J. Transl. Med.* **2013**, *11*, 264. [CrossRef]
75. Hong, M.; Tan, H.Y.; Li, S.; Cheung, F.; Wang, N.; Nagamatsu, T.; Feng, Y. Cancer Stem Cells: The Potential Targets of Chinese Medicines and Their Active Compounds. *Int. J. Mol. Sci.* **2016**, *17*, 893. [CrossRef]
76. Singh, S.K.; Lillard, J.W.; Singh, R. Reversal of drug resistance by planetary ball milled (PBM) nanoparticle loaded with resveratrol and docetaxel in prostate cancer. *Cancer Lett.* **2018**, *427*, 49–62. [CrossRef]
77. Cottet, M.; Rivière-Honegger, A.; Vaudor, L.; Colombain, L.; Dommanget, F.; Evette, A. The end of a myth: Solving the knotweeds invasion "problem". *Anthropocene* **2020**, *30*, 100240. [CrossRef]
78. Martin, F.-M.; Dommanget, F.; Evette, A. Improving the management of Japanese knotweed s.l.: A response to Jones and colleagues. *NeoBiota* **2020**, *63*, 147–153. [CrossRef]
79. El-Readi, M.Z.; Eid, S.Y.; Al-Amodi, H.S.; Wink, M. *Fallopia japonica*: Bioactive Secondary Metabolites and Molecular Mode of Anticancer. *J. Tradit. Med. Clin. Naturop.* **2016**, *5*, 1–20. [CrossRef]
80. Metličar, V.; Vovk, I.; Albreht, A. Japanese and Bohemian Knotweeds as Sustainable Sources of Carotenoids. *Plants* **2019**, *8*, 384. [CrossRef] [PubMed]
81. Chen, H.; Tuck, T.; Ji, X.; Zhou, X.; Kelly, G.; Cuerrier, A.; Zhang, J. Quality Assessment of Japanese Knotweed (*Fallopia Japonica*) Grown on Prince Edward Island as a Source of Resveratrol. *J. Agric. Food Chem.* **2013**, *61*, 6383–6392. [CrossRef] [PubMed]
82. Dassonville, N.; Vanderhoeven, S.; Vanparys, V.; Hayez, M.; Gruber, W.; Meerts, P. Impacts of alien invasive plants on soil nutrients are correlated with initial site conditions in NW Europe. *Oecologia* **2008**, *157*, 131–140. [CrossRef] [PubMed]
83. Rahmonov, O.; Czylok, A.; Orczewska, A.; Majgier, L.; Parusel, T. Chemical composition of the leaves of *Reynoutria japonica* Houtt. and soil features in polluted areas. *Open Life Sci.* **2014**, *9*, 320–330. [CrossRef]
84. Chu, X.; Sun, A.; Liu, R. Preparative isolation and purification of five compounds from the Chinese medicinal herb *Polygonum cuspidatum* Sieb. et Zucc by high-speed counter-current chromatography. *J. Chromatogr. A* **2005**, *1097*, 33–39. [CrossRef]
85. Vrchotová, N.; Bozena, S.; Tříska, J. The Stilbene and Catechin Content of the Spring Sprouts of Reynoutria Species. *Acta Chromatogr.* **2007**, *19*, 21.
86. Yi, T.; Zhang, H.; Cai, Z. Analysis of Rhizoma *Polygoni cuspidati* by HPLC and HPLC-ESI/MS. *Phytochem. Anal.* **2007**, *18*, 387–392. [CrossRef]

87. Bensa, M.; Glavnik, V.; Vovk, A.I. Leaves of Invasive Plants—Japanese, Bohemian and Giant Knotweed—The Promising New Source of Flavan-3-ols and Proanthocyanidins. *Plants* **2020**, *9*, 118. [CrossRef]
88. Alperth, F.; Melinz, L.; Fladerer, J.-P.; Bucar, F. UHPLC Analysis of *Reynoutria japonica* Houtt. Rhizome Preparations Regarding Stilbene and Anthranoid Composition and Their Antimycobacterial Activity Evaluation. *Plants* **2021**, *10*, 1809. [CrossRef]
89. Bunea, A.; Andjelkovic, M.; Socaciu, C.; Bobis, O.; Neacsu, M.; Verhé, R.; Van Camp, J. Total and individual carotenoids and phenolic acids content in fresh, refrigerated and processed spinach (*Spinacia oleracea* L.). *Food Chem.* **2008**, *108*, 649–656. [CrossRef]
90. Chandra-Hioe, M.V.; Rahman, H.H.; Arcot, J. Lutein and β-Carotene in Selected Asian Leafy Vegetables. *J. Food Chem. Nanotechnol.* **2017**, *3*, 93–97. [CrossRef]
91. Glavnik, V.; Vovk, I. High performance thin-layer chromatography–mass spectrometry methods on diol stationary phase for the analyses of flavan-3-ols and proanthocyanidins in invasive Japanese knotweed. *J. Chromatogr. A* **2019**, *1598*, 196–208. [CrossRef]
92. Ristivojević, P.; Trifković, J.; Vovk, I.; Milojković-Opsenica, D. Comparative study of different approaches for multivariate image analysis in HPTLC fingerprinting of natural products such as plant resin. *Talanta* **2017**, *162*, 72–79. [CrossRef]
93. Song, J.-H.; Kim, S.-K.; Chang, K.-W.; Han, S.-K.; Yi, H.-K.; Jeon, J.-G. In vitro inhibitory effects of *Polygonum cuspidatum* on bacterial viability and virulence factors of *Streptococcus mutans* and *Streptococcus sobrinus*. *Arch. Oral Biol.* **2006**, *51*, 1131–1140. [CrossRef]
94. Balouiri, M.; Sadiki, M.; Ibnsouda, S.K. Methods for in vitro evaluating antimicrobial activity: A review. *J. Pharm. Anal.* **2015**, *6*, 71–79. [CrossRef]
95. Magacz, M.; Oszajca, M.; Nawrot-Hadzik, I.; Drożdż, R.; Jurczak, A.; Hadzik, J.; Smakosz, A.; Krzyściak, W. Phenolic Compounds of *Reynoutria* sp. as Modulators of Oral Cavity Lactoperoxidase System. *Antioxidants* **2021**, *10*, 676. [CrossRef]
96. Tong, L.; Chuang, C.-C.; Wu, S.; Zuo, L. Reactive oxygen species in redox cancer therapy. *Cancer Lett.* **2015**, *367*, 18–25. [CrossRef]
97. Jug, U.; Naumoska, K.; Vovk, I. (−)-Epicatechin—An Important Contributor to the Antioxidant Activity of Japanese Knotweed Rhizome Bark Extract as Determined by Antioxidant Activity-Guided Fractionation. *Antioxidants* **2021**, *10*, 133. [CrossRef]
98. Lazurca, M.; Lazurca, D.; Fetea, F.; Ranga, F.; Socaciu, C. Evaluation of the Phenolic Content in the Buds of *Polygonum cuspidatum* Sieb. et Zucc. *Bull. UASVM Agric.* **2012**, *69*, 273–280.
99. Wenzel, E.; Soldo, T.; Erbersdobler, H.; Somoza, V. Bioactivity and metabolism oftrans-resveratrol orally administered to Wistar rats. *Mol. Nutr. Food Res.* **2005**, *49*, 482–494. [CrossRef]
100. Hsu, C.-Y.; Chan, Y.-P.; Chang, J. Antioxidant activity of extract from *Polygonum cuspidatum*. *Biol. Res.* **2007**, *40*, 13–21. [CrossRef]
101. Pan, Y.; Zhang, X.; Wang, H.; Liang, Y.; Zhu, J.; Li, H.; Zhang, Z.; Wu, Q. Antioxidant potential of ethanolic extract of *Polygonum cuspidatum* and application in peanut oil. *Food Chem.* **2007**, *105*, 1518–1524. [CrossRef]
102. Lin, Y.-W.; Yang, F.-J.; Chen, C.-L.; Lee, W.-T.; Chen, R.-S. Free radical scavenging activity and antiproliferative potential of *Polygonum cuspidatum* root extracts. *J. Nat. Med.* **2010**, *64*, 146–152. [CrossRef]
103. Mikulski, D.; Molski, M. Quantitative structure–antioxidant activity relationship of trans-resveratrol oligomers, trans-4,4′-dihydroxystilbene dimer, trans-resveratrol-3-O-glucuronide, glucosides: Trans-piceid, cis-piceid, trans-astringin and trans-resveratrol-4′-O-β-D-glucopyranoside. *Eur. J. Med. Chem.* **2010**, *45*, 2366–2380. [CrossRef]
104. Lee, G.; Choi, T.W.; Kim, C.; Nam, D.; Lee, S.-G.; Jang, H.-J.; Lee, J.; Um, J.-Y.; Jung, S.H.; Shim, B.S.; et al. Anti-inflammatory activities of *Reynoutria elliptica* through suppression of mitogen-activated protein kinases and nuclear factor-κB activation pathways. *Immunopharmacol. Immunotoxicol.* **2011**, *34*, 454–464. [CrossRef]
105. Obrenovich, M.E.; Li, Y.; Parvathaneni, K.; Yendluri, B.B.; Palacios, H.H.; Leszek, J.; Aliev, G. Antioxidants in Health, Disease and Aging. *CNS Neurol. Disord. Drug Targets* **2011**, *10*, 192–207. [CrossRef]
106. Jiao, Y.; Wu, Y.; Du, D. Polydatin inhibits cell proliferation, invasion and migration, and induces cell apoptosis in hepatocellular carcinoma. *Braz. J. Med Biol. Res.* **2018**, *51*, 1–9. [CrossRef]
107. Li, W.; Zhang, Q.; Chen, K.; Sima, Z.; Liu, J.; Yu, Q.; Liu, J. 2-Ethoxystypandrone, a novel small-molecule STAT3 signaling inhibitor from *Polygonum cuspidatum*, inhibits cell growth and induces apoptosis of HCC cells and HCC Cancer stem cells. *BMC Complement. Altern. Med.* **2019**, *19*, 38. [CrossRef]
108. Xie, Q.; Yang, Y.; Wang, Z.; Chen, F.; Zhang, A.; Liu, C. Resveratrol-4-O-D-(2′-galloyl)-glucopyranoside Isolated from *Polygonum cuspidatum* Exhibits Anti-Hepatocellular Carcinoma Viability by Inducing Apoptosis via the JNK and ERK Pathway. *Molecules* **2014**, *19*, 1592–1602. [CrossRef]
109. Liu, Z.; Wei, F.; Chen, L.-J.; Xiong, H.-R.; Liu, Y.-Y.; Luo, F.; Hou, W.; Xiao, H.; Yang, Z.-Q. In Vitro and in Vivo Studies of the Inhibitory Effects of Emodin Isolated from *Polygonum cuspidatum* on Coxsakievirus B4. *Molecules* **2013**, *18*, 11842–11858. [CrossRef]
110. Yiu, C.-Y.; Chen, S.-Y.; Yang, T.-H.; Chang, C.-J.; Yeh, D.-B.; Chen, Y.-J.; Lin, T.-P. Inhibition of Epstein-Barr Virus Lytic Cycle by an Ethyl Acetate Subfraction Separated from *Polygonum cuspidatum* Root and Its Major Component, Emodin. *Molecules* **2014**, *19*, 1258–1272. [CrossRef]
111. Hodgin, K.; Donovan, E.; Kekes-Szabo, S.; Lin, J.; Feick, J.; Massey, R.; Ness, T.; Younger, J. A Placebo-Controlled, Pseudo-Randomized, Crossover Trial of Botanical Agents for Gulf War Illness: Resveratrol (*Polygonum cuspidatum*), Luteolin, and Fisetin (*Rhus succedanea*). *Int. J. Environ. Res. Public Health* **2021**, *18*, 2483. [CrossRef]
112. Nawrot-Hadzik, I.; Zmudzinski, M.; Matkowski, A.; Preissner, R.; Kęsik-Brodacka, M.; Hadzik, J.; Drag, M.; Abel, R. *Reynoutria* Rhizomes as a Natural Source of SARS-CoV-2 Mpro Inhibitors–Molecular Docking and In Vitro Study. *Pharmaceuticals* **2021**, *14*, 742. [CrossRef]

113. Li, L.; Song, X.; Yin, Z.; Jia, R.; Li, Z.; Zhou, X.; Zou, Y.; Li, L.; Yin, L.; Yue, G.; et al. The antibacterial activity and action mechanism of emodin from *Polygonum cuspidatum* against *Haemophilus parasuis* in vitro. *Microbiol. Res.* **2016**, *186–187*, 139–145. [CrossRef]
114. Wang, L.; Qiu, P.; Long, X.-F.; Zhang, S.; Zeng, Z.-G.; Tian, Y.-Q. Comparative analysis of chemical constituents, antimicrobial and antioxidant activities of ethylacetate extracts of *Polygonum cuspidatum* and its endophytic actinomycete, *Streptomyces* sp. A0916. *Chin. J. Nat. Med.* **2016**, *14*, 117–123. [CrossRef]
115. Matkowski, A.; Jamiołkowska-Kozlowska, W.; Nawrot, I. Chinese Medicinal Herbs as Source of Antioxidant Compounds–Where Tradition Meets the Future. *Curr. Med. Chem.* **2013**, *20*, 984–1004.
116. Lee, M.-H.; Kao, L.; Lin, C.-C. Comparison of the Antioxidant and Transmembrane Permeative Activities of the Different *Polygonum cuspidatum* Extracts in Phospholipid-Based Microemulsions. *J. Agric. Food Chem.* **2011**, *59*, 9135–9141. [CrossRef]
117. Pogačnik, L.; Bergant, T.; Skrt, M.; Ulrih, N.P.; Viktorová, J.; Ruml, T. In Vitro Comparison of the Bioactivities of Japanese and Bohemian Knotweed Ethanol Extracts. *Foods* **2020**, *9*, 544. [CrossRef]
118. Wang, Y.; Horng, C.; Hsieh, M.; Chen, H.; Huang, Y.; Yang, J.; Wang, G.; Chiang, J.; Chen, H.; Lu, C.; et al. Autophagy and apoptotic machinery caused by *Polygonum cuspidatum* extract in cisplatin-resistant human oral cancer CAR cells. *Oncol. Rep.* **2019**, *41*, 2549–2557. [CrossRef]
119. Zhang, L.; Li, Y.; Gu, Z.; Wang, Y.; Shi, M.; Ji, Y.; Sun, J.; Xu, X.; Zhang, L.; Jiang, J.; et al. Resveratrol Inhibits Enterovirus 71 Replication and Pro-Inflammatory Cytokine Secretion in Rhabdosarcoma Cells through Blocking IKKs/NF-κB Signaling Pathway. *PLoS ONE* **2015**, *10*, e0116879. [CrossRef]
120. Xing, W.-W.; Wu, J.-Z.; Jia, M.; Du, J.; Zhang, H.; Qin, L.-P. Effects of polydatin from *Polygonum cuspidatum* on lipid profile in hyperlipidemic rabbits. *Biomed. Pharmacother.* **2009**, *63*, 457–462. [CrossRef]
121. Gao, J.P.; Chen, C.X.; Gu, W.L.; Wu, Q.; Wang, Y.; Lü, J. Effects of polydatin on attenuating ventricular remodeling in isoproterenol-induced mouse and pressure-overload rat models. *Fitoterapia* **2010**, *81*, 953–960. [CrossRef]
122. Miao, Q.; Wang, S.; Miao, S.; Wang, J.; Xie, Y.; Yang, Q. Cardioprotective effect of polydatin against ischemia/reperfusion injury: Roles of protein kinase C and mito KATP activation. *Phytomedicine* **2011**, *19*, 8–12. [CrossRef]
123. Yu, H.-R.; Jiang, Y.-L.; Dong, D.-D.; Wang, G.-H. Cardioprotective effects of the total flavonoids of *Polygonum cuspidatum* Sieb. et Zucc. Root extract on experimental myocardial infarction in mice. *Trop. J. Pharm. Res.* **2019**, *18*, 341. [CrossRef]
124. Ge, J.-F.; Xu, Y.-Y.; Qin, G.; Cheng, J.-Q.; Echen, F.-H. Resveratrol Ameliorates the Anxiety- and Depression-Like Behavior of Subclinical Hypothyroidism Rat: Possible Involvement of the HPT Axis, HPA Axis, and Wnt/β-Catenin Pathway. *Front. Endocrinol.* **2016**, *7*, 44. [CrossRef]
125. Ge, J.-F.; Peng, L.; Cheng, J.-Q.; Pan, C.-X.; Tang, J.; Chen, F.-H.; Li, J. Antidepressant-like effect of resveratrol: Involvement of antioxidant effect and peripheral regulation on HPA axis. *Pharmacol. Biochem. Behav.* **2013**, *114–115*, 64–69. [CrossRef]
126. Rege, S.; Egeetha, T.; Griffin, G.D.; Broderick, T.L.; Babu, J.R. Neuroprotective effects of resveratrol in Alzheimer disease pathology. *Front. Aging Neurosci.* **2014**, *6*, 218. [CrossRef]
127. Liu, S.; Li, T.; Liu, H.; Wang, X.; Bo, S.; Xie, Y.; Bai, X.; Wu, L.; Wang, Z.; Liu, D. Resveratrol exerts antidepressant properties in the chronic unpredictable mild stress model through the regulation of oxidative stress and mTOR pathway in the rat hippocampus and prefrontal cortex. *Behav. Brain Res.* **2016**, *302*, 191–199. [CrossRef]
128. Moore, A.; Beidler, J.; Hong, M.Y. Resveratrol and Depression in Animal Models: A Systematic Review of the Biological Mechanisms. *Molecules* **2018**, *23*, 2197. [CrossRef]
129. Li, Y.B.; Lin, Z.Q.; Zhang, Z.J.; Wang, M.W.; Zhang, H.; Zhang, Q.W.; Lee, S.M.Y.; Wang, Y.T.; Hoi, P.M. Protective, Antioxidative and Antiapoptotic Effects of 2-Methoxy-6-Acetyl-7-Methyljuglone from *Polygonum cuspidatumin* PC12 Cells. *Planta Med.* **2011**, *77*, 354–361. [CrossRef]
130. Tang, D.; Zhu, J.-X.; Wu, A.-G.; Xu, Y.-H.; Duan, T.-T.; Zheng, Z.-G.; Wang, R.-S.; Li, D.; Zhu, Q. Pre-column incubation followed by fast liquid chromatography analysis for rapid screening of natural methylglyoxal scavengers directly from herbal medicines: Case study of *Polygonum cuspidatum*. *J. Chromatogr. A* **2013**, *1286*, 102–110. [CrossRef]
131. Sohn, E.; Kim, J.; Kim, C.S.; Jo, K.; Kim, J.S. Extract of Rhizoma *Polygonum cuspidatum* reduces early renal podocyte injury in streptozotocin-induced diabetic rats and its active compound emodin inhibits methylglyoxal-mediated glycation of proteins. *Mol. Med. Rep.* **2015**, *12*, 5837–5845. [CrossRef]
132. Sohn, E.; Kim, J.; Kim, C.-S.; Lee, Y.M.; Kim, J.S. Extract of *Polygonum cuspidatum* Attenuates Diabetic Retinopathy by Inhibiting the High-Mobility Group Box-1 (HMGB1) Signaling Pathway in Streptozotocin-Induced Diabetic Rats. *Nutrients* **2016**, *8*, 140. [CrossRef]
133. Zhao, Y.; Chen, M.X.; Kongstad, K.T.; Jäger, A.K.; Staerk, D. Potential of *Polygonum cuspidatum* Root as an Antidiabetic Food: Dual High-Resolution α-Glucosidase and PTP1B Inhibition Profiling Combined with HPLC-HRMS and NMR for Identification of Antidiabetic Constituents. *J. Agric. Food Chem.* **2017**, *65*, 4421–4427. [CrossRef] [PubMed]
134. Hao, J.; Huang, K.; Chen, C.; Liang, Y.; Wang, Y.; Zhang, X.; Huang, H. Polydatin Improves Glucose and Lipid Metabolisms in Insulin-Resistant HepG2 Cells through the AMPK Pathway. *Biol. Pharm. Bull.* **2018**, *41*, 891–898. [CrossRef] [PubMed]
135. Liu, B.; Li, S.; Sui, X.; Guo, L.; Liu, X.; Li, H.; Gao, L.; Cai, S.; Li, Y.; Wang, T.; et al. Root Extract of *Polygonum cuspidatum* Siebold & Zucc. Ameliorates DSS-Induced Ulcerative Colitis by Affecting NF-kappaB Signaling Pathway in a Mouse Model via Synergistic Effects of Polydatin, Resveratrol, and Emodin. *Front. Pharmacol.* **2018**, *9*, 347. [CrossRef] [PubMed]
136. Xu, H.; Qu, J.; Wang, J.; Han, K.; Li, Q.; Bi, W.; Liu, R. Discovery of pulmonary fibrosis inhibitor targeting TGF-β RI in *Polygonum cuspidatum* by high resolution mass spectrometry with in silico strategy. *J. Pharm. Anal.* **2020**. [CrossRef]

137. Du, J.; Sun, L.-N.; Xing, W.-W.; Huang, B.-K.; Jia, M.; Wu, J.-Z.; Zhang, H.; Qin, L.-P. Lipid-lowering effects of polydatin from Polygonum cuspidatum in hyperlipidemic hamsters. *Phytomedicine* **2009**, *16*, 652–658. [CrossRef]
138. Shen, B.-B.; Yang, Y.-P.; Yasamin, S.; Liang, N.; Su, W.; Chen, S.-H.; Wang, X.-J.; Wang, W. Analysis of the Phytochemistry and Bioactivity of the Genus Polygonum of Polygonaceae. *Digit. Chin. Med.* **2018**, *1*, 19–36. [CrossRef]
139. Zhang, C.; Zhang, X.; Zhang, Y.; Xu, Q.; Xiao, H.; Liang, X. Analysis of estrogenic compounds in *Polygonum cuspidatum* by bioassay and high performance liquid chromatography. *J. Ethnopharmacol.* **2006**, *105*, 223–228. [CrossRef]
140. Brown, V.A.; Patel, K.R.; Viskaduraki, M.; Crowell, J.A.; Perloff, M.; Booth, T.D.; Vasilinin, G.; Sen, A.; Schinas, A.M.; Piccirilli, G.; et al. Repeat Dose Study of the Cancer Chemopreventive Agent Resveratrol in Healthy Volunteers: Safety, Pharmacokinetics, and Effect on the Insulin-like Growth Factor Axis. *Cancer Res.* **2010**, *70*, 9003–9011. [CrossRef]
141. Peltz, L.; Gomez, J.; Marquez, M.; Alencastro, F.; Atashpanjeh, N.; Quang, T.; Bach, T.; Zhao, Y. Resveratrol Exerts Dosage and Duration Dependent Effect on Human Mesenchymal Stem Cell Development. *PLoS ONE* **2012**, *7*, e37162. [CrossRef]
142. Carrizzo, A.; Forte, M.; Damato, A.; Trimarco, V.; Salzano, F.; Bartolo, M.; Maciag, A.; Puca, A.A.; Vecchione, C. Antioxidant effects of resveratrol in cardiovascular, cerebral and metabolic diseases. *Food Chem. Toxicol.* **2013**, *61*, 215–226. [CrossRef]
143. Shaito, A.; Posadino, A.M.; Younes, N.; Hasan, H.; Halabi, S.; Alhababi, D.; Al-Mohannadi, A.; Abdel-Rahman, W.M.; Eid, A.H.; Nasrallah, G.K.; et al. Potential Adverse Effects of Resveratrol: A Literature Review. *Int. J. Mol. Sci.* **2020**, *21*, 2084. [CrossRef]
144. Akkol, E.; Tatlı, I.; Karatoprak, G.; Ağar, O.; Yücel, Ç.; Sobarzo-Sánchez, E.; Capasso, R. Is Emodin with Anticancer Effects Completely Innocent? Two Sides of the Coin. *Cancers* **2021**, *13*, 2733. [CrossRef] [PubMed]
145. Davis, E.S.; Kelly, R.; Maggs, C.A.; Stout, J.C. Contrasting impacts of highly invasive plant species on flower-visiting insect communities. *Biodivers. Conserv.* **2018**, *27*, 2069–2085. [CrossRef]
146. Ferrazzi, P.; Marletto, F. Bee Value of *Reynoutria japonica* Houtt. *Apic. Mod.* **1990**, *81*, 71–76.
147. Russo, L.; Nichol, C.; Shea, K. Pollinator floral provisioning by a plant invader: Quantifying beneficial effects of detrimental species. *Divers. Distrib.* **2016**, *22*, 189–198. [CrossRef]
148. Barudanović, S.; Zečić, E.; Macanović, A.; Duraković, B.; Mašić, E. Invasive Alien Plant Species in Global Perspectives with Special References to Bosnia and Herzegovina. *Invasive Alien Species* **2021**, *3*, 215–252. [CrossRef]
149. Șuba, E.E.; Deak, J.; Coroian, I.; Nap, M.E.; Chiorean, S.; Luputț, I.; Ficior, D. The Use of UAV Photogrammetric Technology in Order to Make the Cadastral Documentation of Detachment in Two Lots of a Property Located in Valea Ierii, Cluj County. *Bull. Univ. Agric. Sci. Vet. Med. Cluj-Napoca* **2021**, *78*, 89–94. [CrossRef]
150. Jabłoński, B.; Kołtowski, Z. Nectar Secretion and Honey Potential of Honey-Plants Growing under Poland's Conditions. Part XII. *J. Apicult. Sci.* **2001**, *45*, 29–35.
151. Peter, A.; Žlabur, J.; Šurić, J.; Voća, S.; Purgar, D.; Pezo, L.; Voća, N. Invasive Plant Species Biomass—Evaluation of Functional Value. *Molecules* **2021**, *26*, 3814. [CrossRef]
152. Simberloff, D.; Martin, J.-L.; Genovesi, P.; Maris, V.; Wardle, D.; Aronson, J.; Courchamp, F.; Galil, B.; García-Berthou, E.; Pascal, M.; et al. Impacts of biological invasions: What's what and the way forward. *Trends Ecol. Evol.* **2013**, *28*, 58–66. [CrossRef]
153. European Council Directive 2001/110 Relating to Honey. *Off. J. Eur. Comm.* **2001**, *L 10*, 47–52. Available online: https://eur-lex.europa.eu/legal-content/EN/ALL/?uri=celex%3A32001L0110 (accessed on 5 November 2021).
154. Bobis, O.; Dezmirean, D.S.; Bonta, V.; Moise, A.; Pasca, C.; Domokos, T.E.; Urcan, A.C. Japanese Knotweed (*Fallopia japonica*): Landscape Invasive Plant Versus High Quality Honey Source. *Sci. Pap. Ser. D Anim. Sci.* **2019**, *62*, 231–235.
155. Alvarez-Suarez, J.M.; Gasparrini, M.; Forbes-Hernández, T.Y.; Mazzoni, L.; Giampieri, F. The Composition and Biological Activity of Honey: A Focus on Manuka Honey. *Foods* **2014**, *3*, 420–432. [CrossRef]
156. Solayman, M.; Islam, M.A.; Paul, S.; Ali, Y.; Khalil, M.I.; Alam, N.; Gan, S.H. Physicochemical Properties, Minerals, Trace Elements, and Heavy Metals in Honey of Different Origins: A Comprehensive Review. *Compr. Rev. Food Sci. Food Saf.* **2016**, *15*, 219–233. [CrossRef] [PubMed]
157. Deng, J.; Liu, R.; Lu, Q.; Hao, P.; Xu, A.; Zhang, J.; Tan, J. Biochemical properties, antibacterial and cellular antioxidant activities of buckwheat honey in comparison to manuka honey. *Food Chem.* **2018**, *252*, 243–249. [CrossRef]
158. Kędzierska-Matysek, M.; Florek, M.; Wolanciuk, A.; Barłowska, J.; Litwińczuk, Z. Concentration of Minerals in Nectar Honeys from Direct Sale and Retail in Poland. *Biol. Trace Elem. Res.* **2018**, *186*, 579–588. [CrossRef]
159. Bertoncelj, J.; Dobersek, U.; Jamnik, M.; Golob, T. Evaluation of the phenolic content, antioxidant activity and colour of Slovenian honey. *Food Chem.* **2007**, *105*, 822–828. [CrossRef]
160. Kaškonienė, V.; Maruška, A.; Kornyšova, O.; Charczun, N.; Ligor, M.; Buszewski, B. Quantitative and Qualitative Determination of Phenolic Compounds in Honey. *Chem. Technol.* **2009**, *52*, 74–80.
161. Lachman, J.; Hejtmánková, A.; Sykora, J.; Karban, J.; Orsák, M.; Rygerová, B. Contents of major phenolic and flavonoid antioxidants in selected Czech honey. *Czech J. Food Sci.* **2010**, *28*, 412–426. [CrossRef]
162. Nešović, M.; Gašić, U.; Tosti, T.; Horvacki, N.; Šikoparija, B.; Nedić, N.; Blagojević, S.; Ignjatović, L.; Tešić, Ž. Polyphenol Profile of Buckwheat Honey, Nectar and Pollen: Polyphenolics in Buckwheat. *R. Soc. Open Sci.* **2020**, *7*, 1–15. [CrossRef] [PubMed]
163. Sanchez, A.; Schuster, T.; Kron, K.A. A Large-Scale Phylogeny of Polygonaceae Based on Molecular Data. *Int. J. Plant Sci.* **2009**, *170*, 1044–1055. [CrossRef]
164. Burke, J.M.; Sanchez, A.; Kron, K.; Luckow, M. Placing the woody tropical genera of Polygonaceae: A hypothesis of character evolution and phylogeny. *Am. J. Bot.* **2010**, *97*, 1377–1390. [CrossRef]

165. Johnston, M.; McBride, M.; Dahiya, D.; Owusu-Apenten, R.; Nigam, P.S. Antibacterial activity of Manuka honey and its components: An overview. *AIMS Microbiol.* **2018**, *4*, 655–664. [CrossRef] [PubMed]
166. Mavric, E.; Wittmann, S.; Barth, G.; Henle, T. Identification and quantification of methylglyoxal as the dominant antibacterial constituent of Manuka (*Leptospermum scoparium*) honeys from New Zealand. *Mol. Nutr. Food Res.* **2008**, *52*, 483–489. [CrossRef]
167. Adams, C.J.; Manley-Harris, M.; Molan, P.C. The origin of methylglyoxal in New Zealand manuka (*Leptospermum scoparium*) honey. *Carbohydr. Res.* **2009**, *344*, 1050–1053. [CrossRef]
168. Tuberoso, C.I.G.; Bifulco, E.; Jerkovic, I.; Caboni, P.; Cabras, P.; Floris, I. Methyl Syringate: A Chemical Marker of Asphodel (*Asphodelus microcarpus* Salzm. et Viv.) Monofloral Honey. *J. Agric. Food Chem.* **2009**, *57*, 3895–3900. [CrossRef]
169. Oelschlaegel, S.; Gruner, M.; Wang, P.-N.; Boettcher, A.; Koelling-Speer, I.; Speer, K. Classification and Characterization of Manuka Honeys Based on Phenolic Compounds and Methylglyoxal. *J. Agric. Food Chem.* **2012**, *60*, 7229–7237. [CrossRef] [PubMed]
170. Alvarez-Suarez, J.M.; Giampieri, F.; Cordero, M.; Gasparrini, M.; Forbes-Hernández, T.Y.; Mazzoni, L.; Afrin, S.; Beltrán-Ayala, P.; González-Paramás, A.M.; Santos-Buelga, C.; et al. Activation of AMPK/Nrf2 signalling by Manuka honey protects human dermal fibroblasts against oxidative damage by improving antioxidant response and mitochondrial function promoting wound healing. *J. Funct. Foods* **2016**, *25*, 38–49. [CrossRef]
171. Pasini, F.; Gardini, S.; Marcazzan, G.L.; Caboni, M.F. Buckwheat honeys: Screening of composition and properties. *Food Chem.* **2013**, *141*, 2802–2811. [CrossRef] [PubMed]
172. Gheldof, N.; Wang, X.-H.; Engeseth, N.J. Identification and Quantification of Antioxidant Components of Honeys from Various Floral Sources. *J. Agric. Food Chem.* **2002**, *50*, 5870–5877. [CrossRef]
173. Dżugan, M.; Grabek-Lejko, D.; Swacha, S.; Tomczyk, M.; Bednarska, S.; Kapusta, I. Physicochemical quality parameters, antibacterial properties and cellular antioxidant activity of Polish buckwheat honey. *Food Biosci.* **2020**, *34*, 100538. [CrossRef]
174. Bobis, O.; Bonta, V.; Moise, A.; Pasca, C.; Domokos, T.; Niculae, M.; Marghitas, L.; Dezmirean, D. Is Color of Honey Related to High Polyphenolic Content and Its Bioactive Properties? Case Study: *Fallopia japonica* Honey from Romania. Symposia "Apitherapy" & "Apiquality" Sibiu—Romania. 2018. Available online: https://www.apimondia.com/en/databases/symposia-proceedings/2-all/52-symposia-apitherapy-apiquality-sibiu-romania-2018 (accessed on 5 November 2021).

plants

MDPI

Review

Review of Existing Knowledge and Practices of Tarping for the Control of Invasive Knotweeds

Marie-Anne Dusz [1,*], François-Marie Martin [2], Fanny Dommanget [1], Anne Petit [3], Caroline Dechaume-Moncharmont [4] and André Evette [1]

1 LESSEM, INRAE, Université Grenoble Alpes, 2 rue de la Papeterie-BP 76, 38402 St-Martin-d'Hères, France; fanny.dommanget@inrae.fr (F.D.); andre.evette@inrae.fr (A.E.)
2 Laboratoire Cogitamus, 38000 Grenoble, France; francois-marie.martin@inrae.fr
3 SNCF Réseau, 6 avenue François Mitterrand, 93574 La Plaine Saint Denis, France; anne.petit@reseau.sncf.fr
4 SNCF Réseau, 19 avenue Georges Pompidou, 69003 Lyon, France; caroline.dechaume@reseau.sncf.fr
* Correspondence: marie-anne.dusz@inrae.fr

Abstract: Managing invasive exotic plant species is a complex challenge, especially for Asian knotweeds (*Reynoutria* spp.). Tarping is a regularly cited but poorly documented control method, which consists of covering the ground with a tarp (agricultural tarp, geotextile, geomembrane, etc.) to create a physical barrier to hinder plant growth and deprive the plants of light in order to deplete their rhizomatous reserves. To improve our knowledge of tarping in order to identify the key factors of its success or failure, we reviewed the relevant grey and scientific literature and conducted an international survey among managers to collect feedback on tarping experiments. In the literature, as well as in the field, practices are quite heterogeneous, and the method's effectiveness is highly contrasted. A better consideration of knotweed biology may improve the efficacy of the method. Based on the bibliography and survey work, we propose practical recommendations including covering the entire stand, extending the tarping up to 2.5 m beyond its edges for a period of at least six years, and ensuring regular monitoring. Even though tarping does not seem to be a one-size-fits-all solution to eradicate knotweed, it could still be a useful control method once knotweed has become a critical management issue.

Keywords: ground covering; *Reynoutria*; *Fallopia*; *Polygonum*; knotweed management; geotextile; invasive alien plants

1. Introduction

The number of invasive alien plant species (IAS) is increasing worldwide mainly as a result of human activities [1]. These species are causing various detrimental impacts on biodiversity, ecosystem services, and human health and activities [2], to such an extent that the Invasive Species Specialist Group (ISSG) of the UICN and the Intergovernmental Science-Policy Platform on Biodiversity and Ecosystem Services (IPBES) consider them one of the main threats to biodiversity at the global level [3,4]. As a consequence, invasive species have become a global concern. Managing invasives and the damage they cause is extremely costly: the annual mean cost at the global scale has been estimated at US$26 billion over the past few decades, and may have reached US$162.7 billion in 2017 according to one study [5]; total IAS-related costs are estimated to be at least €12.5 billion per year in the European Union [6], and €38 million in France alone [7]. Therefore, many countries have adopted policies and strategies to deal with biological invasions. The Convention on Biological Diversity took invasive species into account at the international level in 2011 (Aichi Biodiversity Target 9) and the European Union adopted a regulation (EU No 1143/2014) on the prevention and management of the introduction and spread of IASs in 2014.

In the field, IAS control is a complex challenge for land managers (such as transport infrastructure or natural area managers, hereafter referred to as managers). They have to design local-scale strategies with often limited financial and human resources [8,9] and without being certain of the best control options to implement [10,11]. While IAS management literature is extensive, there is still a "knowing-doing gap" (sensu Matzek et al. [12], i.e., a gap between research and practice), as scientific research rarely meets the manager's needs [13,14]. The complexity of invasion processes and the diversity of species involved also make it difficult to develop effective generalized control methods [15]. For many species, managers often face an absence of consensus [10] and even sometimes contradictory management recommendations [16]. They may also face division in the scientific community concerning IAS impacts and the usefulness of costly management programs [17]. In addition, climate change will likely challenge effective IAS management practices [9]. In such a context of uncertainty [18], management choices rely heavily on local managers' own perception of the IAS, their personal experiences, and advice from fellow managers, more than on the most recent scientific knowledge [19]. Japanese knotweed s.l figures among the world's 100 worst invasive alien species [20] and, as such, is a good example of the challenges faced by scientists and managers. The species complex is composed of the Japanese knotweed (*Reynoutria japonica* Houtt.; syn. *Fallopia japonica* (Houtt.) Ronse Decr. or *Polygonum cuspidatum* Siebold & Zucc.), the Giant knotweed (*Reynoutria sachalinensis* (F. Schmidt) Nakai; syn. *Fallopia sachalinensis* (F. Schmidt) Ronse Decr. or *Polygonum sachalinense* F. Schmidt), and the hybrid of the two: the Bohemian knotweed (*Reynoutria* × *bohemica* Chrtek & Chrtková; syn. *Fallopia* × *bohemica* (Chrtek & Chrtková) J. P. Bailey or *Polygonum* × *bohemicum* (Chrtek & Chrtková) P. F. Zika et al. Jacobson) [21]. Japanese and Giant knotweeds originated from Eastern Asia and were introduced into Europe in the 19th century, mainly as ornamental plants. These invasive perennial geophytes [22] are now widely naturalized in Europe, North America, Australia, New Zealand, Chile, and South Africa [23], where they form large monocultures, called "stands" or "patches", in a wide variety of habitats. Riverbanks, roads, and railways, as well as wastelands and manmade habitats are particularly concerned since the plants spread primarily through the dispersal of vegetative propagules (e.g., fragments of stem or rhizome) and this is facilitated by earth-moving activities and floods.

Though still under discussion [24,25], the numerous negative impacts of knotweeds [26–29] and their locally high rate of spread are forcing managers to take action. A growing body of literature proposes many different control methods including cutting, mowing, pulling, digging, tarping, burning, grazing, planting, salting, and using herbicides or biological control [30–33]. While recommendations for all of these techniques may be found in various management practice guides, their effectiveness has rarely been scientifically assessed in readily available, long-term, robust, controlled experiments [34]. Furthermore, to date, no manual, mechanical, or chemical method has been recognized as fully effective against knotweeds, or fully compatible with sustainable management goals [10,31,34,35]. Worse still, some control techniques could even increase the mean lateral spread of knotweed stands [36,37] or favor the dispersal of knotweed propagules [23,38]. While considerable effort and huge amounts of money are being invested, the lack of information and even consensus on the best knotweed control methods does not allow managers to confidently predict the success of their management campaigns [10].

One technique that is regularly cited in the literature, despite little rigorous documentation, is "tarping". A mechanical method, tarping consists of opposing a physical barrier to plant growth over a period of time (Figure 1). This method is designed to deprive the knotweed of light, thus preventing photosynthesis, until complete depletion of the reserves stored in their belowground organs. The rhizomatous system of knotweeds, which may account for two-thirds of the plant's total biomass, allows them to store and share considerable amounts of resources among shoots to maximize the growth of the whole stand and guarantee its resilience in case of disturbances [26,39,40]. The knotweeds' rhizomatous systems can extend 3 m down into the soil. They can expand laterally from aboveground stems over several meters, with most typically less than 2.5 m and rarely up to 4 m [41].

Figure 1. Installation of a synthetic geotextile as part of a tarping operation to control Asian knotweed. © INRAE.

Tarping is also frequently used in production agriculture and by home gardeners as a soil solarization technique. During solarization, the soil covered with a plastic tarp is heated by solar radiation and reaches a temperature high enough to kill weeds and other plant pests [42]. The effects of solarization on weeds is species dependent, as time and temperature requirements for thermal death may vary considerably among weed species [43]. The method's effectiveness is determined by soil moisture, climate, aspect, and weather but also by the different materials used to cover the soil [44]. Various materials, from the simplest to the most sophisticated, are used for knotweed tarping, including geotextiles (permeable fabrics), agricultural tarps, and geomembranes (i.e., waterproof fabrics), either synthetic or biodegradable. More recently, tarps specifically designed to prevent invasive plant growth have also been developed [45].

As with other control methods, robust evidence is lacking to assess the effectiveness of tarping to control knotweed populations. The few documented studies suggest mixed results [46]. To help fill this knowledge gap and favor well-informed management, the practice of tarping needs to be thoroughly investigated. We therefore propose an overview of tarping practices in order to identify some likely factors for the success or failure of the method: (i) we reviewed the scientific and technical literature related to knotweed tarping, and (ii) we conducted an international survey addressed to managers in French- and English-speaking countries where invasive knotweeds are present.

2. Materials and Methods

2.1. Bibliographic Review

As we were interested in both grey and scientific literature, in spring 2020, we searched for references on Google Scholar and the standard Google search engine with the following keywords in French: 'bâchage' or 'géotextile' or 'solarisation' or 'gestion' AND 'renouée' or 'Fallopia' or 'Reynoutria' or 'Polygonum'; and in English: 'tarping' or 'covering' or 'sheeting' or 'geotextile' or 'solarization' or 'management' AND 'Japanese knotweed' or 'Fallopia' or 'Reynoutria' or 'Polygonum'. We screened each French or English document for relevance. As our focus was on pure in situ tarping, we deliberately chose to exclude ex situ tarping operations (i.e., when knotweed rhizomes are extracted and then tarped on another site) and operations where the soil is crushed with a stone crusher prior to tarping, a technique developed in France (cf. [47]). We read each selected document attentively in order to summarize the available relevant information on the method. Particular attention was paid to the objectives pursued, the environments of use, the protocols followed, and the efficacy described by the authors in order to identify the key factors influencing tarping success or failure.

2.2. Survey of Knotweed Control through Tarping

To complete the existing published knowledge on knotweed tarping techniques, we collected feedback on unpublished field trials via an online questionnaire. The questionnaire, in both French and English (Supplementary File S1), was available online from May to July 2020 and was sent to various manager mailing lists in French- and English-speaking countries where invasive Asian knotweeds are present. We targeted managers who already had experience with the tarping method. The survey was divided into nine sections: general information regarding the tarping project, site description, site preparation before tarping, tarping installation, monitoring operations, tarp removal, post-control observations, cost, and the manager's perception. Presumed key technical points were addressed through specific questions. As such, our questionnaire mostly contained check-all-that-apply questions, rating scale questions, and some open-ended response questions. Most questions included an "other" option in order to collect unexpected answers. Since processing the free responses would have taken a significant amount of time and, as there are few documented tarping experiments, we decided to focus on a descriptive analysis of the data.

3. Results

3.1. Results of the Bibliographic Review

We found 63 relevant documents referring to tarping as a control method for invasive knotweeds (Supplementary File S2). The oldest was published in 2004. However, we only found three references that quantitatively assessed the effectiveness of tarping in comparison to other control methods [31,48,49]. Most of the collected documents were Best Management Practice Guides, technical reports, and conference proceedings. We also found numerous references to tarping operations in local newspapers and internet reports, but we did not include them in an exhaustive list as they often lacked rigor. Our review confirmed both stakeholders' and managers' interest in the method, despite the lack of scientific evaluation of its effectiveness, and many of the documents attempt to provide a methodology and guidance for the proper use of tarping (see Table 1).

3.1.1. Management Objectives and Environmental Context of Tarping Application

The literature review showed that tarping has been used on knotweeds to achieve different objectives. The most common objective is to create a physical barrier to plant growth and to prevent photosynthesis [50–55]. Regrowth (i.e., regenerating shoots) is blocked by the fabric, thus depleting the carbohydrate reserves stored in the rhizomes. A solarization effect is also frequently sought according to the literature. In this case, the fabric acts as a medium designed to expose knotweeds to high temperatures [55] that could "cook the root system" [50,52]. More rarely, tarping has been used to deprive the plant of water by covering it with an impermeable membrane, i.e., a geomembrane (Mireille Boyer, pers. comm.). These objectives, depending on the authors, can be pursued separately or in a complementary approach. In any case, the overall goal is to weaken the rhizomes as much as possible [32,50,53,55–57].

Typically, tarping is only recommended in specific environments or contexts, and for small knotweed stands, but recommendations vary according to country and author. The method is particularly recommended when the use of herbicides is prohibited [58] and is sometimes only considered in that case [32]. This is particularly noticeable in the USA. For example, we found about 10 cases of tarping in the State of Washington [56,59–61] that started in the 2000s after herbicide bans. However, these tarping operations stopped and were replaced by chemical treatments as early as 2010 after a change in the law reauthorized the use of herbicides [61]. In the Cedar River municipal watershed, tarping seems to be currently used only as a complement in case of regrowth on areas already treated with herbicides at least three times [61]. Tarping in the UK is a special case. Knotweed is very prevalent throughout the country and can detrimentally affect the price of real estate [41]. Tarping is mainly used in areas where building construction is planned in order to protect structures and hard surfaces: infested areas are sealed horizontally with a tarp before construction or a tarp is laid under the foundations to prevent the knotweed from entering the building [62]. Regarding the use of tarping on riverbanks, there appears to be no consensus among authors. For some, tarping is not suitable for steep or flood-prone sites [63]. Floods can accelerate the degradation of the tarp [48,64], tear off whole tarps or pieces, and disperse them downstream [45]. For others, the tarping method can be applied on a riverbank if suitable fixations are used [51] and if fabrics do not contain leachable chemicals that could pollute watercourses [62]. The method does seem interesting as it prevents stem or rhizome fragments from spreading downstream [53]. According to SPIGEST [55], tarping should be preferred in hard-to-reach and/or sloping areas, where it is difficult to intervene frequently with machinery or to graze animals. However, Guerin and Hedont [53] advise against its use on sites with an aesthetic vocation, where soil life is rich (because of the risks of soil sterilization) or where the site contains too many obstacles. There is a consensus that tarps should be used only on small knotweed stands, although the definition of a small stand varies among authors: "from 50 stems or less" [57] to less than 500 m^2 [55]. The complexity of tarp installation [53] and the cost of the fabrics [46] do not advocate for the use of this method on large surfaces. According to Soll [57], tarping use should focus on small populations in open, accessible terrain.

3.1.2. Methodological State of the Art

Preparatory Work

The first step in setting up a tarping operation, recommended by all authors, is to clear the area and prepare the ground prior to installing the fabric. This step usually consists of mowing down the knotweed stems. It can be completed with manual or mechanical excavation of the rhizomes [65–67], as digging up the rhizome crowns may speed up the control process [68]. Before installing the tarp, it is recommended to clean up the area as much as possible by removing all debris, stones, and shrubby vegetation to prevent the tarp from tearing or ripping [51,53]. Authors also frequently advise flattening the area to facilitate tarp installation [65,69]. A layer of mulch can also be spread over the area to prevent the cut stems from puncturing the tarp [58]. These operations must be carried out with great care to avoid spreading knotweed fragments [70].

Recommendations regarding the appropriate timing of these preparations vary among authors, but the reasons for these choices are not always stated. Anderson [50] and Godmaire and Houbart [52] advocate for an intervention in late spring, as does Cygan [58], who states that at this time of year, the root system is weakened because "the plant has exhausted stored carbohydrates". Others recommend action at the beginning of the year or in spring [32,55,57,71]; according to Branquart et al. [51], intervening during this period limits the volume of waste to be treated after mowing. Interestingly, even though preparation operations are usually carried out just before the tarp is laid, some may mow or excavate the knotweed stand for several years before laying the tarp in order to weaken the plant as much as possible [53,72].

Tarp Installation

The next step is to cover the infested area with one or several strips of tarps. Various types of covering fabric have been used in the literature from cardboard (in an unsuccessful experiment [57]) to the most sophisticated geomembrane [65]. However, geotextiles (typical in landscaping) and agricultural tarps are the most frequently recommended. Sometimes a combination of materials is used: agricultural tarps plus geotextiles to prevent holes [72] or thick cardboard plus heavy black plastic tarps [69]. The choice of the tarp is quite important since the strength and longevity of the fabric is a key factor for success [32]. According to Miller et al. [73], a black tarp is more effective than a clear one because it blocks the sunlight. Two [55] or three layers of tarp [72] are sometimes superimposed to improve resistance and efficiency. The Public Service of Wallonia [51] specifies that it is preferable to use non-woven fabrics with a density above 240 g/m^2.

Table 1. Summary of recommendations for the implementation of tarping and comments on its effectiveness. Here we selected the documents from our bibliographical review that we considered to be the most detailed on the tarping method.

Author	Duration of the Tarping	Size of the Stand to Control	Preparatory Work	Timing	Distance beyond the Stand Edge	Type of Tarp	Monitoring	Removal of the Tarp	Tarping Effectiveness
Soll, 2004	Throughout the growing season	Isolated and smaller patches on open terrain	Cutting (possibly followed by tilling)	Right at the beginning of the year or after cutting the plant a couple of times in the spring	At least 2 m (and preferably 7 m)	Thick black plastic or multiple layers of cardboard	At least for 2 years	n/a	No reliable reports of successful knotweed control with covering
McHugh, 2006	Throughout the growing season	Isolated and smaller patches on open terrain	Cutting (possibly followed by tilling)	Right at the beginning of the year or after cutting the plant a couple of times in the spring	At least 2 m (and preferably 7 m)	Thick black plastic or multiple layers of cardboard	Diligent monitoring	n/a	Promising results. Key to success seems to be the strength and the longevity of the geotextile fabrics. Method to consider where herbicide use is restricted
Hallworth and Sellentin, 2011	Several years	Small knotweed outbreaks (<0.001 hectares and <10 stems per m^2), small patches (e.g., 300 stems or less)	Cutting	n/a	2 up to 10 m	Woven, heavy grade geotextile or thick cardboard + heavy duty black plasti/poly tarp or geotextile fabric	Continously monitoring for several years	n/a	Not effective along stream banks
Anderson, 2012	More than one growing season (up to 3)	Small to medium (0.1 to 2 ha), dense infestations (more than 1000 plants or 30–100% cover)	Cutting	Late spring	Further out	Dark coloured tarp or heavy material (weed barriers or blue poly tarps)	Monitoring for regrowth, rips and tears in the tarp	n/a	This method may need to be used in conjunction with another method to ensure the entire patch is controlled
Godmaire and Houbart, 2014	At least 5 years	Medium (from 10 to 40 m^2) to large (up to 40 m^2)	Mowing or cutting	Late spring	A few meters	Dark coloured tarp (as geotextile, geomembrane or polyethylene tarp)	Regular monitoring	n/a	Not recommended in shaded areas
King County, 2015	At least 5 growing seasons (longer if soil is wet)	Reasonably small (50 stems or less), isolated patches	Cutting	At the beginning of the year or after cutting the plant down several times during the growing season	At least 2 m beyond	Geotextile or heavy duty black plastic	Intensive monitoring: every 2 to 4 weeks during the growing season	n/a	Moderately effective
Branquart et al., 2018	More than 5 years, up to 8 years	Small (less than 50 m^2, because of the high cost of the tarp)	Uprooting or herbicide injection or mowing	Winter or spring (to limit the biomass to be evacuated)	4 to 5 m	Large non-woven geotextile or tarp class 5 (tensile strength greater than 16 kN and density greater than or equal to 240 g/m^2)	Monitoring at least twice a year	Yes for exposed tarps; No for covered tarps	Mitigation method: very rapid weakening of the knotweed, may lead to its long-term elimination
Comité ZIP des Seigneuries et al., 2018	At least 8 years	n/a	Cutting and excavation (manually or with an excavator) up to 2 m beyond the stand	n/a	At least 2 m	Geomembrane: Texel 800 series or 45 mil ethylene-propylene-diene monomer (EPDM) membrane or Georoute 9	Regular monitoring over several years	Yes	n/a
Cygan, 2018	5 years	n/a	Cutting and spreading a layer of mulch, grass clippings or other material over the cut stems (to prevent them from puncturing the tarp)	June (when rhizomes are weakest)	A few feet beyond	7-mil Black plastic or non-woven geotextile or heavy-duty dark colored tarp	n/a	Yes	Very effective alternative if you wish to avoid the use of herbicides/very successful in sensitive areas
Guerin and Hedont, 2019	Several years, at least 10 years for some	Small (from a dozen to a few hundred m^2, as it is heavy to implement)	Uprooting or mowing	All seasons	1 to 2 m	New, black and opaque tarps. Geomembrane, agricultural tarp or biodegradable geotextile (non-woven)	Monitoring once or twice a year	Yes	One of the most appropriate methods for eradicating the outbreak of vegetatively propagated herbaceous plants
Lavoie, 2019	Up to 7 years	Tarps are too expensive to be laid out over large areas	n/a	n/a	n/a	Geomembrane or rugged geotextile	Regular monitoring	Yes	Popular but few efficacy tests. Mixed results
SPIGEST, 2019	At least 3 years	Small (less than 500 m^2)	Mowing in the winter before laying the tarp	Before vegetation regrowth	1 m	Black agricultural tarp (2 layers)	Regular monitoring: at least in the first years, once a month, from May to October.	Yes	Effective method for weakening rhizomial reserves before the establishment of plant competition

Due to knotweed growth dynamics and their capacity for lateral expansion, the tarp must be laid beyond the visible limits of the stand and carefully fixed to the ground. Recommendations regarding the necessary distance beyond the stand edge vary greatly from 1 m [55] to 10 m [69] while some recommendations are quite vague: a "few meters" [52] or "further out" [50]. Once laid, the tarp is fixed to the ground by various means. The tarp can simply be weighed down with heavy materials (rocks, logs, etc.) to prevent the wind from blowing it away and to prevent it from being lifted by the knotweed growth that almost always occurs in the first few years. Staples can also be used, either alone or as a complement to weighing the tarp down, but they may create weak points in the tarp through which the knotweed can sprout [58]. Tarp edges are sometimes buried vertically into a trench up to 1.5 m in depth [52] in order to hold the tarp in place and to create a vertical rhizome barrier. When several tarps are used, it is necessary to solidly bind them together (with glue, tape, staples, and/or heat sealing) to prevent sunlight from penetrating and to provide sufficient overlap (60 cm [58]; 30 cm [51]) so that knotweed shoots cannot grow through. The fabric is sometimes covered with soil or mulch for aesthetic reasons, but this also protects the fabric from UV rays [58] and helps hold the tarp to the ground. Finally, some guidelines recommend planting trees or shrubs of local species at the same time to create competition for the knotweed [74]. However, deliberately piercing the fabric for plantations creates weak points and is sometimes discouraged [51].

3.1.3. Post-Tarping Follow-ups

Monitoring

Frequent monitoring of the treated area is usually advised to ensure that the fabric or fixations are in a good state of repair, to remove any knotweed regrowth as the plant will likely continue to grow if it has any access to sunlight or air [50,65], and to stomp down re-growth under the tarp [55,71]. Consequently, monitoring recommendations range from twice a year [51] to as often as every two to four weeks during the growing season [54,71].

Tarping Duration and Fabric Removal

The literature usually advises removing the tarp after a certain period of time. However, the recommended duration varies greatly from only one growing season [32] to as much as 10 years [53]; the recommended duration seems to be increasing with time (Figure 2). A Norwegian study showed that rhizomes can survive for more than three years under tarps [49] and the current tendency is towards a longer tarping duration. The duration should also be adapted to the characteristics of the area or of the knotweed stand. For example, managers from King County recommend a longer duration if the soil is wet or the population large and well-established [71]. To ensure that the tarp is removed at the right time, it may be useful to open a test window in the tarp for at least one season to ensure that the knotweed has been eradicated before removing the fabric [53]. Although most authors advise removing the tarps [32,57], especially if they are made of synthetic or plastic materials [75], others do not, particularly when the tarp has been covered with soil or mulch [51].

3.1.4. Tarping Effectiveness

The effectiveness of the method is assessed differently depending on the authors. Some consider tarping to be a very effective alternative to herbicides [58], particularly suitable for the control of woody or clonal herbaceous plants [53]. In other (older) publications, there were no reports of successful long-term control with tarping alone [32,57]. Several operations in the USA were considered successful after five or six years of tarping but only on very small patches; on larger patches, despite eight years of continuous tarping, eradication was unsuccessful [56,61]. The scientific studies we reviewed reported mixed results, making it impossible to conclude on tarping effectiveness. While Jones et al. [31] considered geomembrane tarping to be the least effective technique for controlling knotweed (in comparison with herbicides and integrated physiochemical control treatments), Gerber et al. [48] showed that tarping reduced knotweed biomass but advised against using the technique in flooded environments. Kaczmarek-Derda et al. [49] showed that knotweed rhizomatous systems can survive more than three years and that longer tarping results in a substantial reduction of new shoots. Other than direct eradication, the method can also be used as a control technique [72], which, with regular maintenance, may lead to eradication in the long run [51,63], or at least be a first step in depleting the knotweed's rhizomatous reserves before reinforcing competition by restoring native plant communities [55]. Finally, some consider that the success mostly depends on the rigor with which the method is applied and that the tarping method "is only as good as the way in which the [tarp] has been laid" [62].

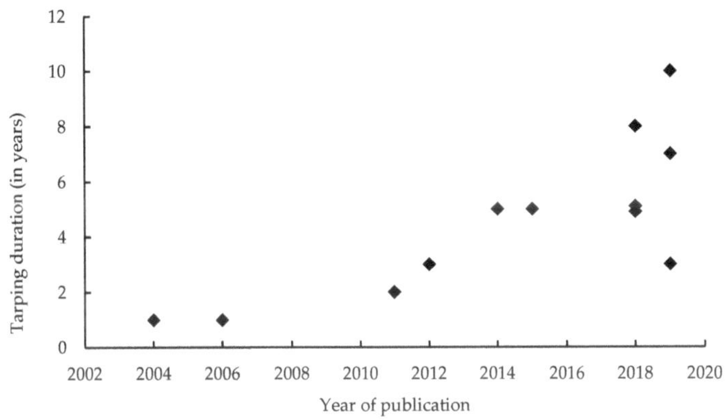

Figure 2. Changes in the recommended tarping duration according to the documents' publication year.

The disadvantages and limitations of the technique have also been discussed, for example, the tarp may have potential non-targeted impacts; the habitat value of tarp on the landscape; the bare soil resulting after the tarp's removal could become a seedbed for other IAS [32,73,76], and the plastic waste from non-biodegradable tarps is difficult to manage [72]; and some authors suggest that the rhizomes may become dormant for several years, rendering the control effort useless [51,52,62,69,77].

3.2. *Questionnaire Results*

We received 81 responses to our questionnaire (Supplementary File S3), 69 of which came from France, while the rest were from the USA (4), Canada (3), Belgium (2), Germany (1), the Netherlands (1), and Slovenia (1).

The respondents worked mainly for local authorities (34%) and watershed management organizations (32%). Most respondents felt that they had a good level of knowledge of knotweeds, with a mean score of 6.42 on a scale from 0 (novice) to 10 (expert). The identified tarping operations were launched between 2003 and 2020 (Figure 3). Three quarters of the operations started more than three years ago.

3.2.1. Management Objectives and Environmental Context of Tarping Application

The reported tarping is mainly being used for the sake of comparison with other methods (53%), and most operations are experimental trials. Only 15% of the respondents had had a previously successful experience with tarping. Bibliographical research (38%) and technical or regulatory constraints (42%) also explained the choice of tarping as a control method. Through this method, the managers' main goal was to eradicate knotweed (69%), to limit knotweed dispersal on the site (54%) and/or to other areas (40%—several answers were possible).

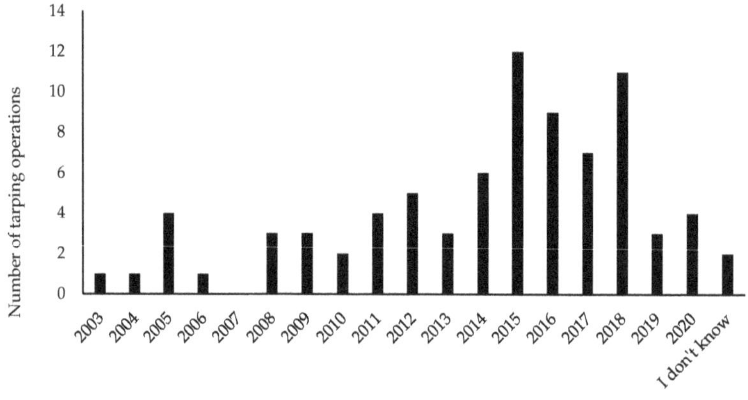

Figure 3. Number of tarping operations identified from the questionnaire according to the starting date of the operation, for a total of 81 tarping operations.

Tarping was mostly used on riverbanks (47%) and on roadsides (37%). Overall, tarped sites were typically easy to access, sunny, sparsely wooded, and rarely flooded (although 21% were flooded regularly). Operations were apparently set up regardless of slope steepness and focused mainly on small- to medium-sized knotweed stands (Figure 4): half of the tarping trials were carried out on stands < 90 m^2 and 75% on stands < 230 m^2 (still, the largest tarped stand covered almost 5000 m^2).

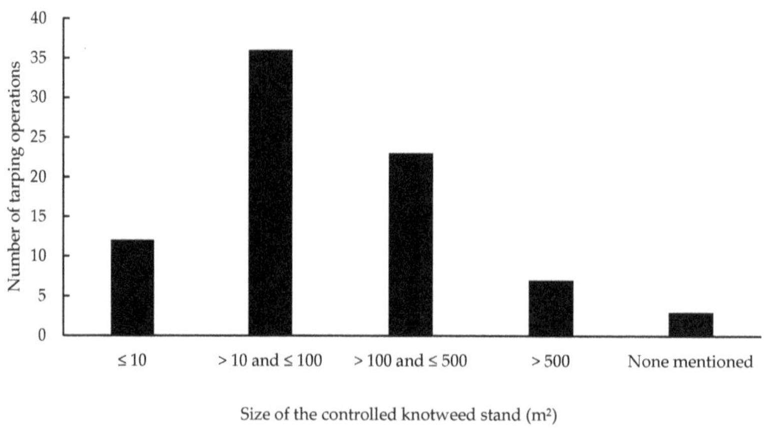

Figure 4. Size of the tarped knotweed stands based on respondents' answers, for a total of 81 tarping operations.

3.2.2. Methodological State-of-the Art

Preparatory Work

In 84% of the cases, managers performed some kind of preparation on the site prior to tarping. Most respondents (56%) applied more than one of the following preparatory measures: mowing (60%), manually or mechanically extracting rhizomes (29%), flattening the area with an excavator (19%), and removing stones (19%). Respondents considered preparation an important step, with an average mark of 7.6 on a scale from 0 (useless) to 10 (essential). Preparation work almost always took place just before the tarp was laid, regardless of the season, since the preparation date was chosen mainly because of organizational constraints (70%). Still, the knotweed biological cycle was also frequently taken into account, as 41% of the respondents favored action before or at the beginning of knotweed emergence in order to limit the amount of waste to treat, 16% wanted to intervene when the rhizomes were the weakest, and 12% intervened before flowering to prevent the formation of seeds.

Tarp Installation

Geotextile was the preferred material for tarping operations (43%). Agricultural and other geomembranes were also used (30% and 22%, respectively). Only 10% of the listed operations were carried out with biodegradable geotextiles. In 78% of the cases, the knotweed stands were entirely covered. When the stands were not entirely covered, it was mainly due to technical constraints, though some managers did so on purpose to test the method on a small plot before considering a larger project. Interestingly, 71% of the managers decided to extend the tarped area beyond the initial stand's edges to cover a buffer zone of at least one meter.

Around three quarters of the projects required the use of several strips of fabric. Various techniques were used to bind the strips together, but staples were most frequently used (50%). On average, the staples were placed every 67 cm (min 6 cm; max 2 m) on strips overlapping by 47 cm on average (min 1 cm; max 1.5 m). Adhesive tape, weighing the fabric down with heavy objects, heat sealing, and glue were also sometimes used, either alone or in conjunction with staples. The tarps were anchored to the ground with staples (52% of the cases, typically a concrete reinforcing bar was pinned down every 71 cm on average) and/or by burying the edges of the sheet in a trench (54%). Fixation was sometimes enhanced by covering the fabric with various materials like gravel or topsoil (37% of the cases, with an average covering layer of 20 cm) or by ballasting with heavy objects (31%). When trenches were dug to fix the edges of the tarp to the ground, they were on average 35 cm deep (with a minimum of 10 cm and a maximum of 1.2 m). The average depth was higher (58 cm with a minimum of 10 cm and a maximum of 2 m) when the objective was also to create a vertical rhizome barrier.

Covered areas usually had few obstacles and 64% of the tarps were installed without any unintentional damage. The few holes identified were mainly due to rocks, rubble, rebar, and branches (45%) present in the ground. Installation was considered moderately difficult, but 80% of the respondents said it was a crucial step in the process. Interestingly, 30% of the respondents carried out planting within the tarped area, either by seeding or by planting cuttings and/or transplants. These managers aimed to create competition between the planted vegetation and the knotweed (67%), to improve the aesthetic aspect of the site (50%) and/or to avoid erosion (38%).

3.2.3. Post Tarping Follow-up

Monitoring

After the tarping was installed, the sites were almost always monitored at least once a year (98%), while many managers monitored 3 times a year or more (52%). Monitoring was generally more frequent during the vegetation period in the first few years after installation. While some limited this monitoring to simple visual inspections, most respondents uprooted the knotweed regrowth (66%) and repaired the fabric or fixations (38%).

Tarping Duration and Tarp Removal

More than two-thirds (69%) of the respondents planned the tarping duration at the beginning of their project. This planned duration ranged from one to 20 years, with most projects planned to last three years or more (69% of the cases). This duration was based on bibliographical research (35%), peer recommendation (32%), and/or personal experience (21%). For those who did not plan a tarping duration, it was either because they intended to leave the tarp in place or because they wanted to determine the duration according to the results obtained over time. Of all respondents, only 27% had removed their tarps;

another 23% expressed their intention to do so when the planned duration had been reached, 33% intended to leave the tarp in place, and finally another 5% said they would leave it because it was too degraded to be removed. Whatever the option planned at the end of the tarping operation, it is mainly synthetic tarps that are used: by 91% of those who had removed their tarps, by 87% of those who expressed their intention to do so, and by 72% of those who intended to leave the tarp in place.

For the operations where the tarps had already been removed (n = 22), in 13 cases, removal was carried out after three years or more, but in seven cases, the managers removed the fabric after only one or two years. Following removal, additional management was performed in 15 cases, usually to restore vegetation cover or to mow down the remaining knotweed shoots.

3.2.4. Tarping Effectiveness

Regrowth of knotweed stems was observed in 78% of the cases, mostly during the first six months after tarp installation (71%). The regrowth appeared mainly within the covered area (54%) and/or in the immediate vicinity (within 2 m of the covered area, 46%). Fragile points in the tarp seem to have been exploited by the knotweed, particularly near obstacles, at holes or where the strips overlapped. The tarping system (the tarps themselves or their anchoring system) deteriorated in 43% of the operations. This degradation was observed shortly after the installation in more than 50% of the cases. Furthermore, 9% of the tarping trials were utterly abandoned.

Of the 22 respondents who proceeded to the removal of the tarp, six said they had successfully eradicated knotweed from the area. Seven managers noted that there were no knotweed plants at the time of the removal, while 10 observed some regrowth under the previously tarped area, and 11 in the immediate vicinity of the previously tarped area.

The 22 respondents who had removed their tarps were able to provide an assessment of the method's effectiveness on a scale of 0 (ineffective) to 10 (very effective), first at the time of the removal and second, at the time of the managers' final observation. This final observation was mostly (54%) made within the months or the year following the tarp's removal. The mean score at the removal was 5.7 (±3.87 standard deviation) and 5.6 (±3.74 standard deviation) at the final observation. Effectiveness ratings at the two dates differed little, though the effectiveness assessment at removal was typically confirmed at final observation. For example, seven respondents deemed the method effective (score of 9 or 10) at the time of removal and nine deemed it effective at their final observation. Conversely, three managers considered the method to be totally ineffective (score of 0 or 1) at the time of removal and five managers gave the lowest score at the time of their final observation.

The respondents considered that tarping was slightly more time-consuming than other knotweed control methods. There was no consensus regarding the cost of tarping, but, on average, they felt that the cost was comparable to other methods. The average cost of tarping in our survey was 34 €/m^2 though it varied greatly, from 1 €/m^2 to 167 €/m^2. However, cost estimation methods differed among managers (monitoring and maintenance were not always included, some added the purchase or hire of equipment needed for installation, others used volunteer labor, etc.), thus partly explaining these differences. However, costs also seem to depend on the type of fabric used. The average cost of an operation with an agricultural tarp (n = 14) was the lowest at 26 €/m^2, compared to 30 €/m^2 for an operation with geotextile (n = 19) and 60 €/m^2 with a geomembrane (n = 9). Finally, 52% of the respondents intended to do new tarping trials at other sites.

3.3. *Consistency of the Results of the Bibliography and the Questionnaire*

3.3.1. Management Objectives and Environmental Context of Tarping Application

The number of responses to our questionnaire and the amount of grey literature found on the subject of tarping shows that the technique is frequently used both in France and abroad, despite a lack of scientific data. Through the results of the questionnaire, we observed an increase in the number of tarping operations over the last few years, suggesting that the method is becoming increasingly popular. The development of more advanced geotextiles specifically dedicated to the control of invasive rhizomatous plants could partially explain this trend. However, these results may also reflect the difficulty of obtaining reliable information about older experiments (e.g., absence of publications, unavailable data).

The method is frequently chosen for comparison with other methods. This confirms that in the 'war' against knotweeds, managers are still struggling to find an effective way to eradicate them. While many authors reckon that eradicating knotweed at the large scale is impossible [10,19,78], 69% of the respondents were still interested in locally eradicating knotweed with tarping. Furthermore, eradication is not necessarily the managers' only goal [79].

Contrary to what is recommended in the literature, managers seem to favor tarping in fairly accessible environments over a wide gradient of slope steepness. Tarping is particularly used on roadsides and riverbanks, even though its use along rivers is controversial in the literature. Theory and practice do agree, however, on the size of the knotweed stands to be controlled. Tarping is mainly practiced on small- to medium-sized stands (75% of the experiments reported through the questionnaire were carried out on stands smaller than 230 m^2), although we found no detailed justifications for this choice. It is unclear, however, whether managers assume tarping's effectiveness decreases with increasing knotweed stand size or, as stated by Guerin and Hedon [53] and Lavoie [46], whether the complexity and/or cost of the method prevent it from being applied on larger surfaces.

3.3.2. On Assessing Tarping Effectiveness

It emerges from the review and the questionnaire that properly assessing tarping's effectiveness is challenging as tarping is implemented in various management and environmental contexts, and the technical characteristics of tarping operations can be extremely diverse (in terms of fabric used, fixing devices, overlap, area covered, etc.), making comparisons precarious at best.

In the literature, the effectiveness of the method is assessed differently depending on the authors and the few scientific studies available reported mixed results.

In our questionnaire, tarping effectiveness was measured through two criteria: the presence of knotweed regrowth after tarping, and an evaluation of the method carried out by the managers themselves. In the first case, the managers were asked whether regrowth had been observed during the tarping period and if so, where (within the tarped area, in its immediate vicinity, and/or in the surrounding area). Seventy-eight percent of the respondents reported regrowth within and/or around the tarped area during the first year after installation. However, it would be premature to deduce the ineffectiveness of the method merely from this information. The data collected did not account for the quantity of regrowth or its trend over time. We know that the rhizomes do not die even with three years of tarping [49], so it is unsurprising that regrowth will persist during the first years. This regrowth does not necessarily invalidate the long-term effectiveness of the tarping operation: regular monitoring, uprooting knotweed regrowth, and repairing the tarp if necessary can still lead to eradication of the knotweed stand. In the second case, the answer rate for the evaluations of the method was low because only managers whose operation had ended and whose tarps had been removed were asked to answer this question. This choice was partly based on literature recommendations and on the idea that the effectiveness of a control operation cannot be properly evaluated when the said operation is still ongoing. However, it turned out that most managers (73%) had not yet removed their tarps or even do not intend to remove them, making their tarping operations, in our sense, endless. Consequently, only 22 respondents gave an evaluation of the method, and their evaluations were, as expected, highly contrasted. If the mean score at the removal was 5.7/10 (+/− 3.87 standard deviation), six managers out of 22 reported the eradication of targeted knotweed stands. Although infrequent, such results indicate that tarping can, in some instances, be objectively "effective". Research efforts should next focus on increasing the number of evaluated operations and elucidating which implementations (i.e., set-ups) or contexts explain the success or failure of tarping operations. Effectiveness assessments should also be based on more quantitative criteria (e.g., density of stem regrowth per square meter) than a simple evaluation by the managers themselves, which remains very subjective and could be influenced by the effort made, the cost of the method, and the manager's perception and knowledge of knotweed's biology [19]. Ideally, rhizomes should also be sampled prior to the removal of the tarp in order to assess their vitality and regenerative capacity.

Prolonged knotweed rhizome dormancy is frequently mentioned as a limiting factor to tarping's effectiveness [31,48,51,52,62,69,77,80–83], yet we were unable to find any scientific basis for that statement. Prolonged dormancy (also known as "vegetative dormancy") is a stage in which mature plants remain belowground during one or more growing seasons instead of emerging to grow and acquire resources [84]. This strategy can be a response to a stressful above-ground environment caused by drought, herbivory, etc., [85,86] and/or a consequence of a lack of stored resources [87,88]. Despite the fact that it is a relatively common phenomenon, prolonged dormancy has never been reported in scientific studies on knotweeds. The reference to knotweed rhizome dormancy seems to have originated in "The Knotweed Code of Practice", published in 2006 by the Environment Agency [62]. In this publication, which itself refers to "unconfirmed observations", it is stated that the repeated use of herbicides could induce a rhizome dormancy lasting for up to 20 years. Tarping was therefore not originally linked to such observations. Still, by preventing photosynthesis and depriving the plant of water (if an impermeable geomembrane is used), tarping could certainly create the stressful conditions required to induce prolonged dormancy. However, to date, none of the 81 respondents to our questionnaire have reported such a phenomenon and, although it cannot be totally excluded, prolonged dormancy cannot be considered as a limiting factor for tarping's effectiveness on knotweed.

4. Recommendations

The results from our review and questionnaire revealed a great heterogeneity among knotweed tarping practices (e.g., in terms of context, duration, covered area, chosen materials), sometimes reflecting insufficient knowledge of knotweed's biology and ecology (in particular, on clonal integration and on the expansion capacities of rhizomes). Such variability explains, to a large extent, the difficulty to find comparable operations and the mixed results reported in the literature. Nonetheless, the synthesis presented in this article enable us to propose some practical recommendations to managers.

4.1. Cover the Entire Stand

In a number of operations, not all of the knotweed stand was covered. Yet, since knotweeds share resources through clonal integration [26,89], the uncovered part of the plant may compensate for the lack of photosynthesis within the tarped part, as was shown for partial mowing [37]. Therefore, we feel that it is particularly important to implement a tarping operation only when covering the entire stand is feasible, except perhaps in the case (mentioned by a few respondents) where the aim is not to eradicate the knotweed population but to prevent its growth in a specific area (e.g., when the stand impedes visibility or access).

4.2. Extend up to 2.5 m

Although doubts remain on the horizontal distance over which knotweed rhizomes can grow, we know that the rhizomatous system can extend several meters beyond the edge of the visible stand [30,41]. Yet, 55% of the respondents only covered the ground 1 m beyond the edge of the stands. According to the latest study by Fennell et al. [41], covering at least 2.5 m, and preferably up to 4 m, beyond the edge of the stand is recommended. Similarly, while rhizomes are found mostly in the first 50 cm of the soil, they can reach a depth 3 m depending on the type of soil, the site characteristics, and the presence of obstacles to circumvent [41]. A quarter of the respondents tried to create an anti-rhizome barrier by burying the edge of the tarp in a trench, but the average depth of this trench was only 58 cm (with trench depth ranging from 10 cm to 2 m). Given the depth sometimes reached by knotweed rhizomes, it may be necessary to install tarps deeper in the soil to achieve a more efficient vertical barrier effect, especially if the distance covered beyond the stand is small.

4.3. *Maintain the Tarping for at Least Six Years*

The recommended tarping duration has steadily increased over time, reflecting a lack of hindsight and empirical development of the method. While it is now quite certain that a short tarping duration cannot kill knotweed rhizomes [49], the most appropriate tarping duration is not yet known. Experiments carried out in the USA over a period of five to eight years have given mixed results [56,61]. The required duration could also depend on the context, the initial vigor of the patch, and its age (in relation to the amount of biomass it has stored underground) [45]. Based on current knowledge, we believe that it is essential to maintain the tarping over at least six years. However, such a long duration is not necessarily adapted to all managers' objectives and means (financial and human).

4.4. *Use a Durable Synthetic Tarp*

The choice of the fabric should also reflect rhizome lifespan and the duration of the tarping project. We feel it is essential to choose a tarp for its durability, whether it is an agricultural tarp, a geomembrane, or a geotextile. In this respect, we question the relevance of using biodegradable tarps. Their lifespan does not seem sufficient for a successful tarping operation. Therefore, despite the potential environmental impact of synthetic tarps, this choice seems to be the most relevant to truly impede knotweed populations.

4.5. *Ensure Long-Term Management*

Though tarping appears to be a passive technique [46], it is not. Regular monitoring must be carried out to find and repair possible degradations, regrowth must be crushed down under the tarp to prevent it from lifting and piercing the fabric, and sprouts that may have grown through the fabric or in its immediate vicinity must be uprooted. Since most observed deteriorations of the tarps occurred in the year following their installation, regular monitoring makes it possible to detect and fix most problems in due time.

Tarping also requires costly work to remove the tarp at the end of the operation. We found an important discrepancy between theory and practice regarding tarp removal. While guidelines advise removing tarps after a certain period of time, 42% of respondents left the tarp in place at the end of the treatment, or intend to do so. In practice, tarp removal can be difficult because of tarp deterioration (too much fragmentation), the presence of planted vegetation, layers of soil, or simply because of the cost of the operation.

In the literature, it is strongly recommended to revegetate the site after tarp removal [50,51,58,65] since tarping is a non-selective method that makes bare soil prone to invasion by other IAS. Although competition and allelopathy can limit knotweed development [90,91], we recommend revegetation only after tarp removal (unless the aim is to control rather than eradicate the knotweed) as the holes made in the tarp for plantations enable the regrowth of knotweed stems and may limit the effectiveness of the tarping. Though revegetation of the site is obligatory at some point, tarping can be the first step in a more global restoration strategy, implying a long-term investment.

5. Conclusions

Through a review and the feedback collected via a questionnaire, we have shown that the available literature on tarping is mainly grey and that there is a great heterogeneity of practices in the field. This reflects the empirical development of the method and demonstrates the need to improve knowledge on tarping as a method of control for Asian knotweed populations.

For the first time, our article synthesized the scattered and often incomplete information available from the literature and compared it with field practices. This important work of descriptive analysis, by making the first state of the art of the tarping, was an essential prerequisite to improving the method.

Although some tarping operations have been shown to successfully eradicate Asian knotweed in small stands, the bibliography and the results of the survey were insufficient to demonstrate the effectiveness of tarping. More data and further analyses, as well as research and experimentation on the functioning (in particular, the life span) of rhizomes, would be useful. The question of the long-term environmental impact of tarping also deserves further consideration: while it is sometimes presented as an alternative to herbicides, the tarping method, especially with synthetic tarps, is not without consequences for the environment.

However, our article highlighted some important aspects to consider to obtain favorable results and proposed practical recommendations for managers to improve the method. Even though tarping does not seem to be a one-size-fits-all solution to eradicate knotweed, it is still a useful control method once knotweed has become a critical management issue.

Supplementary Materials: The following are available online at https://www.mdpi.com/article/10.3390/plants10102152/s1, Supplementary File S1: Survey on the management of knotweeds through tarping, Supplementary File S2: List of references resulting from the bibliographical research, Supplementary File S3: List of tarping operations identified through the questionnaire and their main characteristics.

Author Contributions: Conceptualization, investigation and writing—original draft preparation, M.-A.D., F.-M.M., F.D. and A.E.; writing—review and editing, M.-A.D., F.-M.M., F.D., A.P., C.D.-M. and A.E. All authors have read and agreed to the published version of the manuscript.

Funding: This research received no external funding.

Institutional Review Board Statement: Not applicable.

Informed Consent Statement: Our research project was conducted in compliance with the current regulations applicable to the processing of personal data and, in particular, Regulation (EU) 2016/679 of the European Parliament and of the Council of 27 April 2016 on the protection of natural persons with regard to the processing of personal data and on the free movement of such data (General Data Protection Regulation).

Data Availability Statement: Not applicable.

Acknowledgments: We thank the SNCF Réseau and INRAE for their financial support. We are also grateful to Marylise Cottet for her feedback on the questionnaire, and to all the managers who took the time to complete the questionnaire.

Conflicts of Interest: The authors declare no conflict of interest.

References

1. Kueffer, C. Plant invasions in the Anthropocene. *Science* **2017**, *358*, 724–725. [CrossRef] [PubMed]
2. Vilà, M.; Espinar, J.L.; Hejda, M.; Hulme, P.E.; Jarošík, V.; Maron, J.L.; Pergl, J.; Schaffner, U.; Sun, Y.; Pyšek, P. Ecological impacts of invasive alien plants: A meta-analysis of their effects on species, communities and ecosystems. *Ecol. Lett.* **2011**, *14*, 702–708. [CrossRef]
3. IPBES. *Summary for Policymakers of the Global Assessment Report on Biodiversity and Ecosystem Services of the Intergovernmental Science-Policy Platform on Biodiversity and Ecosystem Services*; IPBES Secretariat: Bonn, Germany, 2019.

4. ISSG. Invasive Species Specialist Group Gateway—UICN. Available online: http://www.issg.org/index.html (accessed on 3 August 2020).
5. Diagne, C.; Leroy, B.; Vaissière, A.-C.; Gozlan, R.E.; Roiz, D.; Jarić, I.; Salles, J.-M.; Bradshaw, C.J.A.; Courchamp, F. High and rising economic costs of biological invasions worldwide. *Nat. Cell Biol.* **2021**, *592*, 571–576. [CrossRef]
6. Kettunen, M.; Genovesi, P.; Gollasch, S.; Pagad, S.; Starfinger, U. *Technical Support to EU Strategy on Invasive Species (IAS)—Assessment of the Impacts of IAS in Europe and the EU (Final Module Report for the European Commission)*; Institute for European Environmental Policy (IEEP): Brussels, Belgium, 2008.
7. Wittmann, A.; Flores-Ferrer, A. *Analyse Économique Des Espèces Exotiques Envahissantes en France*; Etudes et documents du du Commissariat Général au Développement Durable (CGDD); Ministère de l'Écologie, Service de L'Économie, de L'Évaluation et de L'Intégration du Développement Durable: Paris, France, 2015.
8. Andreu, J.; Vilà, M.; Hulme, P.E. An assessment of stakeholder perceptions and management of noxious alien plants in Spain. *Environ. Manag.* **2009**, *43*, 1244–1255. [CrossRef]
9. Beaury, E.M.; Fusco, E.J.; Jackson, M.R.; Laginhas, B.B.; Morelli, T.L.; Allen, J.; Pasquarella, V.J.; Bradley, B.A. Incorporating climate change into invasive species management: Insights from managers. *Biol. Invasions* **2019**, *22*, 233–252. [CrossRef]
10. Delbart, E.; Mahy, G.; Weickmans, B.; Henriet, F.; Crémer, S.; Pieret, N.; Vanderhoeven, S.; Monty, A. Can land managers control Japanese knotweed? Lessons from control tests in Belgium. *Environ. Manag.* **2012**, *50*, 1089–1097. [CrossRef]
11. Pluess, T.; Cannon, R.J.C.; Jarošík, V.; Pergl, J.; Pyšek, P.; Bacher, S. When are eradication campaigns successful? A test of common assumptions. *Biol. Invasions* **2012**, *14*, 1365–1378. [CrossRef]
12. Matzek, V.; Covino, J.; Funk, J.L.; Saunders, M. Closing the knowing–doing gap in invasive plant management: Accessibility and interdisciplinarity of scientific research. *Conserv. Lett.* **2013**, *7*, 208–215. [CrossRef]
13. Bayliss, H.R.; Wilcox, A.; Stewart, G.B.; Randall, N.P. Does research information meet the needs of stakeholders? Exploring evidence selection in the global management of invasive species. *Evid. Policy A J. Res. Debate Pr.* **2012**, *8*, 37–56. [CrossRef]
14. Matzek, V.; Pujalet, M.; Cresci, S. What managers want from invasive species research versus what they get. *Conserv. Lett.* **2014**, *8*, 33–40. [CrossRef]
15. Cottet, M.; Piola, F.; Le Lay, Y.-F.; Rouifed, S.; Rivière-Honegger, A. How environmental managers perceive and approach the issue of invasive species: The case of Japanese knotweed s.l. (Rhône River, France). *Biol. Invasions* **2015**, *17*, 3433–3453. [CrossRef]
16. Robinson, B.S.; Inger, R.; Crowley, S.L.; Gaston, K.J. Weeds on the web: Conflicting management advice about an invasive non-native plant. *J. Appl. Ecol.* **2016**, *54*, 178–187. [CrossRef]
17. Kuebbing, S.E.; Simberloff, D. Missing the bandwagon: Nonnative species impacts still concern managers. *NeoBiota* **2015**, *25*, 73–86. [CrossRef]
18. Liu, S.; Sheppard, A.; Kriticos, D.; Cook, D. Incorporating uncertainty and social values in managing invasive alien species: A deliberative multi-criteria evaluation approach. *Biol. Invasions* **2011**, *13*, 2323–2337. [CrossRef]
19. Cottet, M.; Rivière-Honegger, A.; Vaudor, L.; Colombain, L.; Dommanget, F.; Evette, A. The end of a myth: Solving the knotweeds invasion "problem". *Anthropocene* **2020**, *30*, 100240. [CrossRef]
20. Lowe, S.; Browne, M.; Boudjelas, S.; De Poorter, M. *100 of the World's Worst Invasive Alien Species: A Selection from the Global Invasive Species Database*; The Invasive Species Specialist Group, Species Survival Commission of the IUCN: Auckland, New Zealand, 2000.
21. Bailey, J.P.; Bímová, K.; Mandák, B. Asexual spread versus sexual reproduction and evolution in Japanese knotweed s.l. sets the stage for the "Battle of the Clones". *Biol. Invasions* **2008**, *11*, 1189–1203. [CrossRef]
22. Barney, J.N.; Tharayil, N.; DiTommaso, A.; Bhowmik, P.C. The biology of invasive alien plants in Canada. 5. *Polygonum cuspidatum* Sieb. & Zucc. [= *Fallopia japonica* (houtt.) Ronse Decr.]. *Can. J. Plant Sci.* **2006**, *86*, 887–906. [CrossRef]
23. Martin, F.-M. The Study of the Spatial Dynamics of Asian Knotweeds (*Reynoutria* spp.) across Scales and Its Contribution for Management Improvement. Ph.D. Thesis, Université Grenoble Alpes: Grenoble, France, 2019.
24. Lavoie, C. The impact of invasive knotweed species (*Reynoutria* spp.) on the environment: Review and research perspectives. *Biol. Invasions* **2017**, *19*, 2319–2337. [CrossRef]
25. Staentzel, C.; Rouifed, S.; Beisel, J.-N.; Hardion, L.; Poulin, N.; Combroux, I. Ecological implications of the replacement of native plant species in riparian systems: Unexpected effects of *Reynoutria japonica* Houtt. leaf litter. *Biol. Invasions* **2020**, *22*, 1917–1930. [CrossRef]
26. Aguilera, A.G.; Alpert, P.; Dukes, J.; Harrington, R. Impacts of the invasive plant *Fallopia japonica* (Houtt.) on plant communities and ecosystem processes. *Biol. Invasions* **2010**, *12*, 1243–1252. [CrossRef]
27. Dassonville, N.; Vanderhoeven, S.; Gruber, W.; Meerts, P. Invasion by *Fallopia japonica* increases topsoil mineral nutrient concentrations. *Ecoscience* **2007**, *14*, 230–240. [CrossRef]
28. Gerber, E.; Krebs, C.; Murrell, C.; Moretti, M.; Rocklin, R.; Schaffner, U. Exotic invasive knotweeds (*Fallopia* spp.) negatively affect native plant and invertebrate assemblages in European riparian habitats. *Biol. Conserv.* **2008**, *141*, 646–654. [CrossRef]
29. Wilson, M.J.; Freundlich, A.E.; Martine, C.T. Understory dominance and the new climax: Impacts of Japanese knotweed (*Fallopia japonica*) invasion on native plant diversity and recruitment in a riparian woodland. *Biodivers. Data J.* **2017**, *5*, e20577. [CrossRef]
30. Child, L.; Wade, M. *The Japanese Knotweed Manual: The Management and Control of an Invasive Alien Weed*; Packard Publishing Ltd.: Chichester, UK, 2000.

31. Jones, D.; Bruce, G.; Fowler, M.S.; Law-Cooper, R.; Graham, I.; Abel, A.; Street-Perrott, F.A.; Eastwood, D. Optimising physio-chemical control of invasive Japanese knotweed. *Biol. Invasions* **2018**, *20*, 2091–2105. [CrossRef]
32. McHugh, J.M. *A Review of Literature and Field Practices Focused on the Management and Control of Invasive Knotweed*; The Nature Conservancy: West Haven, CT, USA, 2006.
33. Rouifed, S.; Cottet, M.; de Battista, M.; Le Lay, Y.-F.; Rateau, P.; Rivière-Honegger, A.; Piola, F. Inefficiency of cutting stems once during the vegetative growth of *Fallopia* spp. *Manag. Biol. Invasions* **2020**, *11*, 399–405. [CrossRef]
34. Kabat, T.J.; Stewart, G.B.; Pullin, A.S. *Are Japanese Knotweed (Fallopia japonica) Control and Eradication Interventions Effective?* CEE review 05-015 (SR21); Collaboration for Environmental Evidence: Birmingham, UK, 2006.
35. Bashtanova, U.B.; Beckett, K.P.; Flowers, T. Review: Physiological approaches to the improvement of chemical control of Japanese knotweed (*Fallopia japonica*). *Weed Sci.* **2009**, *57*, 584–592. [CrossRef]
36. Beerling, D.J.; Palmer, J.P. Status of *Fallopia japonica* (Japanese knotweed) in Wales. In *Ecology and Management of Invasive Riverside Plants*; John Wiley and Sons: Chichester, UK, 1994; pp. 199–212.
37. Martin, F.-M.; Dommanget, F.; Lavallée, F.; Evette, A. Clonal growth strategies of *Reynoutria japonica* in response to light, shade, and mowing, and perspectives for management. *NeoBiota* **2020**, *56*, 89–110. [CrossRef]
38. Meier, E.S.; Dullinger, S.; Zimmermann, N.E.; Baumgartner, D.; Gattringer, A.; Hülber, K. Space matters when defining effective management for invasive plants. *Divers. Distrib.* **2014**, *20*, 1029–1043. [CrossRef]
39. Price, E.A.C.; Gamble, R.; Williams, G.G.; Marshall, C. Seasonal patterns of partitioning and remobilization of 14C in the invasive rhizomatous perennial Japanese knotweed (*Fallopia japonica* (houtt.) Ronse decraene). In *Ecology and Evolutionary Biology of Clonal Plants*; Stuefer, J.F., Erschbamer, B., Huber, H., Suzuki, J.-I., Eds.; Springer: Dordrecht, The Netherlands, 2002; pp. 125–140. ISBN 978-94-017-1345-0.
40. Suzuki, J.-I. Growth dynamics of shoot height and foliage structure of a Rhizomatous perennial herb, *Polygonum cuspidatum*. *Ann. Bot.* **1994**, *73*, 629–638. [CrossRef]
41. Fennell, M.; Wade, M.; Bacon, K.L. Japanese knotweed (*Fallopia japonica*): An analysis of capacity to cause structural damage (compared to other plants) and typical rhizome extension. *PeerJ* **2018**, *6*, e5246. [CrossRef]
42. Stapleton, J. Soil solarization in various agricultural production systems. *Crop. Prot.* **2000**, *19*, 837–841. [CrossRef]
43. Linke, K.-H. Effects of soil solarization on arable weeds under Mediterranean conditions: Control, lack of response or stimulation. *Crop. Prot.* **1994**, *13*, 115–120. [CrossRef]
44. Al-Shammary, A.A.G.; Kouzani, A.; Gyasi-Agyei, Y.; Gates, W.; Rodrigo-Comino, J. Effects of solarisation on soil thermal-physical properties under different soil treatments: A review. *Geoderma* **2020**, *363*, 114137. [CrossRef]
45. Evette, A.; Breton, V.; Petit, A.; Dechaume-Moncharmont, C.; Brasier, W. Les techniques de bâchage pour le contrôle de la renouée. *Revue Science Eaux Territoires* **2019**, *27*, 62–67. [CrossRef]
46. Lavoie, C. *50 Plantes Envahissantes: Protéger la Nature et L'agriculture*; Publications du Quebec: Québec, QC, Canada, 2019; ISBN 978-2-551-26390-5.
47. Boyer, M. Une nouvelle technique d'éradication mécanique des renouées du japon testée avec succès au bord de l'ain et de l'isère. *Revue Science Eaux Territoires* **2009**, *57–58*, 17–31.
48. Gerber, E.; Murrell, C.; Krebs, C.; Bilat, J.; Schaffner, U. *Evaluating Non-Chemical Management Methods against Invasive Exotic Knotweeds, Fallopia spp.*; CABI: Egham, UK, 2010.
49. Kaczmarek-Derda, W.; Holm, A.-K.; Brandsæter, L.; Solhaug, K.; Fløistad, I. Survival Time of Rhizomes of Invasive Reynoutria Taxa when Above-Ground Shoot Production is Prevented by Covering with Geotextile. In Proceedings of the 15th International Conference on Ecology and Management of Alien Plant Invasions, Prague, Czech Republic, 9–13 September 2019.
50. Anderson, H. *Invasive Japanese knotweed (Fallopia japonica (houtt.) Best Management Practices in Ontario*; Ontario Invasive Plant Council: Peterborough, ON, Canada, 2012.
51. Branquart, E.; Trojan, M.; Vanparys, V. *La Gestion des Renouées Asiatiques. Technique d'atténuation. Gestion Par Bâchage*; Service Public de Wallonie, Cellule Interdépartementale Espèces invasives: Gembloux, Belgium, 2018.
52. Godmaire, H.; Houbart, C. *Renouée du Japon Guide Technique de Contrôle Mécanique*; Ville de Granby Conseil Québécois des Espèces Exotiques Envahissantes; CQEEE Fondation pour la Sauvegarde des Écosystèmes du Territoire de la Haute-Yamaska; Fondation SÉTHY: Granby, CO, USA, 2014.
53. Guerin, M.; Hedont, M. *Plantes Envahissantes: Pratiques des Gestionnaires d'Espaces Verts. Recueil d'Expériences*; Plante & Cité: Angers, France, 2019.
54. LUPIN (Lutte Contre les Plantes Invasives). *Lupin Se Mobilise Pour La Biodiversité, Lutte Contre Les Plantes Invasives*; GDON Flandre Maritime: Bourbourg, France, 2014.
55. SPIGEST (Synergie Plantes Invasives Grand Est). *Guide Pratique. Pour Une Gestion Raisonnée des Renouées Asiatiques*; SPIGEST: Pierre-La-Treiche, France, 2019.
56. Nickelson, S. *Taylor Townsite Habitat Restoration Project As-Built Document, 2007–2017*; Watershed Management Division Seattle Public Utilities: Seattle, DC, USA, 2017.
57. Soll, J. *Controlling Knotweed (Polygonum cuspidatum, P. sachalinense, P. polystachyum and Hybrids) in the Pacific Northwest*; The Nature Conservancy: Portland, OR, USA, 2004.
58. Cygan, D. *Preventing the Spread of Japanese Knotweed (Reynoutria japonica) Best Management Practices*; New Hampshire Department of Agriculture, Markets & Food: Concord, CA, USA, 2018.

59. King County. *Hatchery Natural Area. Site Management Guidelines*; King County Department of Natural Resources and Parks, Water and Land Resources Division: Seattle, DC, USA, 2004.
60. Nickelson, S. *Webster Creek and Walsh Lake Riparian Habitat Restoration Project As-Built Document*; Watershed Management Division Seattle Public Utilities: Seattle, DC, USA, 2014.
61. Nickelson, S. *Knotweed Treatment through 2012 Cedar River Municipal Watershed. Annual Report Libraries, Utilities, and Center Committee Seattle City Council*; Seattle Public Utilities, Watershed Services Division: Seattle, DC, USA, 2013.
62. Environment Agency. *Managing Japanese Knotweed on Development Sites: The Japanese Knotweed Code of Practice*; Environment Agency: Bristol, UK, 2006.
63. King County. *Shadow Lake Integrated Aquatic Vegetation Management Plan*; King County Department of Natural Resources and Parks Water and Land Resources Division Noxious Weed Control Program: Seattle, DC, USA, 2016.
64. Terrin, E. *Les Espèces Végétales Exotiques Envahissantes Dans l'Ain: Fiches de Synthèse Par Espèce*; Conservatoire Botanique National Alpin: Gap, France, 2011.
65. Comité ZIP des Seigneuries; Comité ZIP Jacques Cartier; Conseil Québécois des Espèces Exotiques Envahissantes. *Renouée Du Japon Protocole de Contrôle: Comité ZIP des Seigneuries*; Comité ZIP Jacques Cartier; Conseil Québécois des Espèces Exotiques Envahissantes: Montréal, QC, Canada, 2018.
66. Louboutin, B. *Les Plantes Invasives de Brennilis Situation et Moyens de Lutte*; Brennilis, France, 2010.
67. Syndicat mixte de la Vallée de l'Orge Aval. *Bilan Du Programme de Lutte Contre La Renouée Du Japon 2005–2009*; Syndicat Mixte de la Vallée de l'Orge Aval: Viry-Chatillon, France, 2009.
68. Derickx, L.M.; Antunes, P.M. *A Guide to the Identification and Control of Exotic Invasive Species in Ontario's Hardwood Forests*; Algoma University: Sault Ste. Marie, ON, Canada, 2013; ISBN 978-0-929100-21-0.
69. Hallworth, J.; Sellentin, E. *Understanding and Controlling Invasive Knotweeds in BC.*; BC Ministry of Forests, Lands and Natural Resource Operations: Victoria, BC, Canada, 2011.
70. Association Rivière Rhône Alpes (ARRA). *Renouées du Japon: Gestion et Lutte. Actes des Journées Techniques*; Association Rivière Rhône Alpes (ARRA): Grenoble, France, 2011.
71. King County. *Best Management Practices. Invasive Knotweeds*; King County Noxious Weed Control Program: Seattle, DC, USA, 2015.
72. Association des Gestionnaires d'Espaces Naturels Bretons. *Journée d'Échanges Sur la Gestion Des EEE Végétales Terrestres 7 Novembre 2017/Landes de Ploumanac'h, Compte Rendu*; Association des Gestionnaires d'Espaces Naturels Bretons: Perros-Guirec, France, 2017.
73. Miller, J.H.; Manning, S.T.; Enloe, S.F. A Management Guide for Invasive Plants in Southern Forests. *Gen Tech Rep SRS–131 Asheville NC US Dep. Agric. For. Serv. South. Res. Stn.* **2013**, *131*, 1–120. [CrossRef]
74. Association ECHEL. *Régulation des Renouées Asiatiques (Fallopia japonica, Fallopia sachalinensis, Fallopia x bohemica)*; Fiche Technique N° 1: Nans-sous-Sainte-Anne, France, 2018.
75. UICN France. *Les Espèces Exotiques Envahissantes Sur Les Sites d'entreprises. Livret 2: Identifier et Gérer Les Principales Espèces*; UICN France: Paris, France, 2015.
76. Haury, J.; Matrat, R.; Hudin, S.; Lambert, E.; Anras, L.; Alain Dutartre, A.; Bottner, B.; Gentil, E.; Gressette, S.; Loriot, S.; et al. *Manuel de Gestion Des Plantes Exotiques Envahissant Les Milieux Aquatiques et Les Berges du Bassin Loire-Bretagne*; Fédération des Conservatoires D'espaces Naturels: Olivet, France, 2010.
77. INNSA (Invasive Non-Native Specialists Association). *Code of Practice. Managing Japanese Knotweed*; INNSA: Thornby, UK, 2017.
78. Dommanget, F.; Evette, A.; Martin, F.-M.; Piola, F.; Thiébaut, M.; Rouifed, S.; Dutartre, A.; Sarat, E.; Lavoie, C.; Cottet, M. Les renouées asiatiques, espèces exotiques envahissantes. *Revue Science Eaux Territoires* **2019**, *27*, 8–13. [CrossRef]
79. Martin, F.-M.; Dommanget, F.; Evette, A. Improving the management of Japanese knotweed s.l.: A response to Jones and colleagues. *NeoBiota* **2020**, *63*, 147–153. [CrossRef]
80. Coeur Emeraude. *Lutte Contre La Renouée Du Japon, Espèce Invasive. Réunion d'information Organisée Par COEUR Emeraude Le 1er Juillet 2013 à St Lunaire, Dans le Cadre de la Mise en Place Du Futur PNR*; Coeur Emeraude: Saint-Lunaire, France, 2013.
81. ENSAIA. *Lutte Contre Des Plantes Invasives, Les Renouées Asiatiques. Projet Professionnel 2015–2016*; ENSAIA: Vandœuvre-lès-Nancy, France, 2016.
82. Parkinson, H.; Mangold, J. *Biology, Ecology and Management of the Knotweed Complex (Polygonum spp.)*; Montana State University Extension: Bozeman, MT, USA, 2010.
83. Spiegelberger, T.; Gerber, E.; Schaffner, U. *Reynoutria 2006—Synthese. Ecologie, Impact Sur L'environnement et Gestion Des Renouées Envahissantes*; CABI: Wallingford, UK, 2006.
84. Lesica, P.; Steele, B.M. Prolonged dormancy in vascular plants and implications for monitoring studies. *Nat. Areas J.* **1994**, *14*, 209–212.
85. Miller, M.T.; Allen, G.A.; Antos, J.A. Dormancy and flowering in two mariposa lilies (*Calochortus*) with contrasting distribution patterns. *Can. J. Bot.* **2004**, *82*, 1790–1799. [CrossRef]
86. Morrow, P.A.; Olfelt, J.P. Phoenix clones: Recovery after long-term defoliation-induced dormancy. *Ecol. Lett.* **2003**, *6*, 119–125. [CrossRef]
87. Gremer, J.R.; Sala, A.; Crone, E.E. Disappearing plants: Why they hide and how they return. *Ecology* **2010**, *91*, 3407–3413. [CrossRef]

88. Shefferson, R.P.; Kull, T.; Hutchings, M.J.; Selosse, M.; Jacquemyn, H.; Kellett, K.M.; Menges, E.S.; Primack, R.B.; Tuomi, J.; Alahuhta, K.; et al. Drivers of vegetative dormancy across herbaceous perennial plant species. *Ecol. Lett.* **2018**, *21*, 724–733. [CrossRef] [PubMed]
89. Callaghan, T.V.; Scott, R.; Whittaker, H.A. *The Yield, Development and Chemical Composition of Some Fast-Growing Indigenous and Naturalised British Plant Species in Relation to Management as Energy Crops*; Institute of Terrestrial Ecology: Cumbria, UK, 1981.
90. Christina, M.; Rouifed, S.; Puijalon, S.; Vallier, F.; Meiffren, G.; Bellvert, F.; Piola, F. Allelopathic effect of a native species on a major plant invader in Europe. *Naturwissenschaften* **2015**, *102*, 12. [CrossRef] [PubMed]
91. Dommanget, F.; Evette, A.; Spiegelberger, T.; Gallet, C.; Pacé, M.; Imbert, M.; Navas, M.-L. Differential allelopathic effects of Japanese knotweed on willow and cottonwood cuttings used in riverbank restoration techniques. *J. Environ. Manag.* **2014**, *132*, 71–78. [CrossRef] [PubMed]

plants MDPI

Article

UHPLC Analysis of *Reynoutria japonica* Houtt. Rhizome Preparations Regarding Stilbene and Anthranoid Composition and Their Antimycobacterial Activity Evaluation

Fabian Alperth, Lena Melinz, Johannes-Paul Fladerer and Franz Bucar *

Institute of Pharmaceutical Sciences, University of Graz, Beethovenstraße 8, 8010 Graz, Austria; fabian.alperth@uni-graz.at (F.A.); lena.melinz@gmail.com (L.M.); johannes.fladerer@uni-graz.at (J.-P.F.)
* Correspondence: franz.bucar@uni-graz.at; Tel.: +43-316-380-5531

Abstract: *Reynoutria japonica* Houtt. is a critical invasive alien plant in Europe and North America with a drastic impact on native flora. However, *R. japonica* has medicinal potential, especially as a source of stilbenes. In order to explore the potential of simple extractions of *R. japonica*, we conducted qualitative and quantitative analyses of fresh *R. japonica* rhizome infusion, decoction, and macerates with ethanol by UHPLC-DAD-ESI-MS^n and UHPLC-DAD, with a focus on major constituent groups of stilbenes and anthranoids. Since *R. japonica* rhizome extracts showed antimicrobial potential in the past, we also evaluated the antimycobacterial effect of raw *R. japonica* extracts for the first time against *Mycobacterium smegmatis*. Of thirty-four characterized substances, six were stilbenes and twelve anthranoids. The main constituents, four trans-stilbenes and eight anthranoids, were quantified in a validated UHPLC-DAD method. The 38% ethanol macerate showed high stilbene (155.078 mg/100 g fluid extract) and low anthranoid content (5.420 mg/100 g fluid extract), while decoction showed the highest anthranoids. Antimycobacterial testing gave good results for all macerates (MIC 256 µg/mL) and *trans*-resveratrol (64 µg/mL). Extraction and enrichment of stilbenes from fresh plant material by simple extraction methods with food-grade solvents might encourage consideration of wild harvest of rhizomes over classic means of eradication of *R. japonica*.

Keywords: *Reynoutria japonica*; anthranoids; stilbenes; UHPLC-MS; UHPLC-DAD; MIC; antimicrobial; antimycobacterial; *Mycobacterium smegmatis*

1. Introduction

Reynoutria japonica Houtt. (Syn. *Fallopia japonica* (Houtt.) Ronse Decr., *Polygonum cuspidatum* Siebold and Zucc.) is an herbal perennial plant in the family of Polygonaceae [1]. The plant is native to eastern Asian regions of China, Japan, and Korea, but was introduced to the west as a garden plant in the 19th century [2]. It grows to heights of up to three metres in various conditions, and possesses thickened rhizomes, which may extend 4.5 m deep and up to 20 m away from parent plants [3,4]. Rhizomes regenerate easily even from small fragments, and therefore asexual dispersal forms the main way of reproduction in Europe and North America, often following natural disturbances or disposal of rhizome contaminated soil. Due to *R. japonica* being a fast-growing competitor early in the season, capturing space and resources, it is an invasive alien plant which can significantly diminish native flora [3,4]. However, in its native regions of China, *R. japonica* is also grown and harvested for long-lasting medicinal use in traditional Chinese medicine (TCM) [5]. Dried rhizome is called "Hu Zhang" in Chinese and listed in the Pharmacopeia of the People's Republic of China, as well as the European Pharmacopeia [6,7]. Indications in TCM comprise jaundice, hepatitis, amenorrhoea and cough, among others [6]. In Korean folk medicine, *R. japonica* rhizome is used to support dental hygiene. Antibacterial effects against *Streptococcus mutans* and *Streptococcus sanguinis* are reported and considered to prevent dental plaque formation and therefore tooth decay, as well as accelerate wound healing on gums [8]. Major secondary metabolites in the constituent profile of *R. japonica* rhizomes are stilbenes, such as polydatin and resveratrol, and anthranoids, such as emodin. Emodin and polydatin serve as quality indicators according to the Chinese Pharmacopoeia, with respective minimum contents of 0.6% and 0.15% [5]. According to the European Pharmacopoeia, emodin content of 1.0% and polydatin content of 1.5% are mandatory for quantitative quality assurance of dried plant material [7]. Furthermore, naphthalene derivates such as torachrysone, gallates, catechins and proanthocyanidins are reported [9,10].

R. japonica rhizome is a major source of resveratrol for the dietary supplement industry [11]. Resveratrol is used for its antioxidative properties. It can support the therapy of liver ailments following oxidative stress or improve skin condition in combination with oligomeric proanthocyanidins [12,13]. Other stilbenes found in *R. japonica* rhizomes, polydatin and piceatannol also show high antioxidative potential and cytotoxic effects against various cancer cells [14,15]. Resveratrol also exhibits antimicrobial effects against various bacteria, including *Mycobacterium smegmatis, Helicobacter pylori, Vibrio cholerae, Neisseria gonorrhoeae, Campylobacter coli* and *Arcobacter cryaerophilus* [16]. For piceatannol, effects against *Staphylococcus aureus* and *Pseudomonas aeruginosa* are reported [17]. The anthranoid emodin is a known stimulant laxative used in treatment of constipation. It also shows anticancer, antibacterial, diuretic and vasorelaxant effects [18]. However, in order to estimate the biological activities of glycosidic plant constituents, the impact of gut microbiota has to be taken into account. The interindividual differences in gut microbiota composition, which is also influenced by individual physiological or pathophysiological conditions, can lead to differences in absorption rates of glycosides as well as their metabolization [19]. In this study, we asked the question of whether traditional preparations like infusion, decoction, and maceration with different ethanol concentrations are suitable methods for extracting the main constituent groups of stilbenes and anthranoids in favorable composition from fresh *R. japonica* rhizome material. We hypothesise that different extraction methods, especially variations in ethanol concentration of macerates, have major influences on yield and composition of extracts and on their antimycobacterial activities. Extracts were evaluated in detail by UHPLC-DAD-ESI-MS^n for qualitative, and UHPLC-DAD for quantitative analysis, in validated methods. Since *R. japonica* rhizome extracts and its major constituents showed antimicrobial potential in the past, including good results for pure resveratrol against *M. smegmatis* achieved in our own working group [20], we also evaluated the antimycobacterial effect of raw *R. japonica* extracts for the first time, by investigating their minimum inhibitory concentration (MIC) against *Mycobacterium smegmatis* mc^2 155.

2. Results

2.1. Drug Extract Ratio (DER)

Drug extract ratios were determined for all investigated *R. japonica* rhizome extracts after freeze-drying. Subsequently used abbreviations are as follows: RI for infusion, RD for decoction, and RM for macerates of different ethanol concentrations. Results for DERs are given in Table 1.

Table 1. Drug extract ratios for investigated *Reynoutria japonica* rhizome extracts.

Extract [1]	DER	Yield [%]
RI	25.75:1	3.88%
RD	8.24:1	12.13%
RM96	8.29:1	12.07%
RM70	9.49:1	10.54%
RM38	9.69:1	10.32%

[1] RI = infusion, RD = decoction, RM = macerates prepared with 96, 70, 38% (*v/v*) ethanol.

Within macerates, yields rose with higher alcohol content of extraction solvent ($p < 0.01$). The extract yield of 12.13% for RD was comparable to the highest macerate yield in RM96 (12.07%) ($p > 0.05$). RM70 and RM38 were similar in extraction efficiency ($p > 0.05$), and both yielded over 10%, while RI gave, by far, the lowest yield of all investigated extraction methods at 3.88% ($p < 0.01$).

2.2. Qualitative Analysis via UHPLC-DAD-ESI-MS^n

Interpretation of DAD-UV and MS^n data in negative ionization mode led to the characterization of thirty-four substances in *R. japonica* rhizome extracts by comparison with literature data and available reference compounds. UV spectra of *cis*-resveratrol and *cis*-polydatin obtained by photoisomerization of reference substances aided identification. With the exception of substance **3** (procyanidin dimer Type B), which only seemed to be present in RD, traces of all substances could be found in every extract, confirmed by spectral data and matching retention times (RT). Results are given in Table 2, with threshold for listed fragment ions in MS^n set as ≥10% relative intensity.

Table 2. Qualitative analysis of *Reynoutria japonica* rhizome extracts.

Substance	RT [min]	Molecular Ion [m/z]	MS^n [m/z], Relative Intensity (%)	Molecular Weight [g/mol]	Tentative Identification
1	5.15	577 [M-H]$^-$	MS^2[577]: 425 (100), 470 (35), 289 (15) MS^3[425]: 407 (100), 273 (10) MS^4[407]: 285 (100), 281 (70), 398 (53), 297 (50)	578	Procyanidin dimer Type B [9,10]
2	5.35	335 [M+HCOO]$^-$ 289 [M-H]$^-$	MS^2[289]: 245 (100), 205 (38), 179 (15) MS^3[245]: 203 (100), 227 (30), 187 (25), 161 (20) MS^4[203]: 175 (100), 188 (65), 161(35), 157 (20)	290	Catechin [9,10,21]
3	6.15	577 [M-H]$^-$	MS^2[577]: 425 (100), 470 (35), 289 (15) MS^3[425]: 407 (100), 273 (10) MS^4[407]: 285 (100), 281 (70), 398 (53), 297 (50)	578	Procyanidin dimer Type B [9,10]
4	6.81	577 [M-H]$^-$	MS^2[577]: 425 (100), 470 (35), 289 (15) MS^3[425]: 407 (100), 273 (10) MS^4[407]: 285 (100), 281 (70), 398 (53), 297 (50)	578	Procyanidin dimer Type B [9,10]
5	7.46	577 [M-H]$^-$	MS^2[577]: 425 (100), 470 (35), 289 (15) MS^3[425]: 407 (100), 273 (10) MS^4[407]: 285 (100), 281 (70), 398 (53), 297 (50)	578	Procyanidin dimer Type B [9,10]
6	7.72	335 [M+HCOO]$^-$ 289 [M-H]$^-$	MS^2[289]: 245 (100), 205 (38), 179 (15) MS^3[245]: 203 (100), 227 (30), 187 (25), 161 (20) MS^4[203]: 175 (100), 188 (65), 161(35), 157 (20)	290	Epicatechin [9,10,21]
7	7.97	383 [M-H]$^-$	MS^2[383]: 303 (100), 285 (13) MS^3[303] 285 (100) MS^4[285]: 241 (100), 175 (50), 243 (20), 199 (18), 257 (15), 217 (15)	384	Undefined
8	8.77	186 [M-H]$^-$	MS^2[186]: 125 (100) MS^3[125]: 97 (100), 125 (35), 57 (18)	187	Undefined
9	9.88	865 [M-H]$^-$	MS^2[865]: 695 (100), 577 (68), 739 (65), 575 (35), 425 (25), 847 (22), 287 (20) MS^3[695]: 543 (100), 451 (38), 677 (35), 405 (32), 243 (30), 525 (28), 289 (17)	866	Procyanidin trimer Type B [10]
10	9.90	405 [M-H]$^-$	MS^2[405]: 243 (100) MS^3[243]: 225 (100), 201 (50), 199 (35), 175 (30) MS^4[225]: 157 (100), 197 (60), 181 (45), 225 (30)	406	*trans*-Piceatannol-hexoside [9,10,21]
11	10.13	435 [M+HCOO]$^-$	MS^2[435]: 289 (100), 227 (13) MS^3[289]: 227 (100) MS^4[227]: 185 (100), 183 (50), 157 (40), 159 (40)	390	Resveratroloside [10,21]
12	10.40	729 [M-H]$^-$	MS^2[729]: 559 (100), 407 (95), 577 (93), 441 (75), 603 (70), 451 (55), 711 (30), 289 (28)	730	Procyanidin dimer monogallate [10]
13	11.86	435 [M+HCOO]$^-$	MS^2[435]: 289 (100) MS^3: 227 (100) MS^4[227]: 185 (100), 183 (50), 157 (40), 159 (40)	390	*trans*-Polydatin [1]
14	12.08	441 [M-H]$^-$	MS^2[441]: 289 (100), 169 (22), 331 (18) MS^3[289]: 245 (100), 205 (38), 179 (18) MS^4[245]: 203 (100), 127 (25), 187 (22), 161 (20), 217 (10)	442	Catechin/Epicatechin monogallate [9]

Table 2. *Cont.*

Substance	RT [min]	Molecular Ion [m/z]	MS^n [m/z], Relative Intensity (%)	Molecular Weight [g/mol]	Tentative Identification
15	14.86	435 [M+HCOO]$^-$	MS^2[435]: 289 (100) MS^3: 227 (100) MS^4[227]: 185 (100), 183 (50), 157 (40), 159 (40)	390	*cis*-Polydatin [2]
16	15.12	393 [M-H]$^-$ 505 [M-H]$^-$	MS^2[393]: 231 (100) MS^2[505]: 289 (100), 215 (38)	394 (1) 506 (2)	Demethylated torachrysone-hexoside [9,10] (1), Undefined (2)
17	15.43	445 [M-H]$^-$	MS^2[445]: 283 (100) MS^3[283]: 240 (100), 265 (70), 268 (35)	446	Emodin methyl ether hexoside [22]
18	15.58	227 [M-H]$^-$	MS^2[227]: 185 (100), 183 (58), 157 (50), 159 (47)	228	*trans*-Resveratrol [1]
19	15.97	431 [M-H]$^-$	MS^2[431]: 269 (100), 311 (10) MS^3[269]: 225 (100), 241 (18) MS^4[225]: 181 (100), 210 (80), 225 (65), 197 (25)	432	Emodin-1-*O*-β-glucopyranoside [9,10]
20	16.25	511 [M-H]$^-$	MS^2[511]: 269 (100), 431 (43) MS^3[296]: 225 (100), 241 (25), 269 (15)	512	Emodin-*O*-(sulfonyl)-hexoside [9,10]
21	16.87	299 [M-H]$^-$	MS^2[299]: 256 (100), 284(25) MS^3[256]: 227 (100), 228 (70), 226 (65), 239 (26), 211 (20), 256 (20)	300	Undefined anthranoid
22	16.9	273 [M+HCOO]$^-$	MS^2[273]: 227(100) MS^3[227]: 185 (100), 183 (55), 157 (40), 159 (40)	228	*cis*-Resveratrol [2]
23	17.49	431 [M-H]$^-$	MS^2[431]: 269 (100), 311 (10) MS^3[269]: 225 (100), 241 (18) MS^4[225]: 181 (100), 210 (83), 225 (63), 197 (25)	432	Emodin-hexoside [9,10]
24	18.20	517 [M-H]$^-$ 1035 [2M-H]$^-$	MS^2[517]: 473 (100) MS^3[473]: 269 (100), 311 (10), MS^4[269]: 225 (100), 241 (18)	518	Emodin-8-*O*-(6′-*O*-malonyl)-hexoside [9,10]
25	18.46	325 [M-H]$^-$	MS^2[325]: 245 (100) MS^3[245]: 230 (100) MS^4[245]: 215 (100)	326	Sulfonyl-torachryson [9]
26	18.7	285 [M-H]$^-$	MS^2[285]: 241 (100), 257 (45), 285 (30), 211 (10) MS^3[241]: 211 (100), 195 (52), 212 (35), 224 (35)	286	Hydroxyemodin [23]
27	18.87	779 [M-H]$^-$	MS^2[779]: 633 (100), 615 (12), 487 (10) MS^3[633]: 487 (100), 453 (25), 469 (20)	780	Hydropiperoside [10]
28	19.57	533 [M+HCOO]$^-$	MS^2[533]: 487 (100), 283 (92), 486 (20) MS^3[487]: 283 (100), 427 (35), 455 (33), 409 (25), 469 (15) MS^3[283]: 240 (100), 268 (40)	488	Emodin methyl ether acetylhexoside [22]
29	19.80	355 [M-H]$^-$	MS^2[355]:311 (100) MS^3[311]: 267 (100), 268 (65), 283 (35), 311 (15)	356	Malonylemodin [9]
30	20.94	283 [M-H]$^-$	MS^2[283]: 240 (100), 268 (30) MS^3[240]: 212 (100), 240 (33), 196 (10), 184 (10) MS^4[212]: 184 (100)	284	Emodin methyl Ether [22]
31	22.38	311 [M-H]$^-$	MS^2[311]: 267 (100), 268 (50), 283 (60), 311 (43) MS^3[267]: 224 (100), 225 (33), 240 (30), 223 (12) MS^4[224]: 196 (100), 195 (15)	312	Acetylemodin [9]
32	22.79	269 [M−H]$^-$	MS^2[269]: 225 (100), 241 (24), 269 (15) MS^3[225]: 281 (100), 210 (73), 225 (63), 197 (30) MS^4[281]: 181 (100), 153 (10)	270	Emodin [1]
33	23.81	509 [M-H]$^-$	MS^2[509]: 254 (100), 491(10) MS^3[254]: 226 (100), 254 (23), 225 (15)	510	Emodin-bianthron [10]
34	24.51	353 [M-H]$^-$	MS^2[353]: 123 (100), 335 (15), 309 (12), MS^3[123]: 123 (100), 95 (65)	354	Undefined anthranoid

[1] identified by comparison with reference compound; [2] identified after UV-isomerization of reference compound.

Of thirty-four characterized substances, six stilbenes, twelve anthranoids, six proanthocyanidins, epicatechin, catechin, one catechin/epicatechin gallate derivative, two torachrysones, and hydropiperoside were tentatively identified or confirmed by reference substances. Two compounds could be attributed to the class of anthranoids without further identification, another two substances remained undefined. Structural examples, which represent the two main constituent groups of stilbenes and anthranoids, are given in Figure 1.

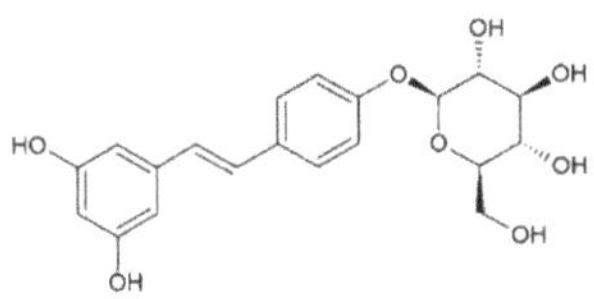

(**11**) Resveratroloside

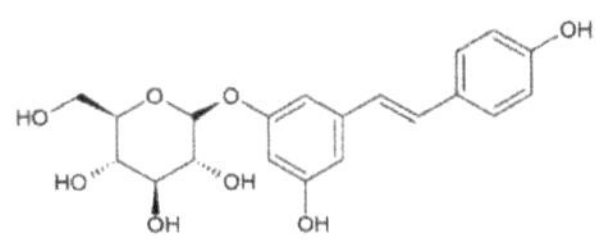

(**13**) *trans*-Polydatin

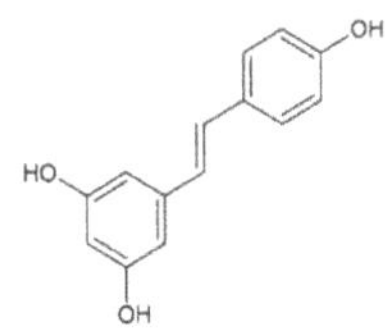

(**18**) *trans*-Resveratrol

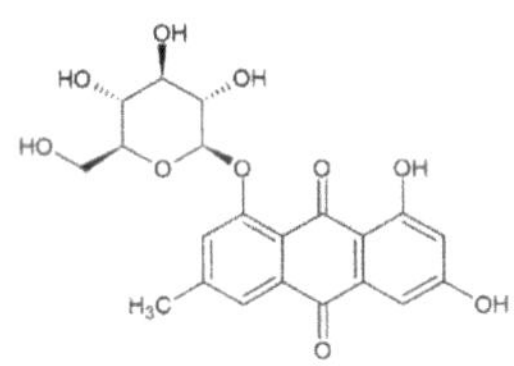

(**19**) Emodin-1-*O*-β-glycopyranoside

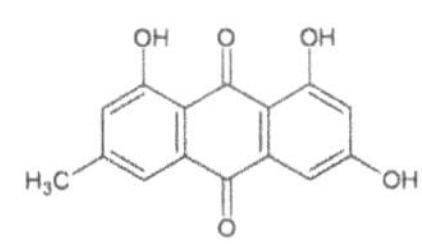

(**32**) Emodin

Figure 1. Molecular structures of several identified compounds in *Reynoutria japonica* rhizome extracts belonging to major constituent groups of stilbenes (**11**, **13**, **18**) and anthranoids (**19**, **32**).

Concerning substances which could not be directly confirmed with available reference substances and where previously reported findings in literature merely showed derivatives, some additional explanations are needed and presented hereafter. For **16**, −162 amu (m/z 393 to 231) in MS^2 was indicative of a loss of one hexose unit, leaving demethylated torachrysone. In **25**, the loss of 80 amu in MS^2 could be attributed to -SO_3 from a sulfonyl moiety to give torachrysone (m/z 325 to 245). Substance **17** showed −162 amu (m/z 445 to 283) in MS^2 and **28** a loss of 204 amu in MS^3 (m/z 487 to 283), which could be attributed to hexose and acetylhexose respectively, both resulting in an aglycone equivalent to emodin methyl ether **30**. In all three substances, this was followed by a loss of 15 amu indicative of -CH_3 stemming from a methoxy group (m/z 283 to 268), and consequently showing deprotonated fragment ions of emodin **32**. MS^2 of substance **29** showed loss of CO_2 at −44 amu (m/z 355 to 311) to give a fragment equivalent to acetylemodin **31**, followed again by deprotonated fragment ions of emodin.

2.3. *Quantitative Analysis of Stilbenes and Anthranoids via UHPLC-DAD*

Quantification was performed for main *trans*-stilbenes and anthranoids in different extracts by UHPLC-DAD. Arithmetic means of triplicate measurements of peak areas (AUC) in UHPLC-DAD chromatograms (λ = 318 nm for stilbenes and 424 nm for anthranoids) were used for calculation. Relative standard deviations (RSD%) for AUCs were <0.60% for stilbenes and < 1.66% for anthranoids, with few exceptions (**2** in RM96, RSD = 4.25%; **6** in RM96, RSD = 2.42%; **6** in RD, RSD = 5.63%). Quantities in freeze-dried extracts, native fluid extracts, and extraction from plant material were calculated from results for sample concentration 2 mg/mL freeze-dried extract in UHPLC analysis, using exact masses of plant material, solvents, and freeze-dried extracts (Supplementary Materials, Table S1). Correction factors (Tables S2 and S3) were applied according to differences in molecular weight of analytes and reference substances. Purity of references was considered and determined to be 99.76% for *trans*-resveratrol and 97.91% for emodin.

Results for stilbenes are given as mg/g dry extract after freeze-drying (Table 3), mg/100 g extract solution after filtration (Table 4), and mg/g plant material used for extraction (Table 5), in the same order, results for anthranoids are presented in Tables 6–8. Glycoside and aglycone contents represent proportions of total amounts.

Table 3. Quantitative analysis of stilbenes in *Reynoutria japonica* rhizome extracts calculated as mg/g freeze-dried extract.

Extract	10	11	13	18	Total	Glycosides	Aglycones
RI	2.669	29.975	47.883	- [1]	80.527	80.527	-
RD	3.618	40.132	67.427	0.236	111.413	111.177	0.236
RM96	4.968	53.561	87.993	2.952	149.474	146.521	2.952
RM70	4.014	42.334	71.068	4.684	122.100	117.416	4.684
RM38	2.024	25.473	29.192	19.926	76.615	56.689	19.926

[1] = not quantifiable.

Table 4. Quantitative analysis of stilbenes in *Reynoutria japonica* rhizome extracts calculated as mg/100 g fluid extract.

Extract	10	11	13	18	Total	Glycosides	Aglycones
RI	0.205	2.302	3.678	- [1]	6.185	6.185	-
RD	0.865	9.590	16.112	0.057	26.623	26.567	0.057
RM96	11.676	125.880	206.802	6.938	351.296	344.358	6.938
RM70	8.293	87.464	146.829	9.678	252.264	242.587	9.678
RM38	4.096	51.561	59.087	40.333	155.078	114.745	40.333

[1] = not quantifiable.

Table 5. Quantitative analysis of stilbenes in *Reynoutria japonica* rhizome extracts calculated as mg/g plant material.

Extract	10	11	13	18	Total	Glycosides	Aglycones
RI	0.104	1.164	1.860	- [1]	3.128	3.128	-
RD	0.439	4.868	8.179	0.029	13.515	13.487	0.029
RM96	0.600	6.464	10.619	0.356	18.039	17.683	0.356
RM70	0.423	4.460	7.488	0.494	12.865	12.371	0.494
RM38	0.209	2.629	3.013	2.057	7.907	5.851	2.057

[1] = not quantifiable.

Table 6. Quantitative analysis of anthranoids in *Reynoutria japonica* rhizome extracts calculated as mg/g freeze-dried extract.

Extract	19	23	24	26	28	30	32	34	Total	Glycosides	Aglycones
RI	2.459	24.433	3.540	0.996	2.919	0.000	1.087	- [1]	35.433	33.350	2.084
RD	4.002	35.318	3.663	2.461	2.923	0.221	1.717	-	50.307	45.907	4.400
RM96	-	1.218	-	-	-	3.104	35.422	1.920	41.664	1.218	40.446
RM70	-	-	-	-	-	2.184	17.170	0.791	20.145	-	20.145
RM38	-	-	-	-	-	1.172	1.505	-	2.678	-	2.678

[1] = not quantifiable.

Table 7. Quantitative analysis of anthranoids in *Reynoutria japonica* rhizome extracts calculated as mg/100 g fluid extract.

Extract	19	23	24	26	28	30	32	34	Total	Glycosides	Aglycones
RI	0.189	1.877	0.272	0.077	0.224	- [1]	0.083	-	2.721	2.561	0.160
RD	0.956	8.440	0.875	0.588	0.698	0.053	0.410	-	12.021	10.970	1.051
RM96	-	2.863	-	-	-	7.294	83.250	4.512	97.919	2.863	95.056
RM70	-	-	-	-	-	4.511	35.475	1.634	41.620	-	41.620
RM38	-	-	-	-	-	2.373	3.047	-	5.420	-	5.420

[1] = not quantifiable.

Table 8. Quantitative analysis of anthranoids in *Reynoutria japonica* rhizome extracts calculated as mg/g plant material.

Extract	19	23	24	26	28	30	32	34	Total	Glycosides	Aglycones
RI	0.095	0.949	0.137	0.039	0.113	- [1]	0.042	-	1.376	1.295	0.081
RD	0.485	4.284	0.444	0.299	0.355	0.027	0.208	-	6.103	5.569	0.534
RM96	-	0.147	-	-	-	0.375	4.275	0.232	5.028	0.147	4.881
RM70	-	-	-	-	-	0.230	1.809	0.083	2.123	-	2.123
RM38	-	-	-	-	-	0.121	0.155	-	0.276	-	0.276

[1] = not quantifiable.

Results for mg/100 g fluid extract (Tables 4 and 7) showed main stilbenes to be **11** and **13** in all extracts. A comparably high amount of **18** could be found in RM38. The main anthranoid was **32** in all macerates, whereas **23** was dominant in RD and RI. Total stilbene and anthranoid extraction followed the trend RM96 > RM70 > RM38 > RD > RI ($p < 0.01$ for total stilbene and $p < 0.037$ for total anthranoid amount), as did stilbene glycoside ($p < 0.01$) and anthranoid aglycone ($p < 0.01$) extraction (Figure 2). An opposite trend was true for stilbene aglycones in macerates ($p < 0.01$), with highest extraction in RM38, while water extractions still gave the lowest results ($p < 0.01$), with no quantifiable stilbene aglycones in RI. Highest amounts of anthranoid glycosides were extracted in RD ($p < 0.01$), none could be quantified in RM70 and RM38.

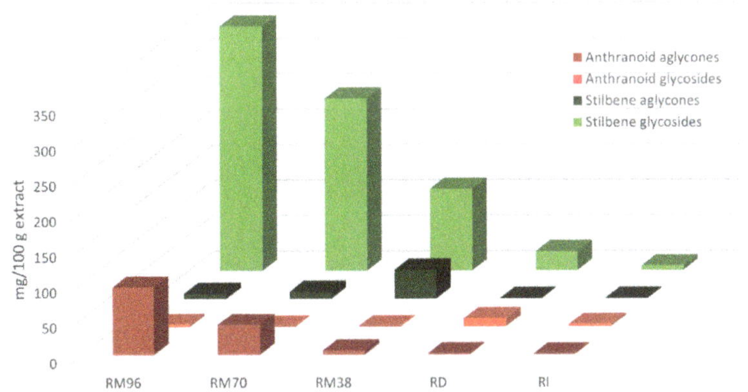

Figure 2. Stilbene and anthranoid glycoside and aglycone content given as mg/100 g fluid extract.

When analysing relative composition of extracts regarding stilbenes and anthranoids, as illustrated in Figure 3, without considering absolute yield, it showed that RM38 had lowest anthranoid content at 3.38% ($p < 0.01$) and a remarkably high fraction of aglycones within stilbenes (25.13%) ($p < 0.01$). RD and RI showed highest anthranoid (31.11 and 30.56%) and total glycoside content (97.13% and 98.20%) in similar relative compositions (Figure 3) ($p < 0.01$ compared to other extraction methods).

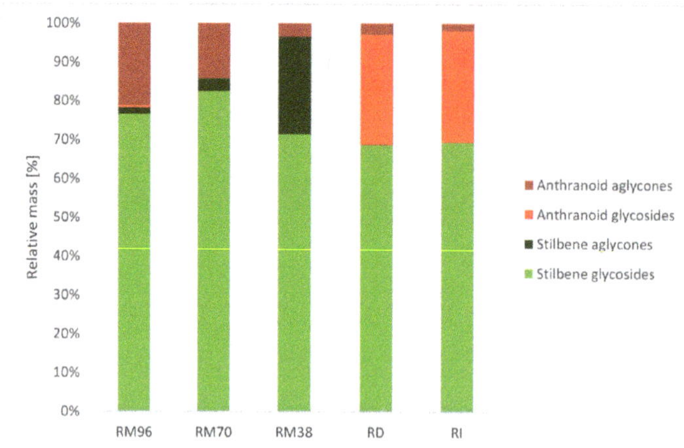

Figure 3. Relative distribution of stilbene and anthranoid glycoside and aglycone content in fluid extracts.

Validation of Quantitative UHPLC-DAD Method

Linearity of measurements was confirmed for the quantitative UHPLC method used in this study. For reference substance, resveratrol in a working range of 2–300 µg/mL and emodin of 2–500 µg/mL separate correlation coefficients showed (R^2) = 1. Intra-day and inter-day precision were investigated for sample RM96. Results are shown in Table 9. Accuracies were found to be 72.44%, 94.99%, and 94.58% for emodin (3, 200, 400 µg/mL) and 82.85%, 93.64%, and 93.83% for resveratrol (3, 100, 200 µg/mL). For emodin, recovery rates at 3, 200, and 400 µg/mL spike concentrations were determined to be 92.54%, 98.76%, and 98.92%, respectively. For resveratrol at 3, 100, and 200 µg/mL, recovery rates were 109.62%, 96.65%, and 96.81%.

Table 9. Intra-day and inter-day precision RSD (%) for several quantified stilbenes (**10–18**) and anthranoids (**30–34**).

Substance	Intra-Day Precision RSD [%]	Inter-Day Precision RSD [%]
10	0.3612	0.9405
11	0.3318	0.8861
13	0.4201	1.0069
18	0.6999	1.1285
30	3.8111	2.6544
32	0.7511	1.5208
34	6.2563	7.7805

2.4. Antimycobacterial Testing of Freeze-Dried Extracts and Reference Substances

MIC testing of freeze-dried extracts showed antimycobacterial effects against *M. smegmatis* for macerates and reference substance resveratrol in the tested concentration ranges (extracts 512–1 µg/mL, reference substances 128–0.25 µg/mL). Infusion and decoction, as well as polydatin and emodin, had no relevant antibacterial effect. Results are summarized in Table 10.

Table 10. Minimum inhibitory concentrations of freeze-dried extracts on *Mycobacterium smegmatis* mc^2 155 growth (n = 4).

Sample	MIC [µg/mL]
RI	>512
RD	>512
RM96	256
RM70	256
RM38	256
trans-Resveratrol	64
trans-Polydatin	>128
Emodin	>128

3. Discussion

In this study, we analysed different extraction methods of fresh *R. japonica* rhizome, which are easily reproducible without specialized instrumental equipment and using food-grade solvents, to compare yields and composition, particularly focusing on main secondary metabolite groups of anthranoids and stilbenes. UHPLC-DAD-ESI-MS^n lead to the characterization of thirty-four substances, of which six were stilbenes, and twelve anthranoids. Substances **16**, **25**, **28**, and **29** (Table 2) can be regarded as new derivatives of previously reported compounds. The main constituents, four *trans*-stilbenes and eight anthranoids, were quantified by means of UHPLC-DAD. Furthermore, we investigated the antimycobacterial potential of extracts and reference compounds against *Mycobacterium smegmatis* in micro-dilution minimum inhibitory concentration (MIC) assays.

Macerates prepared with 38%, 70%, and 96% (*v*/*v*) ethanol (RM38, RM70, RM96) showed yields of 10.32, 10.54, and 12.07% (Table 1), giving the highest ethanol concentration in solvents a slight advantage in yield; however, these did not mark drastic changes in extraction efficiency. Infusion (RI) yields were low at 3.88%, which is to be expected from a short extraction time of 10 min and makes this preparation least suitable for further consideration of usability. Decoction (RD) yielded 12.13% and needs merely 30 min of extraction time compared to three weeks for macerates. However, when extract composition is considered, RD showed the highest anthranoid extraction of all extracts at 31% of quantified substances (Figure 3), which could be considered an undesirable trait of *R. japonica* extracts. Although the European Pharmacopoeia demands minimum contents for polydatin (stilbene) and emodin (anthranoid) in dried *R. japonica* rhizome [7], higher stilbene content and low anthranoid content in extracts seems more desirable to emphasize the antioxidative and antibacterial effects of stilbenes [14–17], while minimizing the laxative effects of emodin and derivatives [18]. In accordance with the reported results for lyophilized and pulverized *R. japonica* rhizome samples [24], the extraction of anthranoids from fresh and cut *R. japonica* rhizome is rising with ethanol concentration. However, a reported decline in extraction efficiency over 80% ethanol concentration was not found to be occurring in fresh rhizome macerates. From this point of view, RM38 seems most promising, with high stilbene and a remarkably low anthranoid content of 3.38% of quantified substances (5.420 mg/100 g fluid extract, Figure 3, Table 7). Absolute stilbene content amounted to 155.078 mg/100 g fluid extract (Table 4) and could easily be improved through scale-up processes like large batch processing and concentration of constituents by evaporation of solvent. A stilbene extraction from fresh *R. japonica* rhizome with liquified dimethyl ether was reported to yield 0.342 mg/g resveratrol and 2.57 mg/g polydatin, calculated per gram of dry rhizome [25]. Extraction of resveratrol from dried and pulverized *R. japonica* rhizome with deep eutectic solvents (DES) yielded up to 9.00 mg/g dry rhizome [26]. In comparison with both methods, we found RM38 to also yield good quantities of 2.057 mg/g resveratrol and 3.013 mg/g polydatin. Maceration with 38% (*v*/*v*) ethanol can, similarly to DES, be employed without the need to dry and grind plant material, but requires neither specialized equipment, nor complex or critical organic solvents.

The highest relative glycoside content among quantified substances was found in hot water extracts RD and RI (97.13% and 98.20%, Figure 3), which could be due to denaturation of glycosidases by hot water, and therefore prohibition of enzymatic hydrolysis. In macerates, high water content encourages hydrolysis and RM38 showed the lowest glycoside content accordingly (71.49%), entirely made up of stilbene glycosides, with no quantifiable anthranoid glycosides. Interestingly, RM70, and not RM96, showed the highest relative glycoside content among macerates. This could be explained by improved glycoside extraction at 30% (*v*/*v*) water content in solvent, while high ethanol of 70% (*v*/*v*) suffices to stabilize glycosides in solution over longer extraction times.

Antimycobacterial testing against *M. smegmatis* showed good results for all freeze-dried macerates at 256 μg/mL, with no visible effects of different extract compositions regarding stilbenes and anthranoids (Table 10). Water extracts RD and RI, however, showed no significant antibacterial effect (>512 μg/mL), which could be attributed to high glycoside content, and therefore less uptake by bacteria due to higher polarity of compounds. Reference compounds *trans*-resveratrol and *trans*-polydatin support this assumption for stilbenes, with promising results for the aglycone *trans*-resveratrol (64 μg/mL), while glycoside the *trans*-polydatin showed no relevant effect (>128 μg/mL). Anthranoid emodin also showed no inhibitory effect (>128 μg/mL).

As *R. japonica* is highly invasive outside its native regions of Eastern Asia and an aggressive competitor in foreign ecosystems, new perspectives on medicinal use in the West show benefits that might encourage consideration of wild harvest over classic means of eradication, which mostly involve heavy use of herbicides [27]. Extraction and enrichment of stilbenes from fresh plant material can be accomplished by simple extraction methods with food-grade solvents which are also suitable for scale-up, most significantly by maceration with 38% (*v*/*v*) ethanol. Stilbene aglycones of natural origin and their derivatives promise to be interesting targets for further antimycobacterial evaluation.

4. Materials and Methods

4.1. Plant Material

Rhizomes of *R. japonica* were harvested in September of 2020 in Graz, Austria (47°04′39.38″ N, 15°27′08.90″ E), and stored at +4 °C. A voucher was deposited at the herbarium of the Department of Pharmacognosy, Institute of Pharmaceutical Sciences, University of Graz (specimen voucher: Alperth 010). The plant material was processed the day after collection by washing with cold water and cutting into approximately 2 mm thick slices shortly before extraction. Only pieces with complete cross-sections were used to facilitate comparability between different extraction methods.

4.2. Solvents and Reference Substances

Reference substances for qualitative and quantitative analysis were trans-resveratrol (99% purity, Sigma-Aldrich, St. Louis, MO, USA), emodin (90% purity, Sigma-Aldrich), and polydatin CRS (trans-polydatin, 99.1% purity, EDQM—Council of Europe, Strasbourg, France). For extract preparation, deionized water and ethanol (AustrAlco, Spillern, Austria) were used. Samples of reference substances were prepared in p.a. grade methanol (VWR, Radnor, PA, USA). HPLC mobile phases were made up of ultrapure water (Barnstead Easypure RF, Barnstead) and HPLC-grade acetonitrile (VWR).

4.3. Photoisomerization of Reference Stilbenes by UV-Exposure

Separate reference solutions of *trans*-resveratrol and *trans*-polydatin were prepared as 1 mg/mL in methanol and treated with UV light (λ = 365 nm) in clear glass vials for 18.5 h to trigger *cis*/*trans*-isomerization for further reference data.

4.4. Extract Preparation

Separate samples of fresh plant material were extracted by infusion, decoction, and maceration before filtration with suction and freeze-drying using a VirTis Sentry freeze-dryer (SP Scientific, Warminster, PA, USA). For infusion, 100 g of boiling, deionized water were added to 2 g of cut rhizome in a beaker and covered for an extraction time of 10 min. For decoction, 100 g of cold, deionized water and 2 g of plant material were brought to a boil and then extracted for 30 min under reflux. Macerates were prepared with 20 g of rhizome and 100 g of 38%, 70% and 96% (*v*/*v*) partly denatured ethanol respectively, with an extraction time of three weeks in the dark.

4.5. UHPLC-DAD-ESI-MS^n for Qualitative Analysis

A Dionex UltiMate 3000 RS Ultra-high performance liquid chromatography (UHPLC) system (Thermo Fisher Scientific, Waltham, MA, USA) comprising pump, autosampler, column compartment and diode array detector (DAD) was used for qualitative analysis of extracts. A Kinetex C18 100 mm × 2.1 mm, 2.6μm column (Phenomenex, Torrance, CA, USA) offered the stationary phase, while the mobile phase consisted of water + 0.1% formic acid (A) and acetonitrile + 0.1% formic acid (B). Runs started at 3% B, increasing to 15% B at 12 min, 42% B at 20 min and 70% B at 23 min, then plateaued until 25 min before dropping back to 3% B at 25.2 min, which was held until 30 min total runtime. Column temperature was kept at 40 °C with flow rate being 0.350 mL/min. DAD recorded spectra in a wavelength range of 190 to 700 nm. For mass spectrometric (MS) detection, the system was coupled to an LTQ XL linear ion-trap mass spectrometer with electrospray ionization (ESI) ion source (Thermo Scientific). Conditions were set to source heater temperature 300 °C, sheath gas flow 40 arb (arbitrary units), auxiliary gas flow 10 arb, source voltage 3.5 kV (ESI neg), and recorded mass range of m/z 50 to 2000 amu.

4.6. UHPLC-DAD for Quantitative Analysis

Quantitative analysis was conducted on a separate Dionex UltiMate 3000 RS UHPLC system (Thermo Fisher Scientific), comprising the same modules as for qualitative analysis without coupling to MS. All chromatographic parameters were identical, with extracted quantification wavelengths from DAD-UV detection being 318 nm for stilbenes and 424 nm for anthranoids. Freeze-dried extracts were redissolved in concentrations of 2 mg/mL, with water as solvent for dry extracts from infusion and decoction, and 50% (*v*/*v*) ethanol for macerates. Reference substances were used in 1 mg/mL methanolic stock solutions and subsequent serial dilutions (1:10, 1:100, 1:1000). Injection volumes were 1 and 5 μL for samples and 1 to 5 μL for references according to desired concentration. All measurements were performed in triplicate and quantified substances calculated for their AUC arithmetic means.

4.7. Validation of UHPLC-DAD Method

4.7.1. Linearity

To establish linearity of measurements in a working range of 2–300 μg/mL for reference substance, resveratrol concentrations 2, 5, 10, 30, 100, 300 μg/mL were evaluated. For emodin in a working range of 2–500 μg/mL, concentrations 2, 5, 10, 30, 100, 300, 500 μg/mL were used.

4.7.2. Precision

Intra-day and inter-day precision were determined for sample RM96 at a sample concentration of 2 mg/mL in 50% (*v*/*v*) ethanol and injection volume of 1 μL. For intra-day precision, the sample solution was injected six consecutive times in one day and relative standard deviation (RSD) in peak areas (area under the curve, AUC) calculated for four stilbenes and three anthranoids. For inter-day precision, the same sample was injected 3 times each on two consecutive days.

4.7.3. Accuracy

For determination of accuracy, serial dilutions of emodin and resveratrol reference stock solutions were measured in different injection volumes to give target concentrations of 3, 200, and 400 μg/mL for emodin and 3, 100, and 200 μg/mL for resveratrol.

4.7.4. Recovery Rate

Recovery rates for emodin and resveratrol were measured by spiking samples of RM38 at a concentration of 2 mg/mL with reference concentrations also used for determination of accuracy (3, 200, and 400 μg/mL for emodin and 3, 100, and 200 μg/mL for resveratrol).

4.8. Antimycobacterial Assay

All freeze-dried extracts and reference compounds were tested in a well-established serial micro-dilution minimum inhibitory concentration (MIC) assay against the bacterial strain *Mycobacterium smegmatis* mc^2 155 (ATCC 700084) [28]. Starting concentrations for dilution were 512 mg/L for extracts and 128 mg/L for reference compounds. In short, samples were dissolved in dimethyl sulfoxide (Roth, Karlsruhe, Germany) before serial dilution with Mueller Hinton Broth medium (Oxoid, Basingstoke, Hampshire, UK) in 96-well plates and incubation for 72 h at 37 °C with a 5×10^5 cfu/mL bacterial inoculum (n = 4). Isoniazid (Sigma, St. Louis, MO, USA) served as control antibiotic. Growth and sterile control were included per plate. Evaluation was performed via Methylthiazolyldiphenyl-tetrazoliumbromide (MTT) (Sigma) colorimetric detection, after additional incubation of 30 min.

4.9. Statistical Analysis

All statistical analyses were performed in IBM SPSS Statistics 26. To test correlations, Spearman Correlation test was employed as correlations might not be linear and extraction methods are transformed to ordinal variables. For mean comparisons, an ANOVA with a Bonferroni-Holm corrected t-test as a post hoc test was used. The α-level was set to 0.05.

5. Conclusions

Reynoutria japonica, as one of the most critical invasive species in Europe and North America, requires constant attention in management. To shed light on benefits beyond eradication can promote the use of invasive species. As hypothesized, differences in simple extraction methods are apparent. Remarkably, a simple macerate with 38% (v/v) ethanol gives good yields of stilbenes while showing low anthranoid content and could therefore be a target for large-scale processing of *R. japonica* rhizomes. However, concerning their antimycobacterial potential, all ethanolic macerates showed similar results, both aqueous preparations were less active.

Optimal harvest times for high stilbene content need to be investigated in future research. Also, antimycobacterial stilbene aglycones, their isolation, derivatization, and testing, proved to be promising for future investigation.

Supplementary Materials: The following are available online at https://www.mdpi.com/article/10.3390/plants10091809/s1, Table S1: Data used for quantitative calculations, Table S2: Correction factors for stilbene quantification relative to reference *trans*-resveratrol, Table S3: Correction factors for anthranoid quantification relative to reference emodin.

Author Contributions: Conceptualization, F.A. and F.B.; methodology, F.A.; validation, L.M. and F.A.; investigation, L.M. and F.A.; resources, F.A. and F.B.; writing—original draft preparation, F.A.; writing—review and editing, F.B., J.-P.F. and L.M.; visualization, J.-P.F.; supervision, F.B. All authors have read and agreed to the published version of the manuscript.

Funding: This research received no external funding.

Data Availability Statement: Data are included in this article.

Acknowledgments: The authors acknowledge the financial support by the University of Graz.

Conflicts of Interest: The authors declare no conflict of interest.

References

1. WFO. *Reynoutria japonica* Houtt. Available online: http://www.worldfloraonline.org/taxon/wfo-0000406106 (accessed on 30 June 2021).
2. Bailey, J.P.; Conolly, A.P. Prize-Winners to Pariahs—A History of Japanese Knotweed s.l. (*Polygonaceae*) in the British Isles. *Watsonia* **2000**, *23*, 93–110.
3. Jones, D.; Bruce, G.; Fowler, M.S.; Law-Cooper, R.; Graham, I.; Abel, A.; Street-Perrott, F.A.; Eastwood, D. Optimising physiochemical control of invasive Japanese knotweed. *Biol. Invasions* **2018**, *20*, 2091–2105. [CrossRef]
4. Global Invasive Species Database. Species Profile: *Polygonum cuspidatum*. Available online: http://www.iucngisd.org/gisd/speciesname/Polygonum+cuspidatum (accessed on 30 June 2021).
5. Peng, W.; Qin, R.; Li, X.; Zhou, H. Botany, phytochemistry, pharmacology, and potential application of *Polygonum cuspidatum* Sieb.et Zucc.: A review. *J. Ethnopharmacol.* **2013**, *148*, 729–745. [CrossRef] [PubMed]
6. Zhang, H.; Li, C.; Kwok, S.-T.; Zhang, Q.-W.; Chan, S.-W. A Review of the Pharmacological Effects of the Dried Root of *Polygonum cuspidatum* (Hu Zhang) and Its Constituents. *Evid. Based Complement. Altern. Med.* **2013**, *2013*, 208349. [CrossRef]
7. *Europäisches Arzneibuch: 10. Ausgabe, Grundwerk. Polygoni Cuspidati Rhizoma et Radix*; Verlag Österreich: Vienna, Austria, 2020; pp. 2085–2087.
8. Nawrot-Hadzik, I.; Hadzik, J.; Fleischer, M.; Choromańska, A.; Sterczała, B.; Kubasiewicz-Ross, P.; Saczko, J.; Gałczyńska-Rusin, M.; Gedrange, T.; Matkowski, A. Chemical Composition of East Asian Invasive Knotweeds, their Cytotoxicity and Antimicrobial Efficacy Against Cariogenic Pathogens: An In-Vitro Study. *Med. Sci. Monit.* **2019**, *25*, 3279–3287. [CrossRef]
9. Fu, J.; Wang, M.; Guo, H.; Tian, Y.; Zhang, Z.; Song, R. Profiling of components of rhizoma et radix polygoni cuspidati by high-performance liquid chromatography with ultraviolet diode-array detector and ion trap/time-of-flight mass spectrometric detection. *Pharmacogn. Mag.* **2015**, *11*, 486–501. [CrossRef]
10. Nawrot-Hadzik, I.; Ślusarczyk, S.; Granica, S.; Hadzik, J.; Matkowski, A. Phytochemical Diversity in Rhizomes of Three Reynoutria Species and their Antioxidant Activity Correlations Elucidated by LC-ESI-MS/MS Analysis. *Molecules* **2019**, *24*, 1136. [CrossRef]
11. Chen, H.; Tuck, T.; Ji, X.; Zhou, X.; Kelly, G.; Cuerrier, A.; Zhang, J. Quality assessment of Japanese knotweed (*Fallopia japonica*) grown on Prince Edward Island as a source of resveratrol. *J. Agric. Food Chem.* **2013**, *61*, 6383–6392. [CrossRef]
12. Ahmad, A.; Ahmad, R. Resveratrol mitigate structural changes and hepatic stellate cell activation in N′-nitrosodimethylamine-induced liver fibrosis via restraining oxidative damage. *Chem. Biol. Interact.* **2014**, *221*, 1–12. [CrossRef]
13. Buonocore, D.; Lazzeretti, A.; Tocabens, P.; Nobile, V.; Cestone, E.; Santin, G.; Bottone, M.G.; Marzatico, F. Resveratrol-procyanidin blend: Nutraceutical and antiaging efficacy evaluated in a placebocontrolled, double-blind study. *Clin. Cosmet. Investig. Dermatol.* **2012**, *5*, 159–165. [CrossRef] [PubMed]
14. Piotrowska, H.; Kucinska, M.; Murias, M. Biological activity of piceatannol: Leaving the shadow of resveratrol. *Mutat. Res.* **2012**, *750*, 60–82. [CrossRef] [PubMed]
15. Su, D.; Cheng, Y.; Liu, M.; Liu, D.; Cui, H.; Zhang, B.; Zhou, S.; Yang, T.; Mei, Q. Comparision of piceid and resveratrol in antioxidation and antiproliferation activities in vitro. *PLoS ONE* **2013**, *8*, e54505. [CrossRef] [PubMed]
16. Vestergaard, M.; Ingmer, H. Antibacterial and antifungal properties of resveratrol. *Int. J. Antimicrob. Agents* **2019**, *53*, 716–723. [CrossRef] [PubMed]
17. Mattio, L.M.; Dallavalle, S.; Musso, L.; Filardi, R.; Franzetti, L.; Pellegrino, L.; D'Incecco, P.; Mora, D.; Pinto, A.; Arioli, S. Antimicrobial activity of resveratrol-derived monomers and dimers against foodborne pathogens. *Sci. Rep.* **2019**, *9*, 19525. [CrossRef]
18. Zheng, Y.-F.; Liu, C.-F.; Lai, W.-F.; Xiang, Q.; Li, Z.-F.; Wang, H.; Lin, N. The laxative effect of emodin is attributable to increased aquaporin 3 expression in the colon of mice and HT-29 cells. *Fitoterapia* **2014**, *96*, 25–32. [CrossRef]
19. Braune, A.; Blaut, M. Bacterial species involved in the conversion of dietary flavonoids in the human gut. *Gut Microbes* **2016**, *7*, 216–234. [CrossRef]
20. Lechner, D.; Gibbons, S.; Bucar, F. Plant phenolic compounds as ethidium bromide efflux inhibitors in Mycobacterium smegmatis. *J. Antimicrob. Chemother.* **2008**, *62*, 345–348. [CrossRef]
21. Fan, P.; Hay, A.-E.; Marston, A.; Lou, H.; Hostettmann, K. Chemical variability of the invasive neophytes *Polygonum cuspidatum* Sieb. and Zucc. and *Polygonum sachalinensis* F. Schmidt ex Maxim. *Biochem. Syst. Ecol.* **2009**, *37*, 24–34. [CrossRef]
22. Feng, J.; Ren, H.; Gou, Q.; Zhu, L.; Ji, H.; Yi, T. Comparative analysis of the major constituents in three related polygonaceous medicinal plants using pressurized liquid extraction and HPLC-ESI/MS. *Anal. Methods* **2016**, *8*, 1557–1564. [CrossRef]
23. Kimura, Y.; Kozawa, M.; Baba, K.; Hata, K. New Constitutents of Roots of *Polygonum cuspidatum*. *Planta Med.* **1983**, *48*, 164–168. [CrossRef]
24. Glavnik, V.; Vovk, I. Extraction of Anthraquinones from Japanese Knotweed Rhizomes and Their Analyses by High Performance Thin-Layer Chromatography and Mass Spectrometry. *Plants* **2020**, *9*, 1753. [CrossRef]
25. Kanda, H.; Oishi, K.; Machmudah, S.; Wahyudiono; Goto, M. Ethanol-free extraction of resveratrol and its glycoside from Japanese knotweed rhizome by liquefied dimethyl ether without pretreatments. *Asia-Pac. J. Chem. Eng.* **2021**, *16*. [CrossRef]

26. Sun, B.; Zheng, Y.-L.; Yang, S.-K.; Zhang, J.-R.; Cheng, X.-Y.; Ghiladi, R.; Ma, Z.; Wang, J.; Deng, W.-W. One-pot method based on deep eutectic solvent for extraction and conversion of polydatin to resveratrol from *Polygonum cuspidatum*. *Food Chem.* **2021**, *343*, 128498. [CrossRef] [PubMed]
27. Delbart, E.; Mahy, G.; Weickmans, B.; Henriet, F.; Crémer, S.; Pieret, N.; Vanderhoeven, S.; Monty, A. Can land managers control Japanese knotweed? Lessons from control tests in Belgium. *Environ. Manag.* **2012**, *50*, 1089–1097. [CrossRef] [PubMed]
28. Stavri, M.; Schneider, R.; O'Donnell, G.; Lechner, D.; Bucar, F.; Gibbons, S. The antimycobacterial components of hops (*Humulus lupulus*) and their dereplication. *Phytother. Res.* **2004**, *18*, 774–776. [CrossRef] [PubMed]

plants

MDPI

Article

The Role of Flavonoids in Invasion Strategy of *Solidago canadensis* L.

Artur Likhanov [1,2,*], Marian Oliinyk [1], Nataliia Pashkevych [3], Andrii Churilov [2] and Mykola Kozyr [1]

1 Institute for Evolutionary Ecology NAS Ukraine, Akademika Lebedeva 37, 03143 Kyiv, Ukraine; MarianOlijnyk@gmail.com (M.O.); geobot2@ukr.net (M.K.)
2 Department of Botany, Dendrology and Forest Tree Breeding, The National University of Life and Environmental Sciences of Ukraine, Henerala Rodimtseva 19, 03041 Kyiv, Ukraine; churilov_konf@ukr.net
3 M.G. Kholodny Institute of Botany of the NAS of Ukraine, Tereshchenkivska 2, 01601 Kyiv, Ukraine; pashkevych.nataly@gmail.com
* Correspondence: likhanov.bio@gmail.com; Tel.: +38-063-619-19-71

Abstract: This study provides data on the problem of potential complexation of phenolic compounds synthesized by the plants *Solidago canadensis* L. and *Solidago gigantea* Ait. with ammonium forms of nitrogen, partly immobilized in the soil. A comparative analysis of secondary metabolites of the studied plants was performed by HPLC. The leaves of invasively active *Solidago canadensis* contain nine times more rutin than the plants of *Solidago gigantea*. Adding to the leaf extracts (v/v1/20) aqueous ammonia solution to pH 8.0 on the chromatograms decreases the intensity or completely causes peaks of flavonoids to disappear; instead, there are peaks of new polar substances (tR 1.5 and 2.0 min). The selective effect of the phenol-ammonium complex on various plant species was revealed. At a concentration of 20 μg/mL, these substances stimulated the formation of lateral roots in soybean seedlings and chrysanthemum cuttings. The suppression of root growth in radish seedlings occurred at a concentration of flavonoids in the extract of 25 μg/mL. In addition, a positive chemotaxis of the *Pseudomonas putida* (PGPR) was detected in the nitrogen-containing complex based on rutin (5 μg/mL). The identified feature allows PGPR colonization of the root system of *Solidago canadensis* with corresponding changes in the structure of the microbial community. The ability of the obtained nitrogen-containing polar complexes to regulate the growth processes of plants at extremely low concentration points to promising research in this direction.

Keywords: allelopathy; ammonia complexes; chemotaxis; flavonoids; growth regulator; PGPR; soil

1. Introduction

The natural biogeographical process of migration of vascular plant species presupposes the presence of at least three conditions: The recipient habitat, the spreading species and the vector of transfer. At the same time, anthropogenic migrations of vascular plants differ, among others, in that they are caused by a significant transformation of one, two or all three components as a result of unintentional or purposeful human activity. Currently, the process of plant migration is accelerating, becoming global. The process of phytoinvasion is the fast distribution of alien plant species, accustomed to changing conditions, in new areas, sometimes intensively increasing the number [1]. In this regard, the study of physiological and biochemical mechanisms of the adaptation strategy of invasive plant species is extremely relevant in order to find out the signs and features of adaptations in the conditions of natural and anthropogenically altered environments that ensure the viability of populations [2]. Thus, the species *Solidago canadensis* L. (kenophyte of North American origin, epecophyte with the European-North American range, mesophyte, sciogeliophyte) and *S. gigantea* Ait. (kenophyte of North American origin with the European-American range, mesophyte, sciogeliophyte) deserve special attention [3–6].

S. canadensis is now actively spreading in the Forest and Forest-Steppe Zones of Ukraine. The species is a transformer of meadow vegetation and forest-fringe communities. It transforms plant communities, creating unbearable conditions of competition [4,7]. It quickly occupies the areas of disturbed phytocenoses through the vegetative propagation by long rhizomes and due to high growth rate, which in combination with the transformation of habitat leads to significant changes in the structure of local vegetation and has negative consequences for phytobiota and zoobiota [1,8,9]. Thus, this species is able to form monodominant communities, maintain its own populations for a long time, lead to collapse in the recovery process, i.e., significantly extend the recovery period of indigenous plant communities of meadow vegetation [10–12].

The reasons for the intensive expansion of this species, despite the close attention of scientists around the world, remain controversial [13–16]. Two provisions are the most discussed. One of those is based on the high allelopathic potency of plants [8,17,18]. The chemical composition of *S. canadensis*, in particular the compounds with allelopathic activity, is well studied [19–22]. The existing metabolic analysis has revealed 122 metabolites, including flavonoids, phenylpropanoids and terpenoids. The synthesis of substances is shown to significantly depend on the cytotype and ploidy of plants [23]. The invasiveness of polyploids is exceptional [24,25], and is primarily due to the increased production of biologically active metabolites [26]. The cytotype and ploidy are often overlooked in plant collection, which may explain the difference in quantitative levels of individual flavonoids and other metabolites in plant material [27]. In addition, plants are characterized by the daily dynamics of the content of phenolic compounds in the vegetative organs [28].

The complex action of metabolites when released into the soil can be accompanied by a significant change in the rhizosphere. The ability of secondary metabolites of *S. canadensis* to inhibit the growth of not only allelochemical-sensitive test cultures, but also highly stress-resistant weeds [29,30] has been experimentally confirmed. However, the complexity of this approach is that the study of allelopathic action of plant metabolites is mainly performed in artificial conditions, where to accurately reproduce the processes that occur in real natural conditions is extremely difficult. This is confirmed by reports that acid rainfall can significantly enhance the allelopathic effects of invasive species, in particular *S. canadensis*, on seed germination and growth of aboriginal species [31]. According to our observations, *S. canadensis* displaces other plant species selectively and with varying intensity. This may be due to the competitive relationship between plants for limited resources. In this regard, the ability of invasive species to compete fiercely for mineral nutrition, mainly nitrogen, is also widely discussed. It is indicated that invasive species actively assimilate ammonium forms of nitrogen. However, the problem is that soil colloids carry positive and negative electrical charges, with the latter being predominant in most soils [32]. This makes it possible to fix NH_4^+ cations by soil colloids in an exchangeable form and to protect them from leaching. The size of the soil particles can also influence the fixation of ammonium nitrogen. However, there is evidence that clay particles and silt are equally good at retaining NH_4^+ [32]. The preferential uptake of NH_4+ over NO_3^- despite the need for additional costs of ATP for pumping protons through PM occurs in plants under conditions of limited energy metabolism [33], for example, at low temperatures [34].

At the same time, aboriginal species usually do not show clear selectivity for the assimilation of certain forms of nitrogen. In addition, it has been shown that many invasive plant species are able to significantly slow down the activity of soil enzymes associated with the C, N and P metabolism [35,36]. This reduces the availability of nutrients in the soil. Such restrictions and imbalances in mineral nutrition may contribute to the rapid spread of invasive species, which are able to use forms of nutrients that are less accessible to other plant species. For example, in clay soils, which are preferred by *S. canadensis*, fixation of NH_4^+ increases with increasing pH and base saturation [32]. In soils with a high pH, the content of flavonoids secreted by the roots of *Solidago canadensis* prevails over the pool of common phenols. At the same time, in acidic soils, these plants predominantly release phenols and less flavonoids [37]. Flavonoids can form complexes with biogenic metal ions and other cations. Of practical interest are organic complexes, which in the presence of appropriate natural conditions, can be formed in plant tissues in the process of mineral nutrition. The barrier functions of secondary metabolites of plants are also important in terms of their potential ability to bind to metal ions when they enter plant tissues in subtoxic concentrations. In addition, ammonium forms of nitrogen are able to form organic complexes with flavonoids. The new complexes have different physico-chemical properties and, accordingly, other molecular targets in physiological processes. Hence, the life strategy of *S. canadensis* plants, in addition to the ability to secrete active metabolites and affect the mineral nutrition of other plant species, also includes the ability to impact the activity of soil enzymes produced by microorganisms [38–40].

Thus, the purpose of our studies was to conduct a comparative phytochemical analysis of plants *S. canadensis* and *S. gigantea*, to determine the most active secondary metabolites that are able to interact with ammonium groups and determine the specifics of their impact on plant growth and microorganism activity in the rhizosphere.

2. Results

2.1. Comparative Analysis of the Secondary Metabolites in Leaves of Solidago Canadensis and Solidago Gigantea

More than 80 components, terpenoids, oxycinnamic and oxybenzoic acids, and 14 flavonoids (Figure 1) were identified by high-performance liquid chromatography (HPLC) in methanolic extracts of leaves and roots of *S. canadensis* and *S. gigantea*, collected in the park "Theofania" (Kyiv, Ukraine). In this work, we also relied on known studies of the biochemical composition of plants of the genus *Solidago* [41,42].

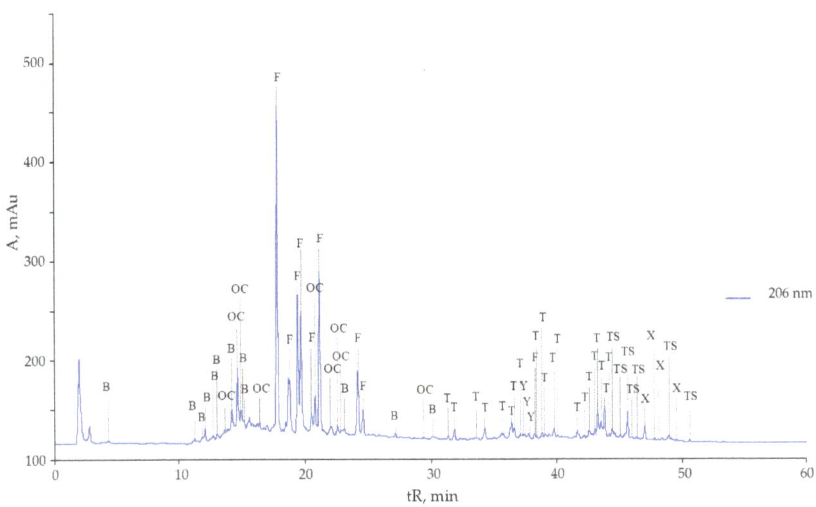

Figure 1. Chromatogram of methanolic extracts of *Solidago canadensis* L. leaves in the flowering phase: B—benzoic and oxybenzoic acids, OC—derivatives of oxycinnamic acids, F—flavonoids, T—terpenoids, TS—triterpenoid saponins, X—chlorophylls and their derivatives, Y—carotenoids and their derivatives.

UV spectra and retention times of the main peaks on the chromatogram in the presence of standards allowed identification of some flavonoids (Figure 2). Their main aglycones were kaempferol and quercetin. The largest share of the total pool of flavonoid glycosides in the leaves of the studied plants belonged to quercetin-3-*O*-beta-rutinoside (rutin).

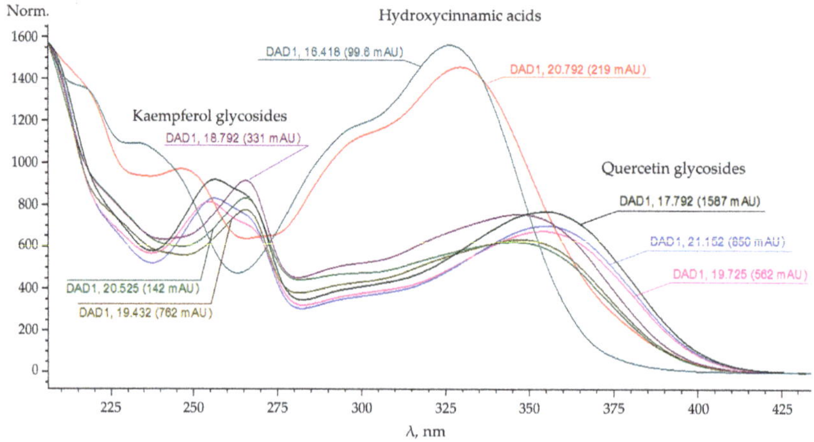

Figure 2. UV spectra of the main flavonoids of *Solidago canadensis* L.

The composition of the secondary metabolites of *S. gigantea* leaves has a very similar biochemical profile to *S. canadensis* (Figure 3). A very diverse composition of terpenoids and triterpenoid saponins was found in the leaves of *S. gigantea* plants in the flowering phase. Those compounds play an important role in the protective reactions of plants, regulation of water regime and interspecific relationships.

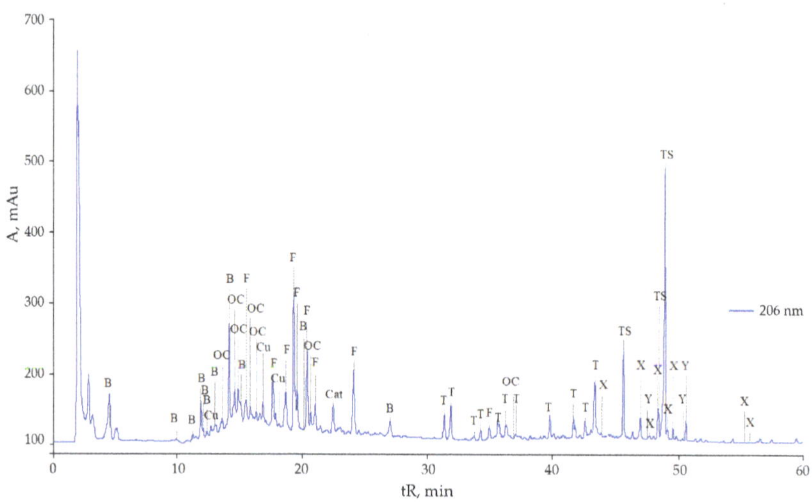

Figure 3. Chromatogram of methanolic extracts of *Solidago gigantea* Ait. leaves in the flowering phase: B—benzoic and oxybenzoic acids, OC—derivatives of oxycinnamic acids, Cu—coumarins, Cat—catechins, F—flavonoids, T—terpenoids, TS—triterpenoid saponins, X—chlorophylls and their derivatives, Y—carotenoids and their derivatives, ?— not identified.

However, it is interesting that the two studied species have an almost identical composition of flavonoids, although their number is less in *S. gigantea* (Table 1). The content of rutin was 9 times, and isoquercitrin was 5.3 times lower in the leaves of *S. gigantea* than in those of *S. canadensis* according to the height of the peaks on the chromatograms (Figures 1 and 3). However, leaves of *S. gigantea* showed a high content of presumably kaempferol-3-*O*-beta-rhamnoside (afselin). Also particularly noteworthy is the relatively high peak (tR 2.5 min) of a rather polar compound.

Table 1. Comparative evaluation of the composition of the main group of flavonoids in the leaves of *S. canadensis* and *S. gigantea*.

Flavonoid	*S. canadensis* Retention Time, min	mAU	*S. gigantea* Retention Time, min	mAU	Absorption Peak, nm	Ref.
Rutin (quercetin-3-*O*-beta-rutinoside)	17.762	1587	17.674	176	256, 355	[43]
Astragalin (kaempferol-3-*O*-beta-glucoside)	18.792	331	18.714	106	266, 350	[43]
Nicotiflorin (kaempferol-3-*O*-beta-rutinoside)	19.432	762	19.354	402	266, 347	[44]
Isoquercitrin (quercetin glycosid)	19.725	562	19.621	107	255, 355	[42]
Afzelin (kaempferol-3-*O*-beta-rhamnoside)	20.525	142	20.407	249	266, 347	-
Quercetin glycoside 1	21.152	850	21.048	70.8	256, 356	-
Kaemperfol glycoside	24.192	323	-	-	265, 347	-
Quercetin glycoside 2	24.592	142	-	-	254, 355	-

2.2. Modeling of Formation Processes of Organic Complexes of Flavonoids with Ammonium Nitrogen

Flavonoids are potentially capable of interacting with the ammonium forms of nitrogen. We simulated this process to study the ecophysiological properties of the generated phenol-ammonia complexes.

Equal volumes of aqueous ammonia solution (10%) were sequentially added to 100 mL of an aqueous extract of *S. canadensis* and *S. gigantea* leaves. When 10 μL of 10% aqueous NH_4OH solution was added to the extracts, they acquired an intense dark brown color (Figure S1). The darkest solutions were at pH 8.0–8.2. Given that the aqueous solution of NH_4OH is a weak base, in water it dissociates into NH_4^+ and OH^- ions. Under the conditions of binding of the ammonium group with 3-*O*-glycosides of flavonols, the balance of ions in the solution shifts towards the increase in OH^-, but not as rapidly as in the case of dissolution of ammonia in water. At the time of maximum complexation, the leaf extract exhibits buffering properties, maintaining alkalinity at pH 8.75. As the titrant volume increases further, the equilibrium gradually shifts toward decreasing pH (Figure 4).

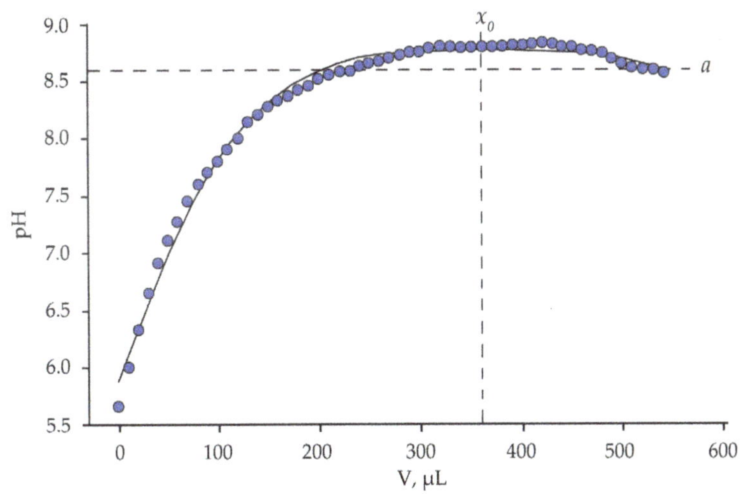

Figure 4. Lognormal model describing the dynamics of changes in the concentration of H^+ ions in the aqueous extract of *Solidago canadensis* L. leaves; x_0—the value corresponds to the volume of the titrant (NH_4OH under appropriate conditions) at which the pH is stabilized, *a*—the calculated coefficient of the lognormal model corresponding to the pH value in a balanced system.

The electrochemical content of the significant coefficients is related to the presence of metabolites in the extract, capable of complexation with NH_4^+ and a corresponding increase in the concentration of OH^- ions in the solution. The assumption that the NH_4^+ groups interact with flavonoids and organic acids with the formation of new complexes under the conditions of adding 10% aqueous solution of ammonia to the plant extract was confirmed by the results of chromatographic profiling.

On the chromatogram, the peaks corresponding to flavonoids completely disappeared or significantly decreased. Instead, two new peaks appeared at the beginning of the chromatogram with a fairly intense signal (Figure 5).

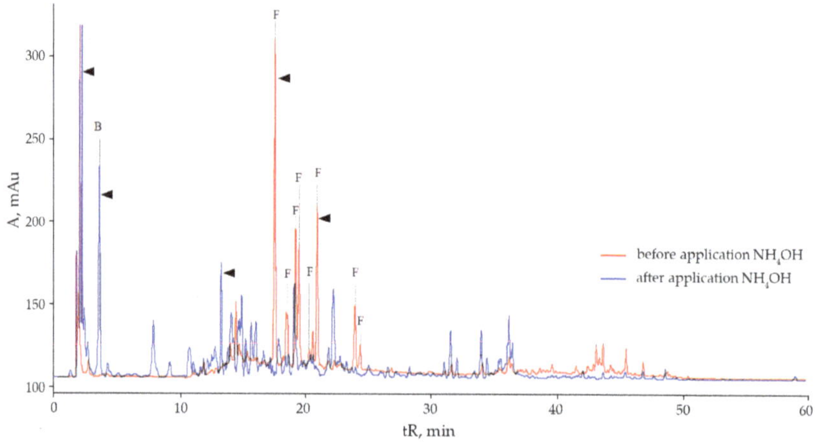

Figure 5. Chromatogram of aqueous extracts of *Solidago canadensis* L. leaves before (marked in red) and after (marked in blue) application of 10% NH_4OH solution; F—flavonoids, B—benzoic and oxybenzoic acids.

This indicates that the formation of new, rather polar compounds in the plant extract occurs mainly from mono- and diglycosides of flavonoids. An intense absorption peak with a retention time of 1.5 min was also found at the beginning of the chromatogram at a wavelength of 250 and 300 nm, and the next one is noted a little later (2.0 min) at a wavelength of 400 and 350 nm. Significant differences in the nature of light energy absorption by substances were detected in the flavonoids detected on chromatograms. For UV spectra, a hypsochromic shift for astragalin from 350 to 335 nm and a batachromic shift from 355 to 373 nm for isoquercitrin were established.

It was experimentally confirmed that under the presence of ammonium cations in plant tissues, the glucosides of kaempferol and quercetin are able to form polar nitrogen-containing complexes. Secondary metabolites of invasive species *S. gigantea* and *S. canadensis* are potentially able to regulate rooting processes. The leaf extracts of the studied species also include substances that can inhibit the development of the root system and growth processes in general. Potentially such compounds may include triterpenoid saponins, which are rich in leaves of the studied plants of the genus *Solidago*.

Principal component analysis confirmed that the biochemical composition of leaf extract is significantly altered by adding NH_4OH. The axis of the first principal component (PC1) reflects the maximum data variance, which was mostly determined by flavonoids, mainly quercetin glycosides (indicated by diamonds and highlighted by an oval on the right along the PC1 axis) (Figure 6).

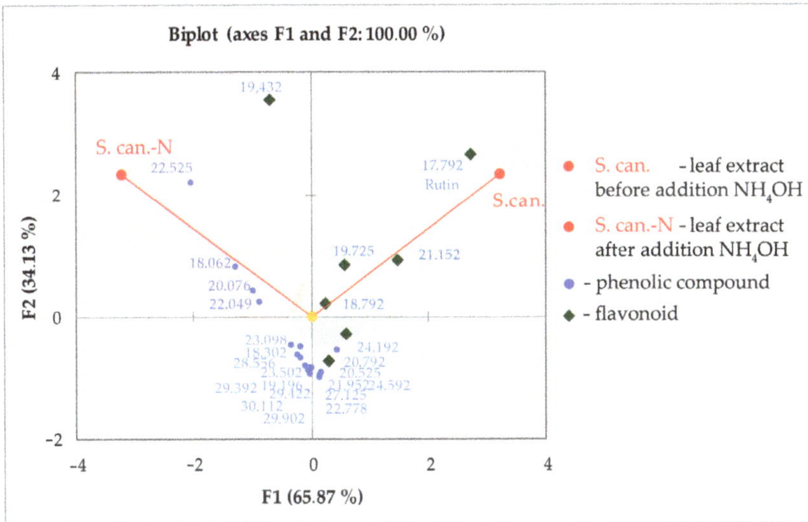

Figure 6. The result of the analysis of the chromatographic profiles of aqueous extracts of *Solidago canadensis* L. leaves by the method of analysis of principal components; numerical values correspond to the retention time of individual components in chromatography.

The largest contribution to the total variance was made by rutin (tR = 17.792) and quercetin glycoside (tR = 21.152). The contribution of kaempferol glycosides (tR = 18.792, 24.192, and 20.525) to the variance was less significant. However, it should be noted that the kaempferol glycoside (presumably nicotiflorin) showed the greatest variance on the PC2 axis. This flavonoid, together with oxycinnamic acid (tR = 22.525), was found in the largest residual amount in the extract after treatment with ammonium. From the point of view of biological activity, three phenolic compounds (indicated by blue circles and highlighted by an oval on the left along the main axis, tR = 18.062, 20.076 and 22.049) are of particular interest. These were detected in the extract (S. can.-N) only after adding the corresponding volume NH_4OH.

The proximity of individual substances on the coordinates of the axes of the principal components to the original composition of the extract (S. can.) indicates that they are contained in a relatively large amount and are easily transformed biochemically.

Comparative analysis of the biochemical composition of leaf extracts of two *Solidago* species and the results of PCA analysis made it possible to consider rutin as the most likely active substance. It was assumed that rutin in the root exudates of *S. canadensis* is able to react with ammonium cations of the soil and then, in the form of polar complexes, actively interact with native plant species and soil microorganisms.

When passing the extracts through the Al_2O_3 column, most flavonoids were retained or irreversibly bound to the sorbent (Figures S1–S2). The total content of kaempferol glycosides in leaf extracts was 2.0 times lower than quercetin. The ratio of the signal of the chromatograph detector at the peaks before and after adsorption decreased on average by 2.1–3.2 times (Table S1). It was found that Al_2O_3 has the highest adsorption capacity in relation to rutin and presumably afzelin. Their content in the extracts decreased more than 3 times. It was assumed that rutin in the root exudates of *S. canadensis* is able to react with ammonium cations of the soil and then, in the form of polar complexes, actively interact with native plant species and soil microorganisms.

2.3. Experimental Determination of Biological Activity of Flavonoids in Complex with Ammonium

The study of the effect of phenol-ammonium complexes revealed significant differences in their effect on the growth processes of different plant species. The polar complex obtained by adding an aqueous solution of ammonia to the extract inhibited the growth of roots of radish seedlings. The degree of growth inhibition (allelopathic effect) was strictly inversely correlated with the concentration of flavonoids (rs = −1.0, significance level $\alpha = 0.05$). The most active effect on root growth was observed at a concentration of flavonoids in the leaf extract of 50–100 μg/mL (Figure 7).

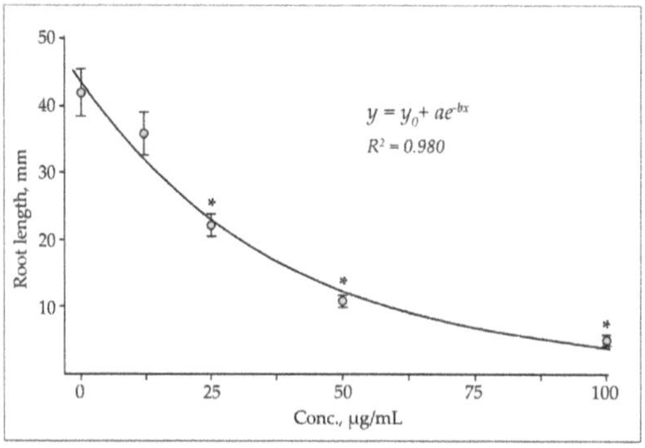

Figure 7. Inhibition of root growth of radish seedlings by phenol-ammonium complex depending on the content of flavonoids in the leaf extract of *Solidago canadensis* L.; the data were compared using Tukey's test (HSD) by one-way ANOVA: *—significant differences to the control at the level $p < 0.01$.

The leveling of the inhibitory effect depending on the concentration of active substances is fairly accurately approximated by an exponential function ($y = 0.7166 + 42.9168e^{-0.256}$, $R^2 = 0.980$). The significance of differences with the control ($p < 0.01$) was confirmed in the variants with the concentration of flavonoids in the initial extract exceeding 25 μg/mL.

The opposite effect was found when growing soybean seeds and processing chrysanthemum shoots. A moderate stimulating effect on root growth was determined for the aqueous extract of *Solidago gigantea,* which increased significantly under the formation of phenol-ammonia complexes (Figure 8a). The difference with the control was significant only in the extracts with the addition of 10% ammonia solution ($p < 0.01$). In the extract of the leaves of this plant species with a reduced content of flavonoids, the stimulating effect was leveled and inhibition of growth processes was observed. A similar but more pronounced effect was found with the use of *S. canadensis* extracts (Figure 8b). Under conditions of partial extraction of flavonoids with aluminum oxide, the effect of the aqueous extract of leaves on the growth of soybean roots was significantly reduced ($p < 0.01$) compared to the original extract and its modification with ammonia solution.

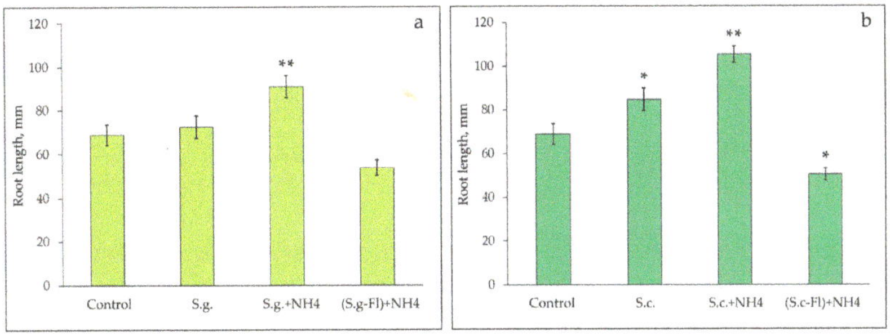

Figure 8. Influence of water extracts of *Solidago giganrea* (**a**) and *Solidago canadensis* (**b**) leaves on the growth of roots of *Glycine max* seedlings (n = 30): S.g., S.c.—initial extracts; S.g.+NH_4, S.c.+NH_4—extracts with added ammonia; (S.g.-Fl)+NH_4, (S.c.-Fl)+NH_4—extracts with added ammonia after adsorption in Al_2O_3 column; the data were compared using Tukey's test (HSD) by one-way ANOVA: *—significant differences to the control at the level $p < 0.05$; **—at the level $p < 0.01$.

An aqueous solution of the phenol-ammonia complex showed high efficiency when used as a root growth stimulator in grafting chrysanthemums. For three weeks of cultivation, the cuttings formed a fairly developed root system (Figure 9a).

Figure 9. Stimulating effect of phenol-ammonia complexes on the growth and development of chrysanthemums: cuttings of the variety 'Opal': a—treatment with rutin-ammonia complex, b—'Kornevin'preparation (control); leaves of chrysanthemum variety 'Valeria'after a single treatment: c—rutin-ammonia complex, d—'Kornevin' (IBA), e—phytocomplex (extract of *Solidago canadensis* L. leaves with aqueous ammonia solution), f—water; g—plants of the variety 'Queen of Autumn' grown as container crops, after a single treatment with phytocomplex and 'Kornevin' (IBA).

The planting material of studied varieties obtained by using the phenol-ammonia complex was significantly larger in terms of morphometric parameters. The plants had longer shoots and a more developed root system (Figure 9g). Notably, cuttings treated with a phenol-ammonium complex obtained on the basis of rutin had an increase in the linear dimensions and leaf area (Figure 9c). The obtained water-soluble complexes, even at a single application, can have a positive effect not only on the development of the root system, but also directly or indirectly stimulate the development of aboveground vegetative mass. Their prolonged stimulating effect indicates sufficient stability in the plant body and the ability to be transported by transport systems in the meristematic tissues. At the same time, the stimulating effect of flavonoid-ammonium complexes was different for different varieties of chrysanthemums. However, during the formation of the final habitus, characteristic of the varieties, the difference in growth rate under the action of stimulants was leveled.

Thus, the assumption that flavonoids play a key role in stimulating growth processes was experimentally confirmed. The active formation of lateral roots in seedlings also deserves special attention. This fact indicates the high activity of pericycle cells. They are probably stimulated by the quite polar phenol-ammonia complexes, which with the flow of water through the apoplast, quite quickly get into the tissues that respond to the initiation of the formation of lateral roots. Stimulation of root system development is an important element of plant survival strategy in conditions of fierce competition for spatial resources and nutrients, as well as in conditions of moisture deficit. For adventitious species with a wide amplitude of adaptive reactions, the active formation of root systems is one of the main features that ensures the successful naturalization of plants in new conditions.

2.4. Positive Chemotaxis of Rhizobial Bacteria to the Rutin-Ammonium Complex

Similar results were obtained when testing the newly formed complex of rutin with an aqueous solution of ammonia (Rutin—NH_4^+) on soybean seeds. The stimulation of root growth was observed at a concentration of 20 μg/mL. Increasing the concentration to 100 μg/mL inhibited seed germination and seedling formation (Figure 10). An essential feature of the action of the studied complex is its ability to stimulate the formation of lateral and adventive roots (Figure 10b,e). This specificity of root formation is characteristic of *S. canadensis* (Figure 10f). The actively forming lateral and adventive roots are located in the surface layer of the soil, which is the most populated by microorganisms. A significant surface area of young roots contributes to the active release of various metabolites into the soil, including phenolic substances.

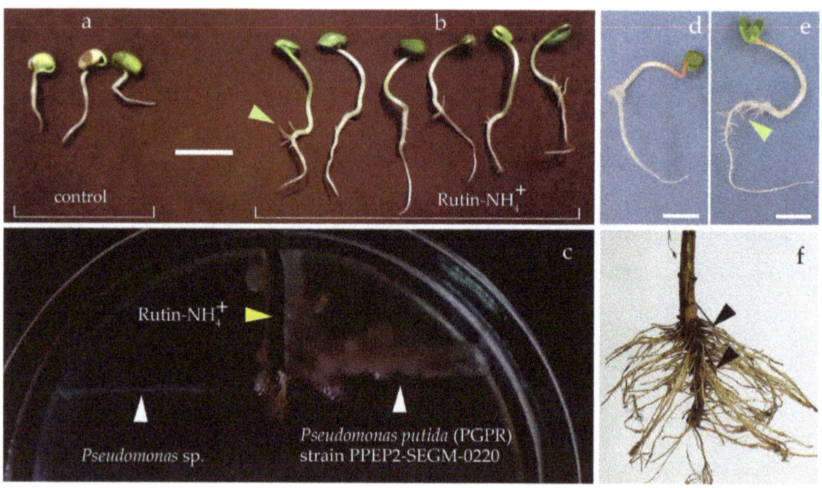

Figure 10. Effects of rutin-ammonia complex on seedlings and bacteria: a—control; b—stimulation of seed germination of test culture (*Glycine max* (L.) Merr.) with a solution of rutin-ammonium complex (20 μg/mL); c—positive chemotaxis of rhizobial bacteria *Pseudomonas putida* (PGPR) strain PPEP2-SEGM-0220 for 48 h; *Pseudomonas* sp.—bacteria that did not show chemotaxis; *Raphanus sativus* var *radicula* Pers. sprouts: d—control; e—after treatment (5 μg/mL); f—root of *Solidago canadensis;* b, e, f—arrows show active formation of lateral roots; scale bar: a, b—3 cm; d, e —1 cm.

Rhizobial bacteria *Pseudomonas putida strain* PPEP2-SEGM-0220, which we isolated from soybean seed skin and identified by 16S RNA, have growth-stimulating activity in relation to other representatives of legumes. In the presence of the Rutin–NH_4+ complex (5 μg/mL) in the nutrient medium, rhizobial bacteria showed positive chemotaxis (Figure 10c). In addition, the presence of this compound contributed to the active growth of the colony. At the same time, other bacteria of the genus *Pseudomonas* did not show explicit chemotaxis.

3. Discussion

In the plant body, phenolic compounds affect the hormonal regulation of morphogenesis [45,46] and perform a wide range of regulatory and protective functions [47–50]. Considering the invasive strategy of plant species, the biotransformation and complexation of secondary metabolites is especially interesting in the rhizosphere under conditions of their entry into the soil. It is known that under conditions of limited nitrogen nutrition in plant tissues, the content of total phenols and flavonoids increases [51]. Since the vast majority of phenylpropanoids and flavonoids are capable of complexation with cations, phenolic compounds can be transformed when they enter the soil. Thus, quercetin aglycones and glycosides actively interact with ammonium groups to form highly polar water-soluble compounds. Additionally, phenol-ammonia complexes are easily soluble in water and become available to the roots. This is important because the availability of nitrogen limits the nutrition of the vast majority of plants [52,53]. As we have shown, the newly created nitrogen-containing complexes can stimulate the growth of lateral roots and increase the total effective area of the root system. Flavonoids are actively secreted by the root system of *Solidago canadensis*. They are extremely reactive towards ammonia complexes. Polar compounds formed as a result of a chemical reaction are biologically active. Their regulating effect on growth processes is observed even at concentrations comparable to those of phytohormones.

However, an increase in the concentration of phenol-ammonia complexes by 3–5 times has a noticeable effect of suppressing the growth of test cultures, which can also be considered as an allelopathic effect. A similar effect was confirmed for kaempferol-3-*O*-β-D-glucoside isolated from the roots of *Solidago canadensis*. An increase in its concentration from 15 μg/mL and higher enhanced the suppression of shoot growth of *Echinochloa colona* (L.) Link. [54].

It is believed that different plant species differ in their ability to absorb various forms of nitrogen [55]. Invasive species are quite competitive for macro- and micronutrient nutrition in the soil, which in turn leads to the suppression of local flora [56]. Compared to aboriginal species, many invasive plants have the ability to effectively absorb nitrogen [57]. Thus, *Flaveria bidentis (L.) Kuntze*, a species invasive for northern China, significantly increases bioproductivity in response to increasing nitrogen content. In addition, the invasive species, compared to local species *Amaranthus retroflexus* L. and *Eclipta prostrata* (L.)L., showed a significant advantage over ammonium forms of nitrogen, while local species did not show a clear preference for certain forms of nitrogen [55].

According to the "novel weapons hypothesis", the root exudates of migratory species can severely suppress aboriginal plants, due to their maladaptation or lack of available nutrients. In this case, it is convincing to view the advantages of new competitive features as an alternative to the compromise of "growing or defending", which underlies the theory of evolution of increased competitiveness [58].

We believe the strategy of *S. canadensis* to be in the development of spatial and other vital resources as a set of features, which is due to the ability of plants to extensively produce secondary metabolites, which are actively deposited in the rhizosphere. Since the intensity of their synthesis directly depends on the ploidy of plants, it is expected that the greatest transformation of the environment will be due to the increase in populations of polyploid cytotypes. It is also important that *Solidago canadensis* plants under the conditions of introduction significantly enhance the synthesis of secondary metabolites with high allelopathic potential, which confirms their direct relationship with plant invasiveness [59].

An important condition for the successful invasion of alien invasive species is the formation of their positive feedback with soil microorganisms [60,61]. Recent studies have shown that content of quercetin is high in root exudates of invasive plants *Triadica sebifera* (L.) Small. The substance, as a part of exudates, is key in signaling for the interaction of plants with mycorrhizal fungi and soil microorganisms in general [62].

Mostly through the root secretions, *Solidago canadensis* plants affect the enzymatic activity of microorganisms, which in stable conditions usually provide the availability of nutrients for other plant species. The potential ability of phenolic compounds to interact with ammonium cations immobilized on the surface of soil particles and the formation of highly polar compounds, creates specific conditions for nitrogen nutrition of plants. This feature of the plant *Solidago canadensis* gives it an extraordinary advantage over other species. In addition, new polar complexes diffuse rapidly in soil solutions, are bioavailable, and at high concentrations (above 25 μg/mL) significantly inhibit the growth of root systems of competing plants.

Based on the experimental data we have obtained, three zones can presumably form around the root system of the *Solidago canadensis*: I—zone of high allelopathic activity (the content of flavonoids is above 100 μg/mL), II—zone of medium activity (50–100 μg/mL) and III—zone of low activity (25–50 μg/mL). In the zone I, the germination of seeds and the development of seedlings can be suppressed, in the zone II—inhibition of the growth of many plant species, in the zone III—inhibition of the growth of susceptible plant species. The spatial position and size of the designated zones are probably largely determined by the composition of the soil, the amount of moisture, the content of ammonium forms of nitrogen, and the intensity of biologically active substances released by plants (Figure 11).

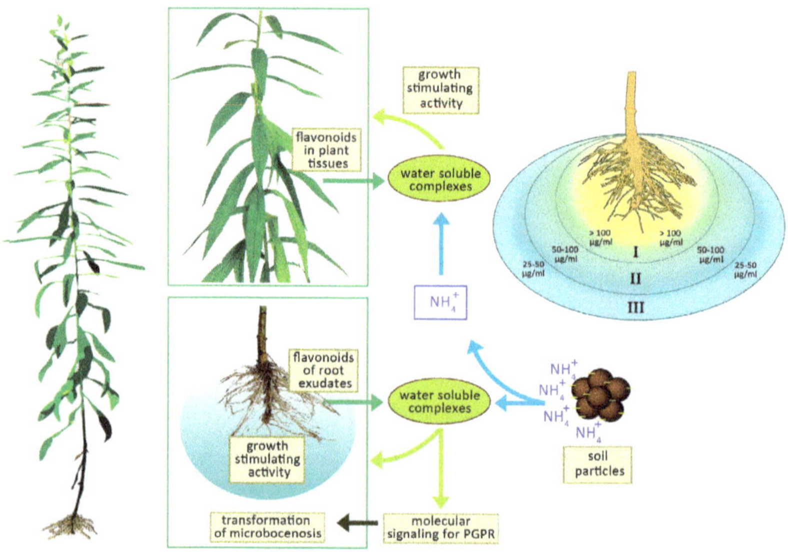

Figure 11. General scheme of the process of formation of water-soluble compounds in the case of interaction of flavonoids of *Solidago canadensis* L. with ammonia complexes that enter through the roots into plant tissues or are formed directly in the soil due to the active release of flavonoids by the root system.

Experimental confirmation that phenol-ammonia complexes cause positive chemotaxis in some species of rhizobial bacteria confirms that the plants are able to selectively change the composition of microorganisms of the rhizosphere and rhizoplan.

In the study of the influence of a complex of microorganisms on the nutrition of *Solidago canadensis* plants, it was shown that plants are positively affected by the higher species diversity of bacteria. It is also interesting to note that inoculation of the soil of model plants *Solidago canadensis* with microorganisms isolated in the zone of invasive distribution of this species in southern China caused a significant increase in plant bioproductivity, which enhances their competitive dominance. The complex of microorganisms from the pastures of Inner Mongolia, where this species has not yet spread, did not cause such an effect.

At the same time, a synergistic effect of a mixed complex of microorganisms on the formation of aboveground mass of plants was determined, which was much better in the presence of sufficient nutrients and moisture [63]. Moisture is a necessary condition for the complexation and transport of nutrients in the soil. Absorbed through the roots, phenol-ammonia complexes move acropetally along the transport system of xylem, as well as laterally through the parenchyma of xylem and bark. Low concentrations of these compounds initiate the formation of lateral roots from the pericycle. It is important to note that the compounds thus obtained are unstable, and in the process of deep oxidation can be converted into other nitrogen-containing organic substances that actively interact with phosphate ions and also form extremely active substances, the chemical structure of which has yet to be determined.

4. Materials and Methods

4.1. Collection and Phytochemical Studies of Plant Material

4.1.1. Source Plant Material

Plant material (ramets) of *Solidago gigantea* and *Solidago canadensis* was collected during the flowering phase in August-September 2018–2020. All ramets for analysis were obtained within the ruderal habitats of sand deposits in the vicinity of park "Theofania" in Kyiv, Ukraine (I 2.242 Ruderal biotopes of sand deposits, EUNIS I1.5 Bare tilled, fallow or recently abandoned arable land). Habitat types are listed according to the biotope classification for the Forest and Forest-Steppe Zones of Ukraine [64]. In the studied habitats, plant communities dominated by *Solidago gigantea* and *Solidago canadensis* form the association *Rudbeckio laciniatae-Solidaginetum canadensis* (Tüxen et Raabe ex Anioł-Kwiatkowska 1974). Geographic coordinates of collection site: 50°20′21.0″ N 30°30′03.1″ E and 50°20′16.0″ N 30°30′47.2″ E.

Leaf extraction was performed with methanol in the ratio of dry mass (DM) to alcohol: *v*/*v*—1/10. The aqueous extract was obtained from the leaves of *Solidago canadensis* and *Solidago gigantea* with double-distilled water (*v*/*v*—1/20). A portion of dry leaves was filled with hot water (T = +80 °C) and kept for 2 h in a water bath. The extract was filtered and stored at T = +4 °C.

4.1.2. Methods of High-Performance Liquid Chromatography

The leaves after drying were thoroughly ground and extracted with methanol at ratio of 100 mg of dry weight per 1 mL, respectively, for a day, in a room protected from light, at room temperature. The material in this state until chromatographic separation was stored in a freezer at −15 ... 20 °C, and immediately before analysis was filtered through a syringe filter 0.2 ... 0.5 µm. Samples were separated using reversed-phase high performance liquid chromatography (HPLC) on an Agilent 1100 system, according to the 2-eluent scheme (eluent I = 0.05 M aqueous solution of orthophosphoric acid; eluent II = acetonitrile) on a column of Thermo Scientific Hypersil ™ BDS C18. Sample volume 5 µL, column temperature 20 °C and after 45 min 40 °C, flow rate 0.3 mL/min and after 30 min −0.6 mL/min; analysis time up to 80 min, elution profile—isocratically 1% eluent II in eluent I for 2 min, then a linear gradient from 1% to 99% II in I for 30 min, and finally, isocrat 99% II in I for 20 min and more. Detection at wavelengths of 206, 254, 300, 350 and 450 nm to determine most organic compounds (including terpenoids), most substances of aromatic nature, phenylpropanoids (oxycinnamic acids and lignans), flavonoids (flavones and flavonols), carotenoids and chlorophylls, respectively. Table 2 shows the values that were used to assign signals to the chromatograph.

Table 2. Symbols of peaks on the chromatogram.

Compounds	Symbols
Derivatives of phenols, benzoic and oxybenzoic acids	B
Catechins	Cat
Coumarins	Cu
Derivatives of oxycinnamic acids	OC
Flavonoids	F
Sterines	S
Terpenoids	T
Triterpenoid saponins	TS
Chlorophylls and their derivatives	X
Carotenoids and their derivatives	Y

Absorption spectra in the ultraviolet and visible ranges were recorded for all substances in order to establish the nature of the secondary metabolites and to assign the chromatographic peaks to certain groups of substances. Standards of chlorogenic acid and rutin were used to verify retention time ranges, spectra, and approximate estimates. The β-sitosterol standard was used as a lipophilic component with low molar extinction.

4.2. *Preparation of Phenol-Ammonium Complexes from Plant Extracts and Flavonoids*

4.2.1. Obtaining Complexes Based on Aqueous Extracts of Leaves with a Solution of Ammonia

To obtain phenol-ammonia complexes of secondary metabolites, 10 μL of an aqueous solution of ammonia (10%) was gradually added to 100 mL of aqueous extracts of leaves of *Solidago canadensis* and *Solidago gigantea* (v/v—1/20) to a pH of 8.0. During the addition of NH_4OH, the extracts acquired an intense dark brown color. The darkest solutions were at pH 8.0–8.2. The dependence of pH value on the volume of NH_4OH is quite accurately described (R^2 = 0.9922) by the modified Gaussian Equation (1):

$$y = ae^{[-0.5(\frac{|x-x_0|}{b})^c]} \quad (1)$$

The coefficients of the regression equation are presented in Table 3.

Table 3. Coefficients of the modified Gaussian function describing the process of complex formation.

Coefficient and Its Value		Standard Error	t	*p*
a	8.7497	0.0138	635.8646	<0.0001
b	390.5112	6.4907	60.1646	<0.0001
c	3.9553	0.1341	29.4951	<0.0001
x_0	369.2351	6.0684	60.8459	<0.0001
R^2	0.9922	-		

In outlining the chemical processes, the value of x_0 corresponds to the maximum volume of titrant (NH_4OH under appropriate conditions), at the introduction of which the pH is stabilized. The coefficient *a* corresponds to the pH value, which is balanced in the system before and after titration. The coefficient *c* determines the steepness of the function (pH value) under the conditions of addition to the extract of NH_4^+ and OH^- ions. A decrease in *c* value in the equation may indicate the presence in the extract of a significant amount of organic acids and phenols capable of reacting with the formation of salts, nitrogen-containing organic compounds and water. At the same time, the buffering properties of the extract will decrease. The coefficient *b* in the equation has the opposite meaning. Its increase may indicate a relatively low pool of organic acids in the aqueous extract, but a much higher content of phenolic compounds, in particular flavonol glycosides, which are able to form complexes with the ammonium group.

4.2.2. Obtaining Polar Complexes Based on Rutin with an Aqueous Solution of Ammonia

A total of 10 mg of rutin (Merck, Darmstadt, Germany) was vortexed in 300 μL of double-distilled water for 60 s in a test tube. Then 100 μL of 10% aqueous ammonia solution was added and stirred again for 30–40 s until complete dissolution of the contents of the tube. The resulting solution was dark brown. The volume of the solution was adjusted with double-distilled water to 1 mL and stored at +4 °C. The resulting stock solution contained 10 mg/mL of polar substances.

4.2.3. Adsorption of Secondary Metabolites from Leaf Extracts of *Solidago Canadensis* and *Solidago Gigantea* on Al_2O_3 and Determination of Flavonoid Content

To determine the most active (against the test cultures) substances contained in the leaves of the test plants, 200 mg of Al_2O_3 powder (Merck, Darmstadt, Germany) for chromatography was added to 2 mL tubes (n = 4) to 1 mL of extract. The contents of the tube were vortexed thoroughly for 1 min and allowed to stand for 5 min at room temperature. The procedure was repeated 3 times. Next, these tubes were centrifuged for 5 min at 6000 rpm. The supernatant was carefully collected without the sorbent, transferred to new tubes and used for testing.

The quantitative content of flavonoids in leaf extracts before and after adsorption on Al_2O_3 was determined by spectrophotometry (Optizen Pop, Mecasys, *Daejeon*, South Korea) at λ = 419 nm. 200 μL of 0.1 M solution of aluminum chloride ($AlCl_3$) and 300 μL of 1 M sodium acetate (CH_3COONa) were added to 300 μL of extract. The calibration graph was based on quercetin (Sigma-Aldrich, Darmstadt, Germany). The phytochemical experiments were repeated 4 times.

4.3. Bioassay

4.3.1. Germination of Seeds of Test Crops and Conditions for Processing and Growing Plants

Seeds of *Raphanus sativus* var *radicula* Pers of the variety 'Rozovo-krasniy s belim konchikom' (bred in 1940 by the All-Russian Research Institute of Vegetable Breeding and Seed Production with the participation of the Biryuchekutskaya Vegetable Breeding Experimental Station) and soybeans *(Glycine max* (L.) Merr.) of the variety 'OAC Strive' (Canada) were used to determine the biological activity of *Solidago canadensis* and *Solidago gigantea* leaf extracts. A total of 30 and 20 seeds of test cultures, respectively, were germinated in Petri dishes on wet filter paper in a thermostat at 25 °C (n = 4). The effect of extracts based on ammonium complexes on growth processes of radish and soybean was assessed on length of seedlings roots on day 5.

To study the effect of phenol-ammonia complexes of *Solidago canadensis* leaf extract on the processes of rhizogenesis, three varieties of chrysanthemums *(Chrysanthemum x koreanum Makai)* 'Opal', 'Valeria' and 'Queen of Autumn' from the collection of the Institute of Evolutionary Ecology of the NAS of Ukraine were used. The lower ends of freshly cut chrysanthemum shoots (n = 30) with 4–5 nodes were immersed for 6 h in water, to which was added the resulting complex (*v*/*v*–1/20). The control batch of shoots (n = 30) was immersed in water for 5 h 40 min. Then the control shoots were soaked for 20 min in a solution of 'Kornevin', which was prepared fresh, according to the manufacturer's recommendations. The active substance in this preparation is indolyl butyric acid (IBA), a well-known phytohormone of the auxin class. An amount of 1 L of prepared solution for rooting of shoots contains 5 mg of IBA. After soaking, the experimental and control batches of shoots were transferred into containers with perlite for 21 days for rooting. Next, the cuttings were transplanted into containers with a soil mixture based on chernozem-peat-perlite (*v*/*v*/*v*–1/1/1) and grown for 1 month.

4.3.2. Study of Chemotaxis of *Pseudomonas* Sp. Bacteria

To study the chemotaxis of bacteria in relation to the complex obtained by the interaction of rutin with an aqueous solution of ammonia, we used the previously isolated strain *Pseudomonas putida* PPEP2-SEGM-0220 (GenBank: MW255059.1), which by a number of indicators belongs to the group PGPR (plant growth promoting rhizobacteria); *Pseudomonas* sp. isolate, which did not show the ability to stimulate plant growth and development, served as a control. A shallow narrow groove was made in King B Medium on Petri dishes to introduce phenol-ammonia complex at a concentration of 5 μg/mL. Bacterial cultures were applied in a dash across the groove. Petri dishes were kept in a thermostat for 28 °C for 2 days. The nature of growth and the distance of bacterial colonies to the complex introduced into the groove were determined.

4.4. Methods of Digital and Statistical Data Processing

Photodocumentation and digital image processing were performed in a specialized program Image-Pro Premier 9.0 (Media Cybernetics, Rockville, MD, USA). The significance of the differences between the values ($p < 0.05$) was determined by the analysis of variance (ANOVA) method in the XLSTAT (Addinsoft Inc., New York, NY, USA, 2010). The data were compared using Tukey's test. The principal components analysis (PCA) was performed in the XLSTAT program. SigmaPlot 12.0 (Systat Software Inc., San Jose, CA, USA, 2011) was used for regression analysis.

5. Conclusions

The analysis of flavonoids in extracts of the invasive species *Solidago canadensis* revealed a significant increase in the content of rutin (quercetin-3-*O*-beta-rutinoside) in its composition compared to *Solidago gigantea*. When entering the rhizosphere with exudates, flavonoids (mainly quercetin glycosides) can interact with ammonium cation, either free or immobilized on negatively charged surfaces of soil particles. As a result, new polar compounds are formed. If there is enough water in the soil, they are able to actively diffuse. The effect of the phenol-ammonia complex on plants of different families (*Asteraceae* Bercht. & J. Presl, *Fabaceae* Lindl., *Brassicaceae* Burnett) differs not only in the strength of the effect, but also in the influence (stimulation or suppression). At low concentrations (20 μg/mL), these substances stimulate the formation of lateral and adventitious roots in soy-bean and chrysanthemum plants, and in concentrations more than 100 μg/mL) inhibit them. Radish sprouts are more susceptible to aqueous extracts of *Solidago canadensis* leaves. The suppression of the growth of their roots is significant at a concentration of flavonoids in the extract of 25 μg/mL. One of the possible explanations for the selective action of the phenol-ammonia complex may be the similarity or difference of metabolic pathways associated with growth-regulation mechanisms in the studied plants. However, to prove this assumption, it is necessary to carry out special studies on a larger number of species from these families.

Rutin, after interacting with ammonium in the form of a new complex, easily dissolves in water and causes positive chemotaxis in symbiotic bacteria of the PGPR group (plant growth promoting rhizobacteria), which stimulate the formation of the root system and enter into competitive relationships with phytopathogenic microorganisms. The formation of a developed system of lateral and adventitious roots contributes to the use of the most nutrient-rich surface layers of the soil inhabited by microorganisms. Enhanced synthesis and release of flavonoids, therefore, is an important element of the strategy for the development and biotransformation of a new habitat.

Supplementary Materials: The following are available online at https://www.mdpi.com/article/10.3390/plants10081748/s1, Figure S1: Color of aqueous extracts of leaves *S. gigantea* and *S. canadensis*, with water (H_2O) and 10% NH_4OH solution; dilution of extracts—*v*/*v*–1/1; Figure S2: Chromatogram of aqueous extracts of *S. canadensis* leaves before adsorption on Al_2O_3; Figure S3: Chromatogram of aqueous extracts of *S. canadensis* leaves after adsorption on Al_2O_3; Table S1: Adsorption capacity of Al_2O_3 for flavonoids from aqueous extract *S. canadensis*.

Author Contributions: Conceptualization, A.L., N.P. and M.O.; methodology, A.L.; software, M.O.; validation, A.L. and N.P.; formal analysis, M.K. and A.C.; investigation, A.L. and M.K.; resources,

M.K.; data curation, M.O.; writing—original draft preparation, A.L. and A.C.; writing—review and editing, A.L. and A.C.; visualization, M.O.; supervision, A.L.; project administration, N.P. All authors have read and agreed to the published version of the manuscript.

Funding: The research was financed by the National Academy of Sciences of Ukraine within the scientific program N 0117U004320 "Strategies of adaptation of species of plants, mushrooms and animals at different levels of anthropogenic transformation of the environment".

Institutional Review Board Statement: Not applicable.

Informed Consent Statement: Not applicable.

Data Availability Statement: The data presented in this study are available on request from the corresponding author.

Acknowledgments: We would like our gratitude to M. Mazura and E. Leshcheniuk for their help in the study in studying the effect of biological activity of the obtained substances on model chrysanthemum plants in greenhouses.

Conflicts of Interest: The authors declare that there is no conflict of interest related to this article.

References

1. Burda, R.I.; Pashkevych, N.A.; Boiko, G.V.; Fitsailo, T.V. *Alien Species of the Protect Floras of Forest-Steppe of Ukraine*; Naukova Dumka: Kyiv, Ukraine, 2015; 120p.
2. Chuj, O.V.; Shumskaja, N.V. Biometric parameters of generative species *Pulsatilla patens* (L.) Mill. In *Cenopopulations of Western Podillia*; Cherkasy University Bulletin: Biological Sciences Series; Cherkasy University: Cherkasy, Ukraine, 2014; Volume 36, pp. 124–128. (In Ukrainian)
3. Kabuce, N.; Priede, N. NOBANIS—Invasive Alien Species Fact Sheet—Solidago Canadensis. From: Online Database of the European Network on Invasive Alien Species—NOBANIS. 2010. Available online: www.nobanis.org (accessed on 18 August 2021).
4. Pashkevych, N.; Burda, R. Distribution of alien species from Poaceae and Asteraceae families in the protected areas of Ukrainian forest-steppe. *Thaiszia J. Bot.* **2017**, *27*, 29–39.
5. Protopopova, V.V.; Shevera, M.V. Ergasiophytes of the Ukrainian flora. *Biodiv. Res. Conserv.* **2014**, *35*, 31–46. [CrossRef]
6. Protopopova, V.V.; Shevera, M.V. Invasive species in the flora of Ukraine. I. The group of highly active species. *Geo Bio.* **2019**, *17*, 116–135. (In Ukrainian) [CrossRef]
7. Abduloyeva, O.S.; Karpenko, N.I. Occurrence of alien invasive plant species in vegetation syntaxa of Ukraine. *Chornomorski Bot. J.* **2009**, *5*, 189–198. (In Ukrainian) [CrossRef]
8. Abhilasha, D.; Quintana, N.; Vivanco, J.; Joshi, J. Do allelopathic compounds in invasive *Solidago canadensis* s.l. restrain the native European flora? *J. Ecol.* **2008**, *96*, 993–1001. [CrossRef]
9. Oitsius, L.V.; Volovyk, P.H.; Doletskyi, S.P.; Lysytsya, A.V. Distribution of adventive species *Solidago canadensis, Phalacroloma annuum, Ambrosia artemisiifolia, Heracleum sosnowsky* in phytocenoses of Volyn' Polissya (Ukraine). *Biosyst. Divers.* **2020**, *28*, 343–349. [CrossRef]
10. Gusev, A.P. The impact of invasive Canadian Goldenrod (*Solidago canadensis* L.) on regenerative succession in old fields (The Southeast of Belarus). *Russ. J. Biol. Invasions* **2015**, *1*, 10–16. (In Russian) [CrossRef]
11. Gusev, A.P. The invasion of Canadian Goldenrod (*Solidago canadensis* L.) into anthropogenic landscapes of Belarus. *Russ. J. Biol. Invasions* **2017**, *4*, 28–35. (In Russian) [CrossRef]
12. Olijnyk, M.P. Variability of morphological signs of populations *Solidago Canadensis* L. and *Solidago Gigantea* Ait. In *Different Ecological Conditions*; Scientific Bulletin of the National Forestry University of Ukraine: Lviv, Ukraine, 2018; Volume 28, pp. 60–63.
13. Keane, R.M.; Crawley, M.J. Exotic plant invasions and the enemy release hypothesis. *Trends Ecol. Evol.* **2002**, *17*, 164–170. [CrossRef]
14. Liao, M.; Xie, X.; Peng, Y.; Chai, J.; Chen, N. Characteristics of soil microbial community functional and structure diversity with coverage of *Solidago canadensis* L. *J. Centr. South Univ.* **2013**, *20*, 749–756. [CrossRef]
15. Inderjit; Callaway, R.M.; Vivanco, J.M. Can plant biochemistry contribute to understanding of invasion ecology? *Trends Plant Sci.* **2006**, *11*, 574–580. [CrossRef]
16. Wang, C.; Jiang, K.; Zhou, J.; Wu, B. *Solidago canadensis* invasion affects soil N-fixing bacterial communities in heterogeneous landscapes in urban ecosystems in East China. *Sci. Total Environ.* **2018**, *631–632*, 702–713. [CrossRef]
17. Zandi, P.; Barabasz-Krasny, B.; Stachurska-Swakon, A.; Pula, J.; Mozdzen, K. Allelopathic effect of invasive Canadian goldenrod (*Solidago canadensis* L.) on early growth of red clover (*Trifolium pratense* L.). *Not. Bot. Horti Agrobot. Cluj-Napoca* **2020**, *48*, 2060–2071. [CrossRef]
18. Xie, L.J.; Zeng, R.S.; Bi, H.H.; Song, Y.Y.; Wang, R.L.; Su, Y.J.; Chen, M.; Chen, S.; Liu, Y.H. Allelochemical mediated invasion of exotic plants in China. *Allelopath. J.* **2010**, *25*, 31–50.
19. Choi, S.Z.; Sang, U.C.; Kang, R.L. Phytochemical constituents of the aerial parts from *Solidago virga-aurea* var. gigantea. *Arch. Pharm. Res.* **2004**, *27*, 164–168. [CrossRef]

20. Iason, G.R.; Dicke, M.; Hartley, S.E. *The Ecology of Plant Secondary Metabolities. From Genes to Global Process*; University Press: Cambridge, UK, 2012; 335p.
21. Radušiene, J.; Marska, M.; Ivanauskas, L.; Jakstas, V.; Karpaviciene, B. Assessment of phenolic compound accumulation in two widespread goldenrods. *Ind. Crop. Prod.* **2015**, *63*, 158–166. [CrossRef]
22. Avertseva, I.N.; Suleymanova, F.S.; Nesterova, O.V.; Reshetnyak, V.Y.; Matveenko, V.N.; Zhukov, P.A. Study of Polyphenolic Compounds in Extracts from Flowers and Leaves of Canadian Goldenrod and Dwarf Goldenrod (*Solidago canadensis* L. and *Solidago nana* Nitt.). *Mosc. Univ. Chem. Bull.* **2020**, *75*, 47–51. [CrossRef]
23. Wu, M.; Ge, Y.; Xu, C.; Wang, J. Metabolome and Transcriptome Analysis of Hexaploid *Solidago canadensis* Roots Reveals its Invasive Capacity Related to Polyploidy. *Genes* **2020**, *11*, 187. [CrossRef] [PubMed]
24. Te Beest, M.; Roux, J.J.L.; Richardson, D.M.; Brysting, A.K.; Suda, J.; Kubešová, M.; Pyšek, P. The more the better? The role of polyploidy in facilitating plant invasions. *Ann. Bot.* **2012**, *109*, 19–45. [CrossRef] [PubMed]
25. Rosche, C.; Hensen, I.; Mráz, P.; Durka, W.; Hartmann, M.; Lachmuth, S. Invasion success in polyploids: The role of inbreeding in the contrasting colonization abilities of diploid versus tetraploid populations of *Centaurea stoebe* s.l. *J. Ecol.* **2017**, *105*, 425–435. [CrossRef]
26. Lavania, U.C.; Sarita, S.; Seshu, L.; Surochita, B.; Nandeesh Kumar, M.; Yasuhiko, M. Autopolyploidy differentially influences body size in plants, but facilitates enhanced accumulation of secondary metabolites, causing increased cytosine methylation. *Plant J.* **2012**, *71*, 539–549. [CrossRef]
27. Zekic, J.; Vovk, I.; Glavnik, V. Extraction and Analyses of Flavonoids and Phenolic Acids from Canadian Goldenrod and Giant Goldenrod. *Forests* **2021**, *12*, 40. [CrossRef]
28. Subin, O.V.; Likhanov, A.F.; Sereda, O.V.; Klyuvadenko, A.A.; Melnychuk, M.D. Dynamics of phenolic compounds in leaves Fragaria ananassa Duch. after treatment with different forms of chitosan. *BioTechnologia* **2020**, *101*, 227–237. [CrossRef]
29. Baležentiene, L. Secondary metabolite accumulation and phytotoxicity of invasive species *Solidago canadensis* L. during the growthperiod. *Allelopath. J.* **2015**, *35*, 217–226.
30. Mozdzen, K.; Barabasz-Krasny, B.; Zandi, P.; Kliszcz, A.; Puła, J. Effect of Aqueous Extracts from *Solidago Canadensis* L. Leaves on Germination and Early Growth Stages of Three Cultivars of *Raphanus sativus* L. var. *Radicula Pers. Plants* **2020**, *9*, 1549.
31. Wang, C.; Xiao, H.; Zhao, L.; Liu, J.; Wang, L.; Zhang, F.; Du, D. The allelopathic effects of invasive plant *Solidago canadensis* on seed germination and growth of *Lactuca sativa* enhanced by different types of acid deposition. *Ecotoxicology* **2016**, *25*, 555–562. [CrossRef]
32. Stevenson, F.J.; Nommik, H.; Vahtras, K. Retention and Fixation of Ammonium and Ammonia in Soils. *Agron. Monograph.* **1982**, *22*, 123–171.
33. Forde, B.G.; Clarkson, D.T. Nitrate and Ammonium Nutrition of Plants: Physiological and Molecular Perspectives. *Adv. Bot. Res.* **1999**, *30*, 1–90.
34. Macduff, J.H.; Jackson, S.B. Growth and Preferences for Ammonium or Nitrate Uptake by Barley in Relation to Root Termperature. *J. Exp. Bot.* **1991**, *42*, 521–530. [CrossRef]
35. Hu, Z.; Li, J.; Shi, K.; Ren, G.; Dai, Z.; Sun, J.; Zheng, X.; Zhou, Y.; Zhang, J.; Li, G.; et al. Effects of Canada Goldenrod Invasion on Soil Extracellular Enzyme Activities and Ecoenzymatic Stoichiometry. *Sustainability* **2021**, *13*, 3768. [CrossRef]
36. Dassonville, N.; Guillaumaud, N.; Piola, F.; Meerts, P.; Poly, F. Niche construction by the invasive Asian knotweeds (species complex Fallopia): Impact on activity, abundance and community structure of denitrifiers and nitrifiers. *Biol. Invasions* **2011**, *13*, 1115–1133. [CrossRef]
37. Zhang, S.; Zhu, W.; Wang, B.; Tang, J.; Chen, X. Secondary metabolites from the invasive *Solidago canadensis* L. accumulation in soil and contribution to inhibition of soil pathogen *Pythium ultimum*. *Appl. Soil Ecol.* **2011**, *48*, 280–286. [CrossRef]
38. Liao, M.; Xie, X.; Peng, Y.; Ma, A. Changes of Soil Microbiological Characteristics after *Solidago canadensis* L. Invasion. *Agric. Sci. China* **2011**, *10*, 1064–1071. [CrossRef]
39. Li, G.; Kim, S.; Han, S.H.; Chang, H.; Du, D.; Son, Y. Precipitation affects soil microbial and extracellular enzymatic responses towarming. *Soil Biol. Biochem.* **2018**, *120*, 212–221. [CrossRef]
40. Ge, Y.; Wang, Q.; Wang, L.; Liu, W.; Christie, P. Response of soil enzymes and microbial communities to root extracts of the alien *Alternanthera philoxeroides*. *Arch. Agron. Soil Sci.* **2018**, *64*, 708–717. [CrossRef]
41. Apati, P.; Szentmihályi, K.; Balázs, A.; Baumann, D.; Hamburger, M.; Kristó, T.S.; Szoke, É.; Kéry, Á. HPLC analysis of the flavonoids in pharmaceutical preparations from Canadian Goldenrod (*Solidago canadensis*). *Chromatographia* **2002**, *56*, S65–S68. [CrossRef]
42. Papp, I.; Apati, P.; Andrasek, V.; Blazovics, A.; Balazs, A.; Kursinszki, L.; Kite, G.C.; Houghton, P.J.; Kery, A. LC-MS Analysis of Antioxidant Plant Phenoloids. *Chromatographia* **2004**, *60*, 93–100. [CrossRef]
43. Moiseev, D.V.; Buzuk, G.N.; Sheluto, V.L. HPLC identification of flavonoids in plants. *Khimiko-Farmatsevticheskii Zhurnal* **2011**, *45*, 35–38. (In Russian) [CrossRef]
44. Kurkina, A.V. *Flavonoids of Pharmacopoeial Plants*; Ophort: Samara, Russia, 2012; 290p. (In Russian)
45. Buer, C.S.; Imin, N.; Djordjevic, M.A. Flavonoids: New roles for old molecules. *J. Integr. Plant Biol.* **2010**, *52*, 98–111. [CrossRef]
46. Volynets, A.P. *Phenolic Compounds in the Life of Plants*; Belaruskaya Navuka: Minsk, Russia, 2013; 283p. (In Russian)
47. Graf, E. Antioxidant potential of ferulic acid. *Free. Radic. Biol. Med.* **1992**, *13*, 435–448. [CrossRef]
48. Zaprometov, M.N. *Phenolic Compounds. Distribution, Metabolism and Function in Plants*; Nauka: Moscow, Russia, 1993; 272p. (In Russian)
49. Apati, P.; Szentmihalyi, K.; Kristo, S.T.; Papp, I.; Vinkler, P.; Szoke, E.; Kery, A. Herbal remedies of Solidago—Correlation of phytochemical characteristics and antioxidative properties. *J. Pharm. Biomed. Anal.* **2003**, *32*, 1045–1053. [CrossRef]

50. Grana, E.; Costas-Gil, A.; Longueira, S.; Celeiro, M.; Teijeira, M.; Reigosa, M.J.; Sanchez-Moreiras, A.M. Auxin-like effects of the natural coumarin scopoletin on *Arabidopsis* cell structure and morphology. *J. Plant Physiol.* **2017**, *218*, 45–55. [CrossRef]
51. Ibrahim, M.H.; Jaafar, H.Z.; Rahmat, A.; Rahman, Z.A. Effects of Nitrogen Fertilization on Synthesis of Primary and Secondary Metabolites in Three Varieties of Kacip Fatimah (*Labisia Pumila Blume*). *Int. J. Mol. Sci.* **2011**, *12*, 5238–5254. [CrossRef] [PubMed]
52. Bobbink, R.; Hicks, K.; Galloway, J.; Spranger, T.; Alkemade, R.; Ashmore, M.; Bustamante, M.; Cinderby, S.; Davidson, E.; Dentener, F.; et al. Global assessment of nitrogen deposition effects on terrestrial plant diversity: A synthesis. *Ecol. Appl.* **2010**, *20*, 30–59. [CrossRef] [PubMed]
53. Boonstra, R.; Krebs, C.J.; Cowcill, K. Responses of key understory plants in the boreal forests of western North America to natural versus anthropogenic nitrogen levels. *For. Ecol. Manag.* **2017**, *401*, 45–54. [CrossRef]
54. Li, J.; Ye, Y.; Huang, H.; Dong, L. Kaempferol-3-O-β-D-glucoside, a potential allelochemical isolated from Solidago Canadensis. *Allelopath. J.* **2011**, *28*, 259–266.
55. Huangfu, C.H.; Li, H.Y.; Chen, X.W.; Liu, H.M.; Wang, H.; Yang, D.L. Response of an invasive plant, Flaveria bidentis, to nitrogen addition: A test of form-preference uptake. *Biol. Invasions* **2016**, *18*, 3365–3380. [CrossRef]
56. Ehrenfeld, J.G. Effects of exotic plant invasions on soil nutrient cycling processes. *Ecosystems* **2003**, *6*, 503–523. [CrossRef]
57. Matzek, V. Trait Values, Not Trait Plasticity, Best Explain Invasive Species Performance in a Changing Environment. *PLoS ONE* **2012**, *7*, e48821. [CrossRef] [PubMed]
58. Callaway, R.M.; Ridenour, W.M. Novel weapons: Invasive success and the evolution of increased competitive ability. *Front. Ecol. Environ.* **2004**, *2*, 436–443. [CrossRef]
59. Yuan, Y.G.; Wang, B.; Zhang, S.S.; Tang, J.J.; Tu, C.; Hu, S.J.; Yong, J.W.H.; Chen, X. Enhanced allelopathy and competitive ability of invasive plant Solidago canadensis in its introduced range. *J. Plant Ecol.* **2013**, *6*, 253–263. [CrossRef]
60. Bever, J.D. Negative feedback within a mutualism: Host specific growth of mycorrhizal fungi reduces plant benefit. *Proc. R. Soc. B Biol. Sci.* **2002**, *269*, 2595–2601. [CrossRef] [PubMed]
61. Callaway, R.M.; Bedmar, E.J.; Reinhart, K.O.; Silvan, C.G.; Klironomos, J. Effects of soil biota from different ranges on *Robinia invasion*: Acquiring mutualists and escaping pathogens. *Ecology* **2011**, *92*, 1027–1035. [CrossRef] [PubMed]
62. Tian, B.; Pei, Y.; Huang, W.; Ding, J.; Siemann, E. Increasing flavonoid concentrations in root exudates enhance associations between arbuscular mycorrhizal fungi and an invasive plant. *ISME J.* **2021**, *15*, 1919–1930. [CrossRef] [PubMed]
63. Adomako, M.O.; Xue, W.; Tang, M.; Du, D.L.; Yu, F.-H. Synergistic Effects of Soil Microbes on *Solidago canadensis* Depend on Water and Nutrient Availability. *Microb. Ecol.* **2020**, *80*, 837–845. [CrossRef]
64. Didukh, Y.P.; Fitsailo, T.V.; Korotchenko, I.A.; Iakushenko, D.M.; Pashkevych, N.A. *Biotopes of Forest and Forest-Steppe Zone of Ukraine*; Didukh, Y.P., Ed.; LLC Macros: Kyiv, Ukraine, 2011; 288p.

plants

Review

A Path Forward: Promoting Microbial-Based Methods in the Control of Invasive Plant Species

Maryam Shahrtash and Shawn P. Brown *

Department of Biological Sciences, The University of Memphis, Memphis, TN 38152, USA; maryam.shahrtash@memphis.edu
* Correspondence: spbrown2@memphis.edu; Tel.: +1-901-678-3596

Abstract: In this review, we discuss the unrealized potential of incorporating plant–microbe and microbe–microbe interactions into invasive plant management strategies. While the development of this as a viable strategy is in its infancy, we argue that incorporation of microbial components into management plans should be a priority and has great potential for diversifying sustainable control options. We advocate for increased research into microbial-mediated phytochemical production, microbial controls to reduce the competitiveness of invasive plants, microbial-mediated increases of herbicidal tolerance of native plants, and to facilitate increased pathogenicity of plant pathogens of invasive plants.

Keywords: microbial consortia; invasive plants; integrated pest management; endophytes; disease facilitation

1. Introduction

Plants serve as habitats for microbes and microbial communities, which can colonize every plant tissue type [1–3]. These includes endophytes, which colonizes inter- and intracellular spaces within leaves, stems, and roots but are asymptomatic on the host [4], as well as epiphytes which colonize external surfaces of plant tissues. These microbes have varied direct and indirect interactions with plants that range from antagonistic (negative), mutualistic (positive) and everything in between [5]. In plant–microbe mutualisms, plants often release compounds that attract and feed the associated microbes, which may in turn secrete compounds that improve plant health and growth, thereby enhancing nutrient acquisition or making plants more resistant to environmental stressors. Prominent examples of mutualistic plant-microbe interactions are the symbioses between plants and nitrogen fixing bacteria and/or mycorrhizal fungi, that help plants grow in soils with low nutrient quality [6]. Conversely, antagonistic microorganisms negatively affect plant growth and/or health, which may be due to direct pathogenicity or a reduction in nutrient uptake rates. However, while we are beginning to better understand plant–microbiome interaction mechanisms [7], there is much about these interactions that remains unresolved [8]. With increased study of these interactions, experimental frameworks are beginning to emerge to investigate how microbiome manipulations can be best done to achieve management goals.

To aid in the control of invasive plants, there has been increased interest in the development and utilization of microbial biocides as targeted direct biocontrol agents [9,10]. Several fungal biocides have shown promise in helping to control invasive plants, including *Fusarium oxysporum*, *Fusarium ploriferatum*, and *Trichoderma koningiopsis* which can partially control the invasive *Euphorbia heterophylla* (Mexican Fire Plant) [11], and *Albifimbria verrucaria* (formally *Myrothecium verrucaria* (Stachybotryaceae) [12]), which has been demonstrated to have biocidal action on numerous invasive plants including Kudzu [10], *Lygodium microphyllum* (old world climb fern) [13] as well as *Salvinia molesta* (floating fern) [14]. Additionally, the fungal genera *Colletotrichum*, *Phoma*, and *Sclerotinia*, as well as bacteria within in the genera *Xanthomonas* and *Pseudomonas* have also been demonstrated to have broad biocidal qualities [15]. However, complex multi-partite microbe–microbe interactions within plants can act in unforeseen ways to limit or modulate targeted goals. Plant-microbiome manipulative investigations are underexplored but have been suggested as a novel tool for invasive plant management [16]. Here, we argue that microbiome manipulations can be a powerful tool for helping to control invasive plants, but this emerging application has hitherto been underutilized and poorly studied.

Given the potential importance of microbial based invasive plant management, we examine existing research on the interactions between invasive plants and their microbiome and posit the impacts of manipulations of these microbial communities to favor invasive plant suppression and control. Together, these studies reflect the need for additional investigations and a broader scope of research into the applications of invasive plant–microbiome and microbe–microbe interactions as a microbial-based management tool. Microbial-informed invasive plant control strategies could include the introduction of plant pathogenic microbes or microbial inhibition of beneficial plant-associated microbes; together, these will act to reduce invasive plant fitness and ecological impacts. Application of synthetic and/or naturally isolated microbial communities or consortia composed of multiple species with different modes of action and various microbe–microbe interactions could be an alternative and complimentary approach in invasive plant management [17].

While biocontrol is still and will likely remain an integral part of invasive plant management, studies into microbiome manipulations to suppress invasive plants are needed to improve efficacy of biocontrol agents as well as provide control opportunities where biocontrol or herbicidal applications are prohibited or otherwise problematic.

Classical biocontrol applications can carry risks; one of the major challenges with a classic biocontrol approach is the adaptation and spread of resistant plant genotypes, which will provide diminishing returns over time [18]. Another drawback is often the lack of biocontrol host specificity [19]. Many plant pathogens used for biocontrol can infect alternative hosts leading to unintended mortality of non-invasive and/or non-target plants. A potentially fruitful frontier in plant management could be the development of microbial consortia that negatively impacts invasive plant fitness through either decreasing their tolerance to biotic or abiotic environmental stress or by improving native plant competitiveness in invaded areas. Microbial consortia are likely to be more effective than individual microbial species introductions as communities tend to be more robust to environmental fluctuations. Also, the probability of plants developing systemic immune responses to consortia is lower than that of individual taxa. In microbial consortia, each member within the consortia interacts and impacts either directly or indirectly with the plant host and/or with one another which ultimately creates an interactive network that impacts host plant fitness and health [20]. We illustrate how these interactive networks may influence invasive plants to achieve control goals (Figure 1). Consortia can be governed by the presence of keystone species (hubs), the major determinants of the microbiome network structure [21], or it may involve tripartite or multipartite interactions [22,23]. Investigations into appropriate network structure for each target invasive plant are needed to develop individual control strategies.

This complex interactive relationship can occur via metabolite or hormone exchange, or through signal transduction pathways [24,25]. Microbial consortia can exhibit complex functionality and their robustness to environmental fluctuations needs to be extensively examined before applications are developed for environmental use. There is an expanding body of literature showing that plant secondary metabolites can alter plant microbiomes and result in differential microbial community assembly [26,27]. Plants release a large proportion of their photosynthates through the soil rhizosphere [28,29] which activates nutrient mobilizing symbionts and/or beneficial plant growth-promoting (PGP) bacteria [30,31]. Plant secondary metabolites impact microbiome structure by acting as signaling molecules, nutrients sources, or as direct toxins [27,32]. Some studies have demonstrated that invasive plants can produce more secondary metabolites than native plants [33,34]. These secondary metabolites facilitate nutrient cycling [35] which may allow invasive plants to outcompete native species. For example, benzoxazinoid indole-derived compounds can function as allelochemicals or protectants against pathogens [36] and act as chemoattractant for (PGP) bacteria in the rhizosphere [37] in invasive plants. Additionally, plant growth promoting rhizobia have been demonstrated to increase scavenged nutrient translocation into legumes, with phosphate additions driving increase nodulation production to facilitate plant growth [38,39]. This, in addition to rhizobia-mediated reduction of ethylene stress associated with degradation of the ethylene precursor molecule 1-aminocyclopropane-1-carboxylic acid (ACC) by ACC deaminase (ACCd) can lead to plant stress reduction [40,41], which facilitates increased resistance to phytopathogens via indirect and direct actions [42]. Together, the utility of considering the integration of plant–microbe and microbe–microbe interactions to alleviate pathogenicity of native plants in the face of biocidal control of proximate invasive plants becomes clear.

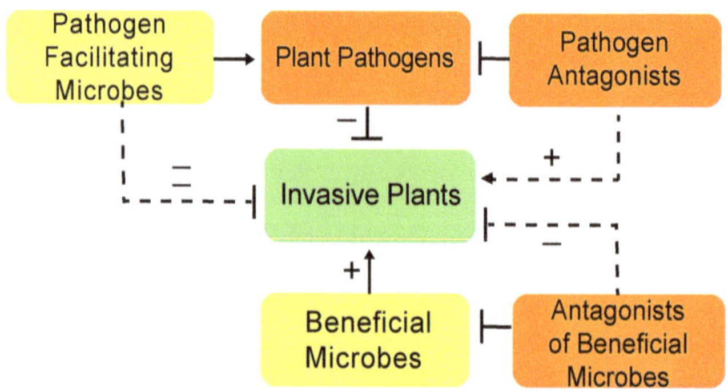

Figure 1. Diagrammatic representation of potential microbial interactions with invasive plants. Direct (solid lines) and indirect (dashed lines) impacts (positive [+] dignified with arrows and negative [−] signified with capped lines) on invasive plant fitness are indicated.

Here, we discuss several scenarios that have been envisaged whereby modification of invasive plant or native plant microbiomes can be considered as a promising sustainable approach in invasive plant control and recovery by native plants. We encourage the research community to incorporate multipartite microbial interactions into the development of the next generation of invasive plant management strategies. These include improving beneficial native plant phytochemical production, reducing the competitiveness of invasive plants, increasing herbicidal tolerance of native plants, and facilitating increased pathogenicity within invasive plants.

2. Research Directions

2.1. Improving Beneficial Native Plant Phytochemical Production

One of the most promising directions of microbial-mediated invasive plant management is perhaps the least well studied. Factors allowing for the success and establishment of invasive species in non-native ranges have been investigated for a long time [43]. According to the novel weapons hypothesis, allelopathic chemicals released by invasive species more effectively inhibit plants from outside of that species' native community than do those of its native range [44]. One major reason is the more 'successful' exotic plants have diverse plant secondary metabolites which may protect against biotic and abiotic stress [45]. These allelopathic effects can accelerate plant invasions [46,47]. Conversely, if the allelopathic potential of native species could be maximized in similar ways, it may provide additional protection against exotic invasions. The allelopathic interactions between native and invasive plants are poorly studied but some studies suggest that allelochemical production by native plants can reduce invasive plants growth [48]. For instance, *Pinus ponderosa* was reported to allelopathically suppress the growth of *Centaurea stoebe*, a noxious weed in the western USA [49]. Further, the presence of pine litter alters soil chemical dynamics, thereby altering the composition soil microbes which can, in turn, further suppress *Centaurea* [50].

More detailed elucidation of plant–plant–microbe interaction mechanisms is needed to fully integrate these interactions into management and restoration strategies, but these interactions are promising in theory. It was reported that the European invasive plant, *Alliaria petiolate* can limit native plants' mycorrhizal fungal richness and colonization rates through releasing secondary metabolites, which in turn negatively impacts native plant health and fitness [51,52]. To provide native plants with a competitive advantage, emphasis needs to be directed toward identifying, selecting and harnessing microbial communities that can improve and/or maximize the native plants' secondary metabolites or allelochemicals production. This could take the shape of exogenous application of selected microbial consortia or as seed coatings used in restoration planting. Unraveling the mechanisms through which microbes control the production of secondary metabolites like allelochemicals and vice versa will help us to pursue the development of management strategies that imitate the structure and function of native plant ecosystems while reducing chemical inputs on the environment. We view this as an important but understudied potential management tool and one that desperately needs additional research to harness this potential.

2.2. *Reducing Competitiveness in Invasive Plants*

Previous studies have found that exotic and/or invasive plants tend to interact differently or more favorably with microbes outside of their native range [53] and can physically alter microbial network structure in invaded ranges [54]. This suggests that soil microbes could be a key component to invasive plant establishment and continued fitness in invaded areas. Many studies reveal that invasive plant microbiomes play a large role in their ability to survive in adverse environmental conditions through mitigation or alleviation of environmental stressors [55]. Understanding the role soil- and plant-associated microbes play in the invasion process will help find strategies to reduce plant fitness in invaded ecosystems. Invasive plants have a competitive advantage over native species [56] and usually have higher net primary productivity (NPP) and greater nitrogen scavenging ability than native plants [57,58]. The rapid radiation of invasive plants can be partially accounted for by co-introduction of pathogens or shifts in abiotic conditions in introduced ranges [46,59], increased abundance and activity of symbiotic microbes [60,61], and higher mineralization rates of nitrogen [62,63], which can be directly influenced by plant endophytes [64]. Soil communities can also be altered following the introduction of invasive species [54], which can account for the higher nitrification rates, a phenomenon that shifts competitive outcomes in favor of invaders and against natives [60]. Invasive plant-mediated shifts in soil properties can further exacerbate microbial community alterations, which can further favor establishment [65].

Manipulating microbial communities within and among invasive plants through the introduction of new microbial populations or providing favorable conditions for shifting established population ratios is a largely uninvestigated option to reduce the competitiveness of invasive plants. This can lead to shifting competitive probabilities in favor of native species, allowing favorable interspecific competition outcomes for native plants. However, it should be noted that utilization of such microbial inoculates, even if favorable outcomes can be achieved, is not without controversy [66], but we need to balance the net benefit with potential ecosystem harm when making these decisions. One avenue towards microbial-mediated invasive plant management is utilizing microbial consortia that can indirectly suppresses invasive plant growth, but development and validation of consortia prior to environmental testing can be difficult [67]. Suppression of invasive plants can occur through inhibition of microbes that mostly benefit the invasive plant, which increases invasive plant fitness. Inhibiting these beneficial microbes will result in net reduction in invasive plant fitness as nutrient acquisition (among other potential mechanisms) capability will be reduced. Evidence for the efficacy of such an approach comes from the counter example of utilizing endophytes and other microbes to increase plant productivity by means of pathogen alleviation [68], whereby endophytes can reduce pathogenicity, thereby benefiting the plant. This other side of the coin is an obvious extension, but one that has been relatively unexplored. Another potential approach is to alter the native plant community near invasive plants to facilitate subsequent cascading effects on soil microbial communities [69]. For instance, planting cover crops in infested areas could affect the quantity and quality of root exudates to the soil, which may in turn affect soil biogeochemical processes and nutrient pools and change the microbial community in invaded areas that favor native plants, although, unfortunately, this type of next-order manipulation is not a major area of active research [70].

2.3. Increasing Herbicide Tolerance in Native Plants

If native or otherwise desirable plants in close proximity to invasive plants can be made more resistant to common herbicides, or less responsive to herbicidal drift, then direct herbicidal application to control invasive plants will produce less ancillary damage. If plant antioxidant content and reactive oxygen species (ROS) scavenging capability could be increased in native plants, they might better tolerate many herbicidal actions. Herbicides can trigger ROS generation in microbes [71] and these ROSs can increase plant cellular damage [72]. It has been shown that certain microbes can function as bio-remediators and convert organic pollutants and xenobiotics into nontoxic products and utilize them a source of carbon, phosphorus, sulfur or nitrogen [73]. Several reports have implicated the significant roles of microbes in degrading the active ingredients of some herbicides [74]. For example, Atrazine can be metabolized by some rhizospheric bacteria including *Arthrobacter* sp. [75], *Pseudomonas aeruginosa*, and *Clavibacter michiganense* [76]. Some *Pseudomonas* strains can metabolize atrazine into cyanuric acid which is then hydrolytically changed to ammonia and carbon dioxide [76]. Manipulating native plant microbiomes in favor of these herbicidal degraders may provide a level of protection to the native plant, but this protective ability is likely to be context-dependent based on the herbicidal mode of action and half-life in soil. Here, we present an incomplete list of taxa that have documented herbicide degradation capabilities. While this is only intended to provide a snapshot of how some microbes can degrade or otherwise transform herbicides, this can serve as a list of potential microbial targets that may have utilization potential for protection against herbicidal action and should be investigated further (Table 1).

Table 1. List of bacterial (top) and fungal (bottom) taxa that have demonstrated herbicidal biodegradation or mineralization capabilities. Presented are species/strain names, herbicides and mode of actions of degradation.

Species/Strain	Herbicide	Mode of Action	Citation
Bacteria			
Pseudomonas sp. ADP.	Atrazine	Mineralization	[77]
Burkholderia (Pseudomonas) cepacia DBO1(pRO101)	2,4-Dichlorophenoxyacetic acid	Biodegradation	[78]
Comamonas sp. SWP-3	Swep	Hydrolysis	[79]
Alicycliphilus sp. PH-34	Swep	Hydrolysis	[79]
Sphingomonas wittichii DC-6	Chlorocetanilide	Mineralization	[80]
Pseudomonas syringae	Triazole	Biotransformation	[81]
Xanthomonas citri	Triazole	Biotransformation	[81]
Enterobacter cloacae K7	Glyphosate	Biodegradation	[82]
Arthrobacter sp. GLP-1	Glyphosate	Biodegradation	[82]
Fungi			
Trichoderma viride	Pirimicarb	Biodegradation	[83]
Trichoderma harzianum	Pirimicarb	Biodegradation	[83]
Nocardioides sp. MFC-A	Mefenacet	Hydrolysis	[84]
Rhodococcus rhodochrous MFC-B	Mefenacet	Hydrolysis	[84]
Stenotrophomonas sp.	Mefenacet	Hydrolysis	[84]
Polyporus tricholoma	Paraquat	Enzymatic Degradation	[85]
Cilindrobasidium leave	Paraquat	Enzymatic Degradation	[85]
Deconica citrospora	Paraquat	Enzymatic Degradation	[85]
Aspergillus terrus	Triazole	Biotransformation	[81]
Penicillium chrysogenum	Triazole	Biotransformation	[81]
Mortierella sp. strain Gr4	Isoproturon	Hydrolysis	[86]
Phoma cf. *eupyrena* Gr61	Isoproturon	Hydrolysis	[86]
Alternaria sp. strain Gr174	Isoproturon	Hydrolysis	[86]
Plectosphaerella cucumerina AR1	Nicosulfuron	Hydrolysis	[87]
Phanerochaete chyrosporium	Atrazine	Biotransformation	[88]

An additional mechanism to confer tolerance to herbicides in native plants could be priming tolerance through microbial-based induction of ROS scavengers within native plants [89,90] or induction of jasmonic acid, oxylipins and salicylic acid production, leading to induction of tolerance responses to herbicide oxidative stress [90]. Another ecologically sound approach to boost native plant tolerance to oxidative stress from herbicides is inoculation with plant growth promoting microbes (PGPM) and certain mycorrhizal fungi. PGPMs can enhance plant growth and resistance to stressors through a wide variety of mechanisms including regulating plant hormones and other phytochemicals, improving nutrition acquisition, siderophore production, enhancing the antioxidant system and activation of induced systemic resistance (ISR) [91]. Most commercially available biofertilizers contain single species inoculants that promote plant growth; however, consortia inoculation might provide higher growth promotion and stronger disease resistance due to cumulative synergistic effects of consortia inoculation over individual inoculations [70].

As discussed earlier, some herbicides cause oxidative damage in plants. For example, glyphosate inhibits the shikimic acid pathway and consequently the production of ROS in tissues [92]. Another mechanism to protect native plants is through the upregulation of phenylpropanoid pathways and boosting the antioxidant system through exogenous treatment of native plants with phytohormones or microbial partners that causes upregulation within the plant [93]. More research into microbial-mediated upregulation of protective pathways needs to be conducted.

2.4. Facilitating Increased Pathogenicity in Invasive Plants

Numerous studies indicate that plant-microbe interactions can improve plant tolerance to biotic stress and/or alleviate pathogenicity effects via multiple mechanisms including secretion of antimicrobial compounds [42,94–96], hyperparasitism [97], and competition for resources such as nutrients or space [98]. However, most research on direct interactions between microbes and pathogens focuses on pathogen mitigation and symptom alleviation [8,99]. Investigations into microbial-mediated pathogen facilitation and increased pathogenicity have not been extensively studied but may have enormous potential to suppress invasive plants [16]. Some plant-associated microbes produce metabolites that can promote pathogen development and facilitate disease [100]. Further, pathogens might exploit specific plant microbes to enhance their pathogenicity or plant susceptibility, and this connection might be driven by production of a plethora of secondary metabolites or hormones by endophytes [99] which may directly or indirectly (via inhibition of a mycoparasite, for instance) facilitate pathogenicity. By developing a framework whereby microbiome manipulations can increase a pathogen's efficacy, invasive plants can be dramatically suppressed via naturally occurring environmental pathogens. Facilitation occurs when one microorganism enhances the development or growth of another. This facilitation may also be due to ecological interactions including competitive exclusion or niche partitioning [101]. By investigating and understanding the dynamics of the invasive plant micro- and mycobiomes, we can develop strategies for modification and manipulation of these communities [16] to favor successful colonization and growth of taxa that facilitate pathogen virulence, or taxa that negatively impact invasive PGPMs, thus resulting in suppression in invasive plants and a reduction in plant fitness. This is an emerging field of study, but one that we feel will be of increasing importance with a growing emphasis on sustainable and non-chemical controls of invasive plants [15]. Targeting invasive plant microbiomes is a novel method of integrated management of invasive plants that deserves to be explored. Identifying, understanding, and the utilization of microorganisms or microbial products to reduce invasive plant fitness are becoming more central parts of sustainable agriculture. To better understand the potential for microbial-mediated facilitation of pathogenicity of invasive plants, and to stimulate the research community into action, it is useful to briefly examine some mechanisms in induction of signaling cascades in plants by microbes.

Plant responses to colonization by microbes can be broadly categorized into one of two main categories SAR (systemic acquired resistance), triggered by plant pathogens, and ISR (induced systemic resistance), triggered by root-colonizing mutualistic microbes (Figure 2) [102,103]. Although both pathways share many common signaling components, their elicitors and regulators are distinct. The conserved microbe-specific elicitors, referred to as microbe- or pathogen-associated molecular patterns (MAMPs or PAMPs), are perceived by the plants' innate immune systems pattern recognition receptors (PRRs). Examples of elicitors include flagellin (Flg), elongation factor Tu (EF-Tu), peptidoglycan (PGN), lipopolysaccharides (LPS), Ax21 (activator of XA21-mediated immunity in rice), fungal chitin, and β-glucans from oomycetes, among others, can be recognized by plant surface localized PRRs [104]. MAMP elicitors, upon perception, can trigger a SAR signaling cascade which is characterized by increased levels of the hormone salicylic acid (SA) which, activates the expression of a large set of pathogenesis-related (PR) genes through the induction of the redox-regulated protein NON-EXPRESSOR OF PR GENES1 (NPR1) leading to the activation of the defense responses [105,106]. The mechanisms underpinning NPR1 action have been well documented [107] and it plays a major role in direct pathogenicity and defense in plants. In SAR, MAMPs elicitors activate ISR signaling pathway which is mediated by an SA-independent pathway where Jasmonic Acid (JA) and ethylene (ET) play major roles, and typically functions without PR gene activation [105,106]. Some studies have suggested that NPR1 may also be required for the ISR triggered by certain rhizospheric microbes [106,108]. Some studies also suggest that ISR is required for SA accumulation in plants [109–111]. ISR eliciting rhizospheric microbes activate plant defense responses which are often effective against a broad spectrum of plant pathogens [112] which could be an avenue to increase native plants fitness. However, there are cases in which harmless or even beneficial microbes can assist pathogen establishment [113], which demonstrates the importance of additional research into these interactions before microbial-mediated disease facilitation and/or protection can be fully developed.

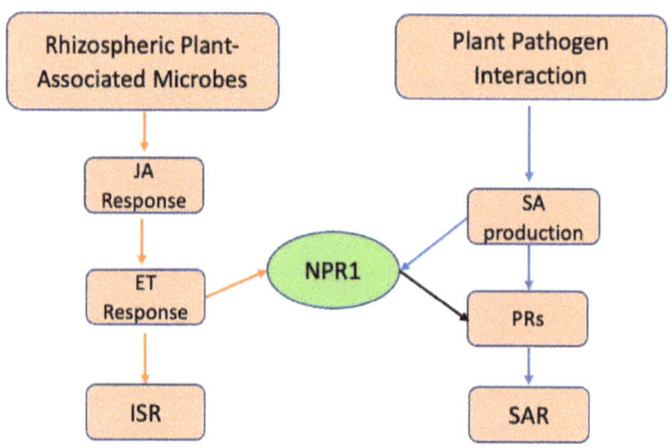

Figure 2. Schematic representation of the common signal-transduction pathways leading to pathogen-induced systemic acquired resistance (SAR) and rhizosphere-mediated induced systemic resistance (ISR) inspired by [114] for *Arabidopsis thaliana* but applicable to plants in general. Crosstalk between the two pathways occurs through the activation of NONEXPRESSOR OF PATHOGENESIS-RELATED GENES1 (NPR1). Non-pathogenic plant-associated microbes, usually from the rhizosphere, can trigger the SAR pathway as well as ISR. In the rhizosphere-mediated ISR pathway, components from the jasmonic acid (JA) and ethylene (ET) responses act in sequence to activate a systemic resistance response (orange arrows). Pathogenic agents could activate the pathogen-induced SAR, through the activation of NPR1 (blue arrows), leading to the expression of PATHOGENESIS-RELATED genes (PRs) (black arrow). NPR1 also mediates crosstalk between the SA signaling pathway.

2.5. Concerns and Potential Problems with Microbial Deployment

To establish a successful microbial-mediated control program, there are many potential concerns that must be taken into account. Introduction of microbes into nature, even if these microbes are already occurring in a particular environment, can carry risk. There is always a risk of non-target associations with applied microbes, which may be problematic particularly for pathogenic applications [115]; extensive host-specificity testing procedures are needed to predict the potential non-target effects. Additionally, development and implementation of microbial applications must be approached from a risk assessment framework [116]. Some have argued that the risk associated with potential unforeseen consequences of microbial inoculants is unacceptable [66] as well as being confronted by too many potential ethical and legal issues [117]. This wariness is understandable, and is largely justified by highlighting the complexity of these systems [118,119] which may occlude potential problems until too late. However, one could argue that the economic and ecological cost of doing nothing [120–122] is far greater than a calculated risk, as long as controlled in planta validations and detailed cost–benefit analyses have been conducted [123]. Extensive work is must be done to study pathogenicity, adaptability, colonization, reproduction, dispersal, and survival efficiency of any potential microbial agents used for biocontrol [124], but this is a desperately needed area of additional research in the future.

3. Research Gaps, Future Directions, and Conclusions

Several studies have explored the role of particular microbes for biocontrol or plant protection services. However, there have only been limited investigations into field-scale investigations and utilization of a consortia approach to either control invasive plants, or benefit native plants to shift competitive outcomes when threatened with invasive plants [79,125–128]. Here, we advocate for additional research to advance sustainability and an integrated microbiological approach to help suppress invasive plant fitness as a potential additional tool for land managers. We identify three main, but not complete, research priorities that need to be investigated to move this field out of its infancy: (1) the development of integrated predictive models to understand the multipartite effects of pathogen–endophyte interactions associated with invasive plants; (2) the definition and elucidation of core pathogen–endophyte combinations on invasive plants to develop targets for additional investigations; (3) the elucidation of how microbial consortia mechanistically interact with hosts, environments, and management strategies in order to develop targeted application plans.

Invasive plant species are one of the challenges facing the world, leading to great economic losses. Inclusion of microbial-based management options for invasive plant management should be investigated with the goal of ultimately reducing invasive plant fitness. Combinations of individual microbes with complementary or synergistic traits may increase the competitive ability of native plants and/or susceptibility of invasive plants which may ultimately reduce invasive plant fitness in the invaded range. Not only does the effectiveness of individual microbes need to be examined for invasive plant control, but so does the interrelation, strength, and directionality of interactions between taxa. These strategies should be incorporated in invasive plant management programs. Here, we implore the invasive plant management research community to incorporate microbial dynamics into explorations of control strategies. The control potential of such methods is promising, but additional investigations are needed to move these strategies into active development.

Author Contributions: Conceptualization, M.S. and S.P.B.; writing, M.S. and S.P.B. All authors have read and agreed to the published version of the manuscript.

Funding: This research received no external funding.

Institutional Review Board Statement: Not applicable.

Informed Consent Statement: Not applicable.

Data Availability Statement: No new data were created or analyzed in this study. Data sharing is not applicable to this article.

Acknowledgments: This was funded in part by the Agriculture and Food Technologies Cluster of the FedEx Institute of Technology and the Center for Biodiversity Research at the University of Memphis.

Conflicts of Interest: The authors declare no conflict of interest. The funders had no role in the design of the study; in the collection, analyses, or interpretation of data; in the writing of the manuscript; or in the decision to publish the results.

References

1. Lindow, S.E.; Leveau, J.H.J. Phyllosphere Microbiology. *Curr. Opin. Biotechnol.* **2002**, *13*, 238–243. [CrossRef]
2. Bonkowski, M.; Villenave, C.; Griffiths, B. Rhizosphere Fauna: The Functional and Structural Diversity of Intimate Interactions of Soil Fauna with Plant Roots. *Plant Soil* **2009**, *321*, 213–233. [CrossRef]
3. Brown, S.P.; Grillo, M.A.; Podowski, J.C.; Heath, K.D. Soil Origin and Plant Genotype Structure Distinct Microbiome Compartments in the Model Legume *Medicago Truncatula*. *Microbiome* **2020**, *8*, 139. [CrossRef]
4. Stone, J.K.; Bacon, C.W.; White, J.F., Jr. An overview of endophytic microbes: Endophytism defined. In *Microbial Endophytes*; Bacon, C.W., White, J.F., Jr., Eds.; Marcel Dekker, Inc.: New York, NY, USA, 2000; pp. 3–30.
5. Hardoim, P.R.; van Overbeek, L.S.; Berg, G.; Pirttilä, A.M.; Compant, S.; Campisano, A.; Döring, M.; Sessitsch, A. The Hidden World within Plants: Ecological and Evolutionary Considerations for Defining Functioning of Microbial Endophytes. *Microbiol. Mol. Biol. Rev.* **2015**, *79*, 293–320. [CrossRef]
6. Goh, C.-H.; Veliz Vallejos, D.F.; Nicotra, A.B.; Mathesius, U. The Impact of Beneficial Plant-Associated Microbes on Plant Phenotypic Plasticity. *J. Chem. Ecol.* **2013**, *39*, 826–839. [CrossRef] [PubMed]
7. Jones, P.; Garcia, B.J.; Furches, A.; Tuskan, G.A.; Jacobson, D. Plant Host-Associated Mechanisms for Microbial Selection. *Front. Plant Sci.* **2019**, *10*. [CrossRef] [PubMed]
8. Busby, P.E.; Soman, C.; Wagner, M.R.; Friesen, M.L.; Kremer, J.; Bennett, A.; Morsy, M.; Eisen, J.A.; Leach, J.E.; Dangl, J.L. Research Priorities for Harnessing Plant Microbiomes in Sustainable Agriculture. *PLoS Biol.* **2017**, *15*, e2001793. [CrossRef]
9. Kowalski, K.P.; Bacon, C.; Bickford, W.; Braun, H.; Clay, K.; Leduc-Lapierre, M.; Lillard, E.; McCormick, M.K.; Nelson, E.; Torres, M.; et al. Advancing the Science of Microbial Symbiosis to Support Invasive Species Management: A Case Study on Phragmites in the Great Lakes. *Front. Microbiol.* **2015**, *6*. [CrossRef]
10. Weaver, M.A.; Boyette, C.D.; Hoagland, R.E. Management of Kudzu by the Bioherbicide, *Myrothecium Verrucaria*, Herbicides and Integrated Control Programmes. *Biocontrol Sci. Technol.* **2016**, *26*, 136–140. [CrossRef]
11. Reichert Júnior, F.W.; Scariot, M.A.; Forte, C.T.; Pandolfi, L.; Dil, J.M.; Weirich, S.; Carezia, C.; Mulinari, J.; Mazutti, M.A.; Fongaro, G.; et al. New Perspectives for Weeds Control Using Autochthonous Fungi with Selective Bioherbicide Potential. *Heliyon* **2019**, 5, e01676. [CrossRef]
12. Lombard, L.; Houbraken, J.; Decock, C.; Samson, R.A.; Meijer, M.; Réblová, M.; Groenewald, J.Z.; Crous, P.W. Generic Hyper-Diversity in Stachybotriaceae. *Persoonia* **2016**, *36*, 156–246. [CrossRef]
13. Clarke, T.C.; Shetty, K.G.; Jayachandran, K.; Norland, M.R. Myrothecium Verrucaria—A Potential Biological Control Agent for the Invasive 'Old World Climbing Fern' (*Lygodium Microphyllum*). *BioControl* **2007**, *52*, 399–411. [CrossRef]
14. Weaver, M.A.; Shearer, J.F.; Grodowitz, M.J.; Boyette, C.D. Potential of *Myrothecium* Species as Bioherbicides for Giant Salvinia (*Salvinia Molesta*). *J. Aquat. Plant Manag.* **2018**, *56*, 120–122.
15. Harding, D.P.; Raizada, M.N. Controlling Weeds with Fungi, Bacteria and Viruses: A Review. *Front. Plant Sci.* **2015**, *6*. [CrossRef] [PubMed]
16. Shahrtash, M.; Brown, S.P. Drivers of Foliar Fungal Endophytic Communities of Kudzu (*Pueraria Montana* Var. *Lobata*) in the Southeast United States. *Diversity* **2020**, *12*, 185. [CrossRef]
17. Lopez-Cervantes, J.; Thorpe, D.T. Microbial Composition Comprising Liquid Fertilizer and Processes for Agricultural Use. U.S. Patent Application 13/829,300, 3 October 2013.
18. Stastny, M.; Sargent, R.D. Evidence for Rapid Evolutionary Change in an Invasive Plant in Response to Biological Control. *J. Evol. Biol.* **2017**, *30*, 1042–1052. [CrossRef]
19. Brodeur, J. Host Specificity in Biological Control: Insights from Opportunistic Pathogens. *Evol. Appl.* **2012**, *5*, 470–480. [CrossRef] [PubMed]
20. Poudel, R.; Jumpponen, A.; Schlatter, D.C.; Paulitz, T.C.; Gardener, B.B.M.; Kinkel, L.L.; Garrett, K.A. Microbiome Networks: A Systems Framework for Identifying Candidate Microbial Assemblages for Disease Management. *Phytopathology* **2016**, *106*, 1083–1096. [CrossRef]
21. Van der Heijden, M.G.A.; Hartmann, M. Networking in the Plant Microbiome. *PLoS Biol.* **2016**, *14*, e1002378. [CrossRef]
22. Bonfante, P.; Anca, I.-A. Plants, Mycorrhizal Fungi, and Bacteria: A Network of Interactions. *Annu. Rev. Microbiol.* **2009**, *63*, 363–383. [CrossRef]

23. Dames, J.F.; Ridsdale, C.J. What We Know about Arbuscular Mycorhizal Fungi and Associated Soil Bacteria. *AJB* **2012**, *11*, 13753–13760. [CrossRef]
24. Keller, L.; Surette, M.G. Communication in Bacteria: An Ecological and Evolutionary Perspective. *Nat. Rev. Microbiol.* **2006**, *4*, 249–258. [CrossRef]
25. Vandenkoornhuyse, P.; Quaiser, A.; Duhamel, M.; Van, A.L.; Dufresne, A. The Importance of the Microbiome of the Plant Holobiont. *New Phytol.* **2015**, *206*, 1196–1206. [CrossRef]
26. Fitzpatrick, C.R.; Copeland, J.; Wang, P.W.; Guttman, D.S.; Kotanen, P.M.; Johnson, M.T.J. Assembly and Ecological Function of the Root Microbiome across Angiosperm Plant Species. *Proc. Natl. Acad. Sci. USA* **2018**, *115*, E1157–E1165. [CrossRef] [PubMed]
27. Jacoby, R.P.; Chen, L.; Schwier, M.; Koprivova, A.; Kopriva, S. Recent Advances in the Role of Plant Metabolites in Shaping the Root Microbiome. *F1000Research* **2020**, *9*. [CrossRef]
28. Jones, D.L.; Nguyen, C.; Finlay, R.D. Carbon Flow in the Rhizosphere: Carbon Trading at the Soil–Root Interface. *Plant Soil* **2009**, *321*, 5–33. [CrossRef]
29. Lakshmanan, V.; Selvaraj, G.; Bais, H.P. Functional Soil Microbiome: Belowground Solutions to an Aboveground Problem. *Plant Physiol.* **2014**, *166*, 689–700. [CrossRef] [PubMed]
30. Egamberdieva, D.; Abdiev, A.; Khaitov, B. Synergistic interactions among root-associated bacteria, rhizobia and chickpea under stress conditions. In *Plant-Environment Interaction*; Azzoz, M., Ahmad, P., Eds.; John Wiley & Sons, Ltd.: Hoboken, NJ, USA, 2016; pp. 250–262. ISBN 978-1-119-08100-5.
31. Barret, M.; Morrissey, J.P.; O'Gara, F. Functional Genomics Analysis of Plant Growth-Promoting Rhizobacterial Traits Involved in Rhizosphere Competence. *Biol. Fertil. Soils* **2011**, *47*, 729. [CrossRef]
32. Pascale, A.; Proietti, S.; Pantelides, I.S.; Stringlis, I.A. Modulation of the Root Microbiome by Plant Molecules: The Basis for Targeted Disease Suppression and Plant Growth Promotion. *Front. Plant Sci.* **2020**, *10*. [CrossRef] [PubMed]
33. Alford, É.R.; Perry, L.G.; Qin, B.; Vivanco, J.M.; Paschke, M.W. A Putative Allelopathic Agent of Russian Knapweed Occurs in Invaded Soils. *Soil Biol. Biochem.* **2007**, *39*, 1812–1815. [CrossRef]
34. Li, Y.-P.; Feng, Y.-L.; Chen, Y.-J.; Tian, Y.-H. Soil Microbes Alleviate Allelopathy of Invasive Plants. *Sci. Bull.* **2015**, *60*, 1083–1091. [CrossRef]
35. Hättenschwiler, S.; Vitousek, P.M. The Role of Polyphenols in Terrestrial Ecosystem Nutrient Cycling. *Trends Ecol. Evol.* **2000**, *15*, 238–243. [CrossRef]
36. de Bruijn, W.J.C.; Gruppen, H.; Vincken, J.-P. Structure and Biosynthesis of Benzoxazinoids: Plant Defence Metabolites with Potential as Antimicrobial Scaffolds. *Phytochemistry* **2018**, *155*, 233–243. [CrossRef] [PubMed]
37. Neal, A.L.; Ahmad, S.; Gordon-Weeks, R.; Ton, J. Benzoxazinoids in Root Exudates of Maize Attract *Pseudomonas Putida* to the Rhizosphere. *PLoS ONE* **2012**, *7*, e35498. [CrossRef] [PubMed]
38. Ditta, A.; Imtiaz, M.; Mehmood, S.; Rizwan, M.S.; Mubeen, F.; Aziz, O.; Qian, Z.; Ijaz, R.; Tu, S. Rock Phosphate-Enriched Organic Fertilizer with Phosphate-Solubilizing Microorganisms Improves Nodulation, Growth, and Yield of Legumes. *Commun. Soil Sci. Plant Anal.* **2018**, *49*, 2715–2725. [CrossRef]
39. Ditta, A.; Muhammad, J.; Imtiaz, M.; Mehmood, S.; Qian, Z.; Tu, S. Application of Rock Phosphate Enriched Composts Increases Nodulation, Growth and Yield of Chickpea. *Int. J. Recycl. Org. Waste Agricult.* **2018**, *7*, 33–40. [CrossRef]
40. Tiwari, G.; Duraivadivel, P.; Sharma, S.; Hariprasad, P. 1-Aminocyclopropane-1-Carboxylic Acid Deaminase Producing Beneficial Rhizobacteria Ameliorate the Biomass Characters of Panicum Maximum Jacq. by Mitigating Drought and Salt Stress. *Sci. Rep.* **2018**, *8*, 17513. [CrossRef] [PubMed]
41. Sagar, A.; Sayyed, R.Z.; Ramteke, P.W.; Sharma, S.; Marraiki, N.; Elgorban, A.M.; Syed, A. ACC Deaminase and Antioxidant Enzymes Producing Halophilic Enterobacter Sp. PR14 Promotes the Growth of Rice and Millets under Salinity Stress. *Physiol. Mol. Biol. Plant.* **2020**, *26*, 1847–1854. [CrossRef]
42. Basu, A.; Prasad, P.; Das, S.N.; Kalam, S.; Sayyed, R.Z.; Reddy, M.S.; El Enshasy, H. Plant Growth Promoting Rhizobacteria (PGPR) as Green Bioinoculants: Recent Developments, Constraints, and Prospects. *Sustainability* **2021**, *13*, 1140. [CrossRef]
43. Theoharides, K.A.; Dukes, J.S. Plant Invasion across Space and Time: Factors Affecting Nonindigenous Species Success during Four Stages of Invasion. *New Phytol.* **2007**, 256–273. [CrossRef] [PubMed]
44. Callaway, R.M.; Ridenour, W.M. Novel Weapons: Invasive Success and the Evolution of Increased Competitive Ability. *Front. Ecol. Environ.* **2004**, *2*, 436–443. [CrossRef]
45. Macel, M.; de Vos, R.C.H.; Jansen, J.J.; van der Putten, W.H.; van Dam, N.M. Novel Chemistry of Invasive Plants: Exotic Species Have More Unique Metabolomic Profiles than Native Congeners. *Ecol. Evol.* **2014**, *4*, 2777–2786. [CrossRef]
46. Reinhart, K.O.; Callaway, R.M. Soil Biota and Invasive Plants. *New Phytol.* **2006**, *170*, 445–457. [CrossRef] [PubMed]
47. Shah, M.A.; Reshi, Z.A.; Rasool, N. Plant Invasions Induce a Shift in Glomalean Spore Diversity. *Trop. Ecol.* **2010**, *51*, 317–323.
48. Christina, M.; Rouifed, S.; Puijalon, S.; Vallier, F.; Meiffren, G.; Bellvert, F.; Piola, F. Allelopathic Effect of a Native Species on a Major Plant Invader in Europe. *Sci. Nat.* **2015**, *102*, 12. [CrossRef] [PubMed]
49. Metlen, K.L.; Aschehoug, E.T.; Callaway, R.M. Competitive Outcomes between Two Exotic Invaders Are Modified by Direct and Indirect Effects of a Native Conifer. *Oikos* **2013**, *122*, 632–640. [CrossRef]
50. Iason, G.R.; Lennon, J.J.; Pakeman, R.J.; Thoss, V.; Beaton, J.K.; Sim, D.A.; Elston, D.A. Does Chemical Composition of Individual Scots Pine Trees Determine the Biodiversity of Their Associated Ground Vegetation? *Ecol. Lett.* **2005**, *8*, 364–369. [CrossRef]

51. Stinson, K.A.; Campbell, S.A.; Powell, J.R.; Wolfe, B.E.; Callaway, R.M.; Thelen, G.C.; Hallett, S.G.; Prati, D.; Klironomos, J.N. Invasive Plant Suppresses the Growth of Native Tree Seedlings by Disrupting Belowground Mutualisms. *PLoS Biol.* **2006**, *4*, e140. [CrossRef]
52. Callaway, R.M.; Cipollini, D.; Barto, K.; Thelen, G.C.; Hallett, S.G.; Prati, D.; Stinson, K.; Klironomos, J. Novel Weapons: Invasive Plant Suppresses Fungal Mutualists in America but Not in Its Native Europe. *Ecology* **2008**, *89*, 1043–1055. [CrossRef]
53. Lu-Irving, P.; Harenčár, J.G.; Sounart, H.; Welles, S.R.; Swope, S.M.; Baltrus, D.A.; Dlugosch, K.M. Native and Invading Yellow Starthistle (*Centaurea Solstitialis*) Microbiomes Differ in Composition and Diversity of Bacteria. *mSphere* **2019**, *4*. [CrossRef]
54. Mamet, S.D.; Redlick, E.; Brabant, M.; Lamb, E.G.; Helgason, B.L.; Stanley, K.; Siciliano, S.D. Structural Equation Modeling of a Winnowed Soil Microbiome Identifies How Invasive Plants Re-Structure Microbial Networks. *ISME J.* **2019**, *13*, 1988–1996. [CrossRef]
55. White, J.F.; Kingsley, K.L.; Zhang, Q.; Verma, R.; Obi, N.; Dvinskikh, S.; Elmore, M.T.; Verma, S.K.; Gond, S.K.; Kowalski, K.P. Review: Endophytic Microbes and Their Potential Applications in Crop Management. *Pest Manag. Sci.* **2019**, *75*, 2558–2565. [CrossRef]
56. MacDougall, A.S.; Gilbert, B.; Levine, J.M. Plant Invasions and the Niche. *J. Ecol.* **2009**, *97*, 609–615. [CrossRef]
57. Ehrenfeld, J.G. Effects of Exotic Plant Invasions on Soil Nutrient Cycling Processes. *Ecosystems* **2003**, *6*, 503–523. [CrossRef]
58. Liao, C.; Peng, R.; Luo, Y.; Zhou, X.; Wu, X.; Fang, C.; Chen, J.; Li, B. Altered Ecosystem Carbon and Nitrogen Cycles by Plant Invasion: A Meta-Analysis. *New Phytol.* **2008**, *177*, 706–714. [CrossRef]
59. Liao, Z.-Y.; Zhang, R.; Barclay, G.F.; Feng, Y.-L. Differences in Competitive Ability between Plants from Nonnative and Native Populations of a Tropical Invader Relates to Adaptive Responses in Abiotic and Biotic Environments. *PLoS ONE* **2013**, *8*, e71767. [CrossRef] [PubMed]
60. Hawkes, C.V.; Wren, I.F.; Herman, D.J.; Firestone, M.K. Plant Invasion Alters Nitrogen Cycling by Modifying the Soil Nitrifying Community. *Ecol. Lett.* **2005**, *8*, 976–985. [CrossRef]
61. Lekberg, Y.; Gibbons, S.M.; Rosendahl, S.; Ramsey, P.W. Severe Plant Invasions Can Increase Mycorrhizal Fungal Abundance and Diversity. *ISME J.* **2013**, *7*, 1424–1433. [CrossRef]
62. Ehrenfeld, J.G.; Kourtev, P.; Huang, W. Changes in Soil Functions Following Invasions of Exotic Understory Plants in Deciduous Forests. *Ecol. Appl.* **2001**, *11*, 1287–1300. [CrossRef]
63. McLeod, M.L.; Cleveland, C.C.; Lekberg, Y.; Maron, J.L.; Philippot, L.; Bru, D.; Callaway, R.M. Exotic Invasive Plants Increase Productivity, Abundance of Ammonia-Oxidizing Bacteria and Nitrogen Availability in Intermountain Grasslands. *J. Ecol.* **2016**, *104*, 994–1002. [CrossRef]
64. Oses-Pedraza, R.; Torres-Díaz, C.; Lavín, P.; Retamales-Molina, P.; Atala, C.; Gallardo-Cerda, J.; Acuña-Rodríguez, I.S.; Molina-Montenegro, M.A. Root Endophytic Penicillium Promotes Growth of Antarctic Vascular Plants by Enhancing Nitrogen Mineralization. *Extremophiles* **2020**, *24*, 721–732. [CrossRef] [PubMed]
65. Gibbons, S.M.; Lekberg, Y.; Mummey, D.L.; Sangwan, N.; Ramsey, P.W.; Gilbert, J.A. Invasive Plants Rapidly Reshape Soil Properties in a Grassland Ecosystem. *mSystems* **2017**, *2*. [CrossRef]
66. Jack, C.N.; Petipas, R.H.; Cheeke, T.E.; Rowland, J.L.; Friesen, M.L. Microbial Inoculants: Silver Bullet or Microbial Jurassic Park? *TIM* **2020**. [CrossRef]
67. Kong, Z.; Hart, M.; Liu, H. Paving the Way From the Lab to the Field: Using Synthetic Microbial Consortia to Produce High-Quality Crops. *Front. Plant Sci.* **2018**, *9*. [CrossRef]
68. Zivanovic, A.; Rodgers, L. The Role of Fungal Endophytes in Plant Pathogen Resistance. *Bios* **2019**, *89*, 192–197. [CrossRef]
69. Pant, P.; Pant, P. Ecological Restoration Techniques for Management of Degraded, Mined-Out Areas and the Role Played by Rhizospheric Microbial Communities. In *Green Technologies and Environmental Sustainability*; Singh, R., Kumar, S., Eds.; Springer: Cham, Switzerland, 2017; pp. 437–453. ISBN 978-3-319-50654-8.
70. Domenech, J.; Reddy, M.S.; Kloepper, J.W.; Ramos, B.; Gutierrez-Mañero, J. Combined Application of the Biological Product LS213 with Bacillus, Pseudomonas or Chryseobacterium for Growth Promotion and Biological Control of Soil-Borne Diseases in Pepper and Tomato. *Biocontrol* **2006**, *51*, 245. [CrossRef]
71. Rutherford, A.W.; Krieger-Liszkay, A. Herbicide-Induced Oxidative Stress in Photosystem II. *Trends Biochem. Sci.* **2001**, *26*, 648–653. [CrossRef]
72. Demidchik, V. Mechanisms of Oxidative Stress in Plants: From Classical Chemistry to Cell Biology. *Environ. Exp. Bot.* **2015**, *109*, 212–228. [CrossRef]
73. Singh, B.K.; Walker, A. Microbial Degradation of Organophosphorus Compounds. *FEMS Microbiol. Rev.* **2006**, *30*, 428–471. [CrossRef]
74. Singh, B.; Singh, K. Microbial Degradation of Herbicides. *Crit. Rev. Microbiol.* **2016**, *42*, 245–261. [CrossRef] [PubMed]
75. Wang, Q.; Xie, S. Isolation and Characterization of a High-Efficiency Soil Atrazine-Degrading *Arthrobacter* sp. Strain. *Int. Biodeterior. Biodegrad.* **2012**, *71*, 61–66. [CrossRef]
76. Popov, V.H.; Cornish, P.S.; Sultana, K.; Morris, E.C.; Popov, V.H.; Cornish, P.S.; Sultana, K.; Morris, E.C. Atrazine Degradation in Soils: The Role of Microbial Communities, Atrazine Application History, and Soil Carbon. *Soil Res.* **2005**, *43*, 861–871. [CrossRef]
77. Govantes, F.; Porrúa, O.; García-González, V.; Santero, E. Atrazine Biodegradation in the Lab and in the Field: Enzymatic Activities and Gene Regulation. *Microb. Biotechnol.* **2009**, *2*, 178–185. [CrossRef] [PubMed]

78. Jacobsen, C.S. Plant Protection and Rhizosphere Colonization of Barley by Seed Inoculated Herbicide Degrading *Burkholderia (Pseudomonas) Cepacia* DBO1(PRO101) in 2,4-D Contaminated Soil. *Plant Soil* **1997**, *189*, 139–144. [CrossRef]
79. Zhang, L.; Hang, P.; Zhou, X.; Dai, C.; He, Z.; Jiang, J. Mineralization of the Herbicide Swep by a Two-Strain Consortium and Characterization of a New Amidase for Hydrolyzing Swep. *Microb. Cell Fact.* **2020**, *19*, 4. [CrossRef] [PubMed]
80. Cheng, M.; Yan, X.; He, J.; Qiu, J.; Chen, Q. Comparative Genome Analysis Reveals the Evolution of Chloroacetanilide Herbicide Mineralization in Sphingomonas Wittichii DC-6. *Arch. Microbiol.* **2019**, *201*, 907–918. [CrossRef]
81. Ahmad, K.S.; Gul, P.; Gul, M.M. Efficient Fungal and Bacterial Facilitated Remediation of Thiencarbazone Methyl in the Environment. *Environ. Res.* **2020**, *188*, 109811. [CrossRef] [PubMed]
82. Kertesz, M.; Elgorriaga, A.; Amrhein, N. Evidence for Two Distinct Phosphonate-Degrading Enzymes (C-P Lyases) in *Arthrobacter* Sp. GLP-1. *Biodegradation* **1991**, *2*, 53–59. [CrossRef] [PubMed]
83. Romeh, A. Biodegradation of Carbosulfan, Pirimicarb and Diniconazole Pesticides by *Trichoderma* Spp. *J. Environ. Res.* **2001**, *3*, 162–172.
84. Harada, N.; Takagi, K.; Harazono, A.; Fujii, K.; Iwasaki, A. Isolation and Characterization of Microorganisms Capable of Hydrolysing the Herbicide Mefenacet. *Soil Biol. Biochem.* **2006**, *38*, 173–179. [CrossRef]
85. Camacho-Morales, R.; Gerardo-Gerardo, J.; Guillén Navarro, K.; Sánchez, H. Ligninolytic Enzyme Production by White Rot Fungi during Paraquat (Herbicide) Degradation. *Rev. Argent Microbiol.* **2017**, *49*, 189–196. [CrossRef]
86. Rønhede, S.; Jensen, B.; Rosendahl, S.; Kragelund, B.B.; Juhler, R.K.; Aamand, J. Hydroxylation of the Herbicide Isoproturon by Fungi Isolated from Agricultural Soil. *Appl. Environ. Microbiol.* **2005**, *71*, 7927–7932. [CrossRef]
87. Carles, L.; Rossi, F.; Besse-Hoggan, P.; Blavignac, C.; Leremboure, M.; Artigas, J.; Batisson, I. Nicosulfuron Degradation by an Ascomycete Fungus Isolated From Submerged Alnus Leaf Litter. *Front. Microbiol.* **2018**, *9*. [CrossRef]
88. Mougin, C.; Laugero, C.; Asther, M.; Dubroca, J.; Frasse, P.; Asther, M. Biotransformation of the Herbicide Atrazine by the White Rot Fungus *Phanerochaete chrysosporium*. *Appl. Environ. Microbiol.* **1994**, *60*, 705–708. [CrossRef]
89. Cummins, I.; Cole, D.J.; Edwards, R. A Role for Glutathione Transferases Functioning as Glutathione Peroxidases in Resistance to Multiple Herbicides in Black-Grass. *Plant J.* **1999**, *18*, 285–292. [CrossRef] [PubMed]
90. Tétard-Jones, C.; Edwards, R. Potential Roles for Microbial Endophytes in Herbicide Tolerance in Plants. *Pest Manag. Sci.* **2016**, *72*, 203–209. [CrossRef] [PubMed]
91. Heil, M. Induced Systemic Resistance (ISR) against Pathogens—A Promising Field for Ecological Research. *PPEES* **2001**, *4*, 65–79. [CrossRef]
92. Ahsan, N.; Lee, D.-G.; Lee, K.-W.; Alam, I.; Lee, S.-H.; Bahk, J.D.; Lee, B.-H. Glyphosate-Induced Oxidative Stress in Rice Leaves Revealed by Proteomic Approach. *Plant Physiol. Biochem.* **2008**, *46*, 1062–1070. [CrossRef]
93. Gao, Z.; Xie, W.; Ashraf, U.; Li, Y.; Ma, L.; Gui, R.; Pan, S.; Tian, H.; Duan, M.; Wang, S.; et al. Exogenous γ-Aminobutyric Acid (GABA) Application at Different Growth Stages Regulates 2-Acetyl-1-Pyrroline, Yield, Quality and Antioxidant Attributes in Fragrant Rice. *J. Plant Interact.* **2020**, *15*, 139–152. [CrossRef]
94. Bailey, B.A.; Bae, H.; Strem, M.D.; Crozier, J.; Thomas, S.E.; Samuels, G.J.; Vinyard, B.T.; Holmes, K.A. Antibiosis, Mycoparasitism, and Colonization Success for Endophytic Trichoderma Isolates with Biological Control Potential in *Theobroma Cacao*. *Biol. Control* **2008**, *46*, 24–35. [CrossRef]
95. Helfrich, E.J.N.; Vogel, C.M.; Ueoka, R.; Schäfer, M.; Ryffel, F.; Müller, D.B.; Probst, S.; Kreuzer, M.; Piel, J.; Vorholt, J.A. Bipartite Interactions, Antibiotic Production and Biosynthetic Potential of the Arabidopsis Leaf Microbiome. *Nat. Microbiol.* **2018**, *3*, 909–919. [CrossRef] [PubMed]
96. Chen, Y.; Wang, J.; Yang, N.; Wen, Z.; Sun, X.; Chai, Y.; Ma, Z. Wheat Microbiome Bacteria Can Reduce Virulence of a Plant Pathogenic Fungus by Altering Histone Acetylation. *Nat. Commun.* **2018**, *9*, 3429. [CrossRef]
97. Parratt, S.R.; Laine, A.-L. The Role of Hyperparasitism in Microbial Pathogen Ecology and Evolution. *ISME J.* **2016**, *10*, 1815–1822. [CrossRef]
98. Zelezniak, A.; Andrejev, S.; Ponomarova, O.; Mende, D.R.; Bork, P.; Patil, K.R. Metabolic Dependencies Drive Species Co-Occurrence in Diverse Microbial Communities. *Proc. Natl. Acad. Sci. USA* **2015**, *112*, 6449–6454. [CrossRef]
99. Busby, P.E.; Ridout, M.; Newcombe, G. Fungal Endophytes: Modifiers of Plant Disease. *Plant Mol. Biol.* **2016**, *90*, 645–655. [CrossRef] [PubMed]
100. Van Wees, S.C.; De Swart, E.A.; Van Pelt, J.A.; Van Loon, L.C.; Pieterse, C.M. Enhancement of Induced Disease Resistance by Simultaneous Activation of Salicylate- and Jasmonate-Dependent Defense Pathways in Arabidopsis Thaliana. *Proc. Natl. Acad. Sci. USA* **2000**, *97*, 8711–8716. [CrossRef]
101. Pan, J.J.; May, G. Fungal-Fungal Associations Affect the Assembly of Endophyte Communities in Maize (*Zea Mays*). *Microb. Ecol.* **2009**, *58*, 668–678. [CrossRef] [PubMed]
102. Verbon, E.H.; Trapet, P.L.; Stringlis, I.A.; Kruijs, S.; Bakker, P.A.H.M.; Pieterse, C.M.J. Iron and Immunity. *Annu. Rev. Phytopathol.* **2017**, *55*, 355–375. [CrossRef]
103. Choudhary, D.K.; Prakash, A.; Johri, B.N. Induced Systemic Resistance (ISR) in Plants: Mechanism of Action. *Indian J. Microbiol.* **2007**, *47*, 289–297. [CrossRef]
104. Newman, M.-A.; Sundelin, T.; Nielsen, J.T.; Erbs, G. MAMP (Microbe-Associated Molecular Pattern) Triggered Immunity in Plants. *Front. Plant Sci.* **2013**, *4*. [CrossRef]

105. Pieterse, C.M.J.; Van der Does, D.; Zamioudis, C.; Leon-Reyes, A.; Van Wees, S.C.M. Hormonal Modulation of Plant Immunity. *Annu. Rev. Cell Dev. Biol.* **2012**, *28*, 489–521. [CrossRef]
106. Pieterse, C.M.J.; Zamioudis, C.; Berendsen, R.L.; Weller, D.M.; Van Wees, S.C.M.; Bakker, P.A.H.M. Induced Systemic Resistance by Beneficial Microbes. *Annu. Rev. Phytopathol.* **2014**, *52*, 347–375. [CrossRef] [PubMed]
107. Backer, R.; Naidoo, S.; van den Berg, N. The NONEXPRESSOR OF PATHOGENESIS-RELATED GENES 1 (NPR1) and Related Family: Mechanistic Insights in Plant Disease Resistance. *Front. Plant Sci.* **2019**, *10*. [CrossRef]
108. Nie, P.; Li, X.; Wang, S.; Guo, J.; Zhao, H.; Niu, D. Induced Systemic Resistance against *Botrytis Cinerea* by *Bacillus Cereus* AR156 through a JA/ET- and NPR1-Dependent Signaling Pathway and Activates PAMP-Triggered Immunity in Arabidopsis. *Front. Plant Sci.* **2017**, *8*, 238. [CrossRef] [PubMed]
109. Alizadeh, H.; Behboudi, K.; Ahmadzadeh, M.; Javan-Nikkhah, M.; Zamioudis, C.; Pieterse, C.M.J.; Bakker, P.A.H.M. Induced Systemic Resistance in Cucumber and *Arabidopsis Thaliana* by the Combination of *Trichoderma Harzianum* Tr6 and *Pseudomonas* Sp. Ps14. *Biol. Control* **2013**, *65*, 14–23. [CrossRef]
110. Ryu, C.-M.; Hu, C.-H.; Reddy, M.S.; Kloepper, J.W. Different Signaling Pathways of Induced Resistance by Rhizobacteria in *Arabidopsis Thaliana* against Two Pathovars of Pseudomonas Syringae. *New Phytol.* **2003**, *160*, 413–420. [CrossRef] [PubMed]
111. Kamle, M.; Borah, R.; Bora, H.; Jaiswal, A.K.; Singh, R.K.; Kumar, P. Systemic Acquired Resistance (SAR) and Induced Systemic Resistance (ISR): Role and Mechanism of Action Against Phytopathogens. In *Fungal Biotechnology and Bioengineering*; Hesham, A.E.-L., Upadhyay, R.S., Sharma, G.D., Manoharachary, C., Gupta, V.K., Eds.; Fungal Biology; Springer: Cham, Switzerland, 2020; pp. 457–470. ISBN 978-3-030-41870-0.
112. de Boer, W.; Wagenaar, A.-M.; Klein Gunnewiek, P.J.A.; van Veen, J.A. In Vitro Suppression of Fungi Caused by Combinations of Apparently Non-Antagonistic Soil Bacteria. *FEMS Microbiol. Ecol.* **2007**, *59*, 177–185. [CrossRef]
113. Venturi, V.; da Silva, D.P. Incoming Pathogens Team up with Harmless "resident" Bacteria. *Trends Microbiol.* **2012**, *20*, 160–164. [CrossRef] [PubMed]
114. Pieterse, C.M.J.; Loon, L.C. van Salicylic Acid-Independent Plant Defence Pathways. *Trends Plant Sci.* **1999**, *4*, 52–58. [CrossRef]
115. Paynter, Q.; Paterson, I.D.; Kwong, R.M. Predicting Non-Target Impacts. *Curr. Opin. Insect Sci.* **2020**, *38*, 79–83. [CrossRef]
116. Molak, V. *Fundamentals of Risk Analysis and Risk Management*, 1st ed.; CRC Press: Boca Raton, FL, USA, 1996.
117. Meena, S.S.; Mohanty, A. Ethical, Patent, and Regulatory Issues in Microbial Engineering. In *Engineering of Microbial Biosynthetic Pathways*; Singh, V., Singh, A.K., Bhargava, P., Joshi, M., Joshi, C.G., Eds.; Springer: Singapore, 2020; pp. 133–142. ISBN 9789811526046.
118. Jamiołkowska, A. Natural Compounds as Elicitors of Plant Resistance Against Diseases and New Biocontrol Strategies. *Agronomy* **2020**, *10*, 173. [CrossRef]
119. Thomloudi, E.-E.; Tsalgatidou, P.C.; Douka, D.; Spantidos, T.-N.; Dimou, M.; Venieraki, A.; Katinakis, P. Multistrain versus Single-Strain Plant Growth Promoting Microbial Inoculants—The Compatibility Issue. *HPPJ.* **2019**, *12*, 61–77. [CrossRef]
120. Welsh, M.J.; Turner, J.A.; Epanchin-Niell, R.S.; Monge, J.J.; Soliman, T.; Robinson, A.P.; Kean, J.M.; Phillips, C.; Stringer, L.D.; Vereijssen, J.; et al. Approaches for Estimating Benefits and Costs of Interventions in Plant Biosecurity across Invasion Phases. *Ecol. Appl.* **2021**, e2319. [CrossRef]
121. Warziniack, T.; Haight, R.G.; Yemshanov, D.; Apriesnig, J.L.; Holmes, T.P.; Countryman, A.M.; Rothlisberger, J.D.; Haberland, C. Economics of Invasive Species. In *Invasive Species in Forests and Rangelands of the United States*; Poland, T.M., Patel-Weynand, T., Finch, D.M., Miniat, C.F., Hayes, D.C., Lopez, V.M., Eds.; Springer: Cham, Switzerland, 2021; pp. 305–320. ISBN 978-3-030-45366-4.
122. Hanley, N.; Roberts, M. The Economic Benefits of Invasive Species Management. *People Nat.* **2019**, *1*, 124–137. [CrossRef]
123. Mozelewski, T.G.; Scheller, R.M. Forecasting for Intended Consequences. *Conserv. Sci. Pract.* **2021**, *3*, e370. [CrossRef]
124. Köhl, J.; Booij, K.; Kolnaar, R.; Ravensberg, W.J. Ecological Arguments to Reconsider Data Requirements Regarding the Environmental Fate of Microbial Biocontrol Agents in the Registration Procedure in the European Union. *BioControl* **2019**, *64*, 469–487. [CrossRef]
125. Yang, Y.; Pratap Singh, R.; Song, D.; Chen, Q.; Zheng, X.; Zhang, C.; Zhang, M.; Li, Y. Synergistic Effect of *Pseudomonas Putida* II-2 and Achromobacter Sp. QC36 for the Effective Biodegradation of the Herbicide Quinclorac. *Ecotoxicol. Environ. Saf.* **2020**, *188*, 109826. [CrossRef]
126. Zhang, L.; Hang, P.; Hu, Q.; Chen, X.-L.; Zhou, X.-Y.; Chen, K.; Jiang, J.-D. Degradation of Phenylurea Herbicides by a Novel Bacterial Consortium Containing Synergistically Catabolic Species and Functionally Complementary Hydrolases. *J. Agric. Food Chem.* **2018**, *66*, 12479–12489. [CrossRef] [PubMed]
127. Pileggi, M.; Pileggi, S.A.V.; Sadowsky, M.J. Herbicide Bioremediation: From Strains to Bacterial Communities. *Heliyon* **2020**, *6*, e05767. [CrossRef]
128. Villaverde, J.; Rubio-Bellido, M.; Lara-Moreno, A.; Merchan, F.; Morillo, E. Combined Use of Microbial Consortia Isolated from Different Agricultural Soils and Cyclodextrin as a Bioremediation Technique for Herbicide Contaminated Soils. *Chemosphere* **2018**, *193*, 118–125. [CrossRef]

MDPI AG
Grosspeteranlage 5
4052 Basel
Switzerland
Tel.: +41 61 683 77 34

Plants Editorial Office
E-mail: plants@mdpi.com
www.mdpi.com/journal/plants

www.ingramcontent.com/pod-product-compliance
Lightning Source LLC
Chambersburg PA
CBHW041950260926
51850CB00019B/2284
9783725855179